D1408371

ALL IN ONE

CompTIA Mobility+™ Certification

EXAM GUIDE

(Exam MB0-001)

ABOUT THE AUTHOR

Bobby E. Rogers is an Information Security Engineer working for a major hospital in southeastern United States. His previous experience includes working as a contractor for Department of Defense agencies, helping to secure, certify, and accredit their information systems. His duties include information system security engineering, risk management, and certification and accreditation efforts. He retired after 21 years in the United States Air Force, serving as a network security engineer and instructor, and has secured networks all over the world. Bobby has a master's degree in Information Assurance (IA), and is pursuing a doctoral degree in IA from Capitol College in Maryland. His many certifications include CompTIA's A+, CompTIA Network+, CompTIA Security+, and CompTIA Mobility+, as well as the CISSP-ISSEP, CEH, and MCSE: Security.

About the Technical Editor

Chris Crayton is an author, technical consultant, and trainer. He has worked as a computer technology and networking instructor, information security director, network administrator, network engineer, and PC specialist. Chris has authored several print and online books on PC repair, Microsoft Windows, CompTIA A+, and CompTIA Security+. He has also served as technical editor and content contributor on numerous technical titles for several of the leading publishing companies, including the *CompTIA A+ Certification All-in-One Exam Guide*, the *CompTIA A+ Certification Study Guide*, and the *CompTIA Security+ All-in-One Exam Guide*. He holds multiple industry certifications, has been recognized with many professional teaching awards, and serves as a state-level SkillsUSA competition judge.

ALL · IN · ONE

CompTIA Mobility+™ Certification

EXAM GUIDE

(Exam MB0-001)

Bobby E. Rogers

New York   Chicago   San Francisco
Athens   London   Madrid   Mexico City
Milan   New Delhi   Singapore   Sydney   Toronto

Cataloging-in-Publication Data is on file with the Library of Congress

McGraw-Hill Education books are available at special quantity discounts to use as premiums and sales promotions, or for use in corporate training programs. To contact a representative, please visit the Contact Us pages at www.mhprofessional.com.

CompTIA Mobility+™ Certification All-in-One Exam Guide (Exam MB0-001)

Figures 1-1, 1-2, 2-1, 3-12, 3-14, 4-1, 4-3, 4-4, 4-8, 4-9, 5-1, 5-2, 5-3, 5-4, 5-11, 5-12, 6-2, 8-3, 9-9, 12-2, 12-4, 12-6, and 12-8 used with permission from Sarah McNeill.
Figures 2-2 and 2-3 used with permission from Dynetics, Inc. Photographed by Michael Thoenes.

1234567890 DOC DOC 10987654

ISBN: Book p/n 978-0-07-182522-1 and CD p/n 978-0-07-182650-1
of set 978-0-07-182532-0

MHID: Book p/n 0-07-182522-3 and CD p/n 0-07-182650-5
of set 0-07-182532-0

Sponsoring Editor Stephanie Evans	**Technical Editor** Chris Crayton	**Production Supervisor** James Kussow
Editorial Supervisor Jody McKenzie	**Copy Editor** Nancy Rapoport	**Composition** Cenveo Publisher Services
Project Manager Vasundhara Sawhney, Cenveo® Publisher Services	**Proofreader** Carol Shields	**Illustration** Cenveo Publisher Services
Acquisitions Coordinator Mary Demery	**Indexer** Jack Lewis	**Art Director** Jeff Weeks

This book is dedicated to all of the information technology professionals
I have both learned from and taught throughout the years.

It Pays to Get Certified

In a digital world, digital literacy is an essential survival skill. Certification demonstrates that you have the knowledge and skills to solve technical or business problems in virtually any business environment. CompTIA certifications are highly valued credentials that qualify you for jobs, increased compensation, and promotion.

IT is Everywhere	IT Knowledge and Skills Get Jobs	Job Retention	New Opportunities	High Pay-High Growth Jobs
IT is mission critical to almost all organizations and its importance is increasing.	Certifications verify your knowledge and skills that qualifies you for:	Competence is noticed and valued in organizations.	Certifications qualify you for new opportunities in your current job or when you want to change careers.	Hiring managers demand the strongest skill set.
• 79% of U.S. businesses report IT is either important or very important to the success of their company	• Jobs in the high growth IT career field • Increased compensation • Challenging assignments and promotions • 60% report that being certified is an employer or job requirement	• Increased knowledge of new or complex technologies • Enhanced productivity • More insightful problem solving • Better project management and communication skills • 47% report being certified helped improve their problem solving skills	• 31% report certification improved their career advancement opportunities	• There is a widening IT skills gap with over 300,000 jobs open • 88% report being certified enhanced their resume

CompTIA Mobility+ Certification Helps Your Career

- The **CompTIA Mobility+ certification** designates an experienced IT professional who can deploy, integrate, support, and manage a mobile environment.

- The designation certifies that the successful candidate has the knowledge and skills to understand and research capabilities of various mobile devices and aspects of over-the-air technologies.
- Certified professionals ensure proper security measures are maintained for devices and platforms to mitigate risks and threats while ensuring usability.
- The exam is geared toward IT professionals with 18 months of experience in a mobile IT environment.
- Relevant job roles include Mobile (Device Management) Administrator, Mobility Engineer, Mobility Architect, and Help Desk Level 1.
- **The market for mobility professionals is growing**   Mobile professional careers rank #4 in an InfoWorld study of the hottest IT jobs.
- **A smart next step**   After CompTIA Network+ certification, in a growing market for mobility expertise, is CompTIA Mobility+.

Five Steps to Getting Certified and Staying Certified

1. **Review exam objectives.** Review the certification objectives to make sure you know what is covered in the exam: http://certification.comptia.org /examobjectives.aspx.
2. **Practice for the exam.** After you have studied for the exam, review and answer the sample questions to get an idea of what type of questions might be on the exam: http://certification.comptia.org/samplequestions.aspx.
3. **Purchase an exam voucher.** You can purchase exam vouchers on the CompTIA Marketplace: www.comptiastore.com.
4. **Take the test!** Go to the Pearson VUE website: www.pearsonvue.com/comptia/, and schedule a time to take your exam.
5. **Stay Certified!** The CompTIA Mobility+ certification is valid for three years from the date of certification. There are a number of ways the certification can be renewed. For more information, go to http://certification.comptia.org/ce.

How to Obtain More Information

- **Visit CompTIA online**   Go to http://certification.comptia.org/home.aspx to learn more about getting CompTIA certified.
- **Contact CompTIA**   Please call 866-835-8020 and choose option 2 or e-mail questions@comptia.org.
- **Connect with CompTIA**   Find CompTIA on Facebook, LinkedIn, Twitter, and YouTube.

Content Seal of Quality

This courseware bears the seal of CompTIA Approved Quality Content. This seal signifies this content covers 100 percent of the exam objectives and implements important instructional design principles. CompTIA recommends multiple learning tools to help increase coverage of the learning objectives.

AUTHORIZED

CompTIA Approved Quality Content Disclaimer

CONTENTS AT A GLANCE

Chapter 1	Basic Networking Concepts	1
Chapter 2	Network Infrastructure and Technologies	39
Chapter 3	Radio Frequency Principles	81
Chapter 4	Cellular Technologies	107
Chapter 5	Wi-Fi Client Technologies	133
Chapter 6	Planning for Mobile Devices	173
Chapter 7	Implementing Mobile Device Infrastructure	225
Chapter 8	Mobile Security Risks	275
Chapter 9	Mobile Security Technologies	313
Chapter 10	Troubleshooting Network Issues	351
Chapter 11	Monitoring and Troubleshooting Mobile Security	387
Chapter 12	Troubleshooting Client Issues	419
Appendix	About the CD-ROM	451
	Glossary	453
	Index	463

CONTENTS

Acknowledgments xix

Introduction xxi

Chapter 1   Basic Networking Concepts 1

Introduction to Networking 2
    How Networks Are Built 2
    Basic Terms and Concepts 2

The OSI Model 8
    How the OSI Model Works 9
    The Physical Layer 12
    The Data Link Layer 13
    The Network Layer 13
    The Transport Layer 14
    The Session Layer 16
    The Presentation Layer 17
    The Application Layer 17

Ports and Protocols 18
    How Ports Work 18
    Common Protocols and Ports Used in Networks 22

Chapter Review 31
    Questions 32
    Answers 36

Chapter 2   Network Infrastructure and Technologies 39

Network Devices 39
    Common Network Devices 40
    Specialized Network Devices 45

Network Topologies 51
    Traditional Topologies 51
    Topology Design 57
    Network Architectures 58

Networking Technologies 62
    IP Addressing 62
    Network Management 69

Chapter Review 71
    Questions 74
    Answers 78

Chapter 3   Radio Frequency Principles 81
                Radio Frequency Fundamentals 81
                        Radio Frequency Concepts and Characteristics 82
                Exercise 89
                        RF Propagation 89
                Antennas 92
                        Basic Concepts 92
                        Antenna Characteristics 92
                        Types of Antennas 96
                Faraday Cages 99
                Chapter Review 99
                        Questions 100
                        Answers 104

Chapter 4   Cellular Technologies 107
                Introduction to Cellular Technologies 107
                        History 108
                        Basic Terms and Concepts 109
                Signaling Technologies and Standards 115
                        Code Division Multiple Access 116
                        Time Division Multiple Access 116
                        Global System for Mobile Communication (GSM) 118
                        Third Generation (3G) Technologies 120
                        Fourth Generation (4G) Technologies 122
                        Roaming and Switching Between Network Types 125
                Chapter Review 126
                        Questions 128
                        Answers 131

Chapter 5   Wi-Fi Client Technologies 133
                Introduction to Wi-Fi 133
                        History 134
                        Basic Terms and Concepts 134
                Wi-Fi Standards 143
                        802.11 Standards 143
                Exercise 149
                        Bluetooth 149
                Authentication and Encryption 153
                        Authentication 153
                        Encryption 154
                        WEP 154
                        WPA 155
                        802.1X 156

Site Surveys 157
Preparation 158
Capacity 159
Coverage 159
Signal Strength 160
Receive Signal Strength 161
Interference 161
Spectrum Analysis 162
Site Survey Documentation 163
Post Site Survey 164
Exercise 164
Chapter Review 164
Questions 166
Answers 170

Chapter 6   Planning for Mobile Devices 173

Basic Mobile Device Concepts 173
History of Mobile Devices in the Enterprise 174
Comparing Apples to Androids 176
MDM Concepts 178
Policy and the Mobile Infrastructure 182
Organizational IT and Security Policies 182
Vendors, Platforms, and Telecommunications 184
Enterprise Mobile Device Infrastructure Requirements 186
Security Requirements 186
Interoperability and Infrastructure Support 190
Location-Based Services 193
Mobile Application Management 194
Disaster Recovery Principles and the Mobile Infrastructure 198
Business Continuity and Disaster Recovery 199
Best Practices for Mobile Device Data Protection 209
Data Recovery 211
New Technologies and the Enterprise Mobile Infrastructure 213
Chapter Review 215
Questions 218
Answers 222

Chapter 7   Implementing Mobile Device Infrastructure 225

Install and Configure Requirements-Based Mobile Solutions 225
Pre-Installation 226
Deployment Best Practices 234
Mobile Device On-Boarding and Off-Boarding 236
Device Activation 237
Provisioning 238
Off-Boarding and De-Provisioning 243

Implementing Mobile Device Operations
and Management 246
   Centralized Content and Application
      Management and Distribution 246
   Implementing Remote Capabilities 249
   Life-Cycle Operations 252
Configuring and Deploying Mobile
Applications and Associated Technologies 256
   Messaging Standards 256
   Network Configuration 258
   Types of Mobile Applications 260
   In-House Application Requirements 262
   Push Notification Technologies 265
Chapter Review 267
   Questions 269
   Answers 272

Chapter 8   **Mobile Security Risks** 275
Mobile Infrastructure Risks 275
   Wireless Risks 275
   Software Risks 281
   Hardware Risks 286
   Organizational Risks 287
   Mitigation Strategies 290
Incident Response 299
   Preparation 299
   Response 303
Chapter Review 305
   Questions 307
   Answers 311

Chapter 9   **Mobile Security Technologies** 313
Encryption 313
   Encryption Basics 314
   Data-at-Rest 315
   Data-in-Transit 318
   Implementing Encryption 326
Access Control 330
   Authentication Concepts 330
   PKI Concepts 334
   Certificate Management 335
   Software-Based Container Access and Data Segregation 340

Chapter Review 343
    Questions 345
    Answers 349

Chapter 10 Troubleshooting Network Issues 351
    Troubleshooting Methodology 351
        Troubleshooting Steps 352
        Identify the Problem 353
        Establish a Theory of Probable Cause 356
        Testing the Theory to Determine Cause 356
        Resolving the Problem 357
        Implement 358
        Verify 359
        Documentation 359
    Troubleshooting Network and Over-the-Air Connectivity 360
        Basic Connectivity Troubleshooting Tools 361
        Common Network Issues 365
        Over-the-Air Connectivity Issues 374
    Chapter Review 378
        Questions 380
        Answers 384

Chapter 11 Monitoring and Troubleshooting Mobile Security 387
    Monitoring and Reporting 388
        Device Compliance and Audit Information Reporting 388
    Common Security Problems 394
        Certificate Problems 394
        Authentication Issues 399
        Network Device Problems 403
        False Positives and Negatives 405
        Other Security Issues 407
    Chapter Review 411
        Questions 412
        Answers 416

Chapter 12 Troubleshooting Client Issues 419
    Troubleshooting Device Problems 419
        Power Issues 420
        Device Issues 425
    Troubleshooting Common Application Problems 433
        Application Installation and Configuration 434
    Chapter Review 443
        Questions 444
        Answers 448

Appendix About the CD-ROM 451
        System Requirements 451
        Installing and Running Total Tester 451
            About Total Tester 451
        Free PDF Copy of the Book 452
        Technical Support 452

        Glossary 453

        Index 463

ACKNOWLEDGMENTS

I'd like to thank all the good folks at McGraw-Hill Education and their associates for their guidance throughout the writing of this book, and for helping me to improve my writing and ensure a quality product. Meghan Manfre and Mary Demery were awesome to work with: they made sure I stayed on track and they did everything they could to make this a wonderful experience. I'm very grateful to them for giving me the chance to do this book and for believing in me every step of the way.

Chris Crayton deserves some special thanks because, as the technical editor, he had a much harder job than I did, which was to make sure that I did my job the right way. Chris significantly contributed to the clarity, readability, and technical accuracy of the text. Thanks for all your help, Chris; this book is a direct testament to your extraordinary competence and professionalism.

I'd also like to take this opportunity to thank some of the folks I've worked with who gave me great ideas, sanity checks, and even provided some insight into the topics in the book, often correcting my misconceptions and patiently explaining to me "how things really work." In particular, I'd like to thank the good folks at Dynetics, Inc., whom I worked for during the majority of the time I was writing this book. They are some of the smartest and most professional folks I've ever worked with. Special thanks goes to some Dynetics folks who helped me get some really good pictures of some cool network equipment for Figures 2-2 and 2-3 in Chapter 2, including Tom Gates, Janet Felts, Steve Lee, Eddy Hammond, and Michael Thoenes, the photographer for those shots. I'd also like to thank my good friend Kelly Sparks for helping me out in the early stages of this book, and for keeping me sane in the later stages. Same goes to my other three partners in crime, Tim Turrell, Mark Teresin, and Pete Frambach. Thanks for being great friends, guys.

Of special note is my daughter, Sarah McNeill, who took almost all of the photos for the book (except for the networking equipment) and endured a great many re-shoots and special requests for taking photos of strange equipment, cell towers, and other assorted objects. Each one of her pictures is better than a thousand of my words in describing some of the concepts covered in this book. Thanks, Sarah, for the fantastic work.

Finally, and most importantly, I would like to thank my family, who sacrificed a great deal to give me the time to write this book. To my wife, Barbara, who endured many nights of going to bed without me while I stayed up writing to meet a deadline, as well as doing my share of the housework so I could write, I give my deepest thanks and love. To my kids, Greg, Sarah, and AJ, thanks for putting up with the old man all these years, especially during this past year when I could get a little bit grumpy while writing. I'd also like to thank my two grandsons, Sam and Ben, who were constantly told that I couldn't play because I was writing. I have some time to play now, guys! Go get the bat and gloves.

INTRODUCTION

Welcome to the *All-in-One Exam Guide* for CompTIA's Mobility+ certification exam! I'm Bobby Rogers, and I'm going to be your facilitator throughout this book. This book is designed to help you study for, and successfully pass, one of CompTIA's coolest new certification exams, the Mobility+ exam. This exam went live in November 2013, and is designed to test your knowledge of a wide variety of topics related to the fast-growing field of mobile devices. The exam focuses on mobile devices used in an enterprise infrastructure, versus from an individual user perspective. In this book, the real focus is on how to integrate mobile devices into your organization's network and manage them for the best balance of functionality and security possible.

Mobile devices have proliferated the consumer markets and now, finally, the workplace. Conservative estimates indicate that over 1 billion devices will be sold annually over the next several years. A great many of those devices are being bought for inclusion in an organization's network infrastructure, as we are seeing the differences between processing capabilities of traditional desktops and mobile devices become insignificant. Additionally, mobile devices have become ubiquitous in our society and people are naturally taking them to work and using them to process and access corporate data. This presents a challenge both to management and IT professionals because they have an obligation to protect organizational data. The old paradigm of traditional desktops that stay in the office whenever a user leaves for the day is pretty much over and has been replaced by one in which employees use laptops, smartphones, and tablets to receive, process, transmit, and store sensitive organizational data.

This book covers topics such as networking technologies, cellular and Wi-Fi communications technologies, mobile device management, and security. I also cover troubleshooting issues associated with mobile device use, including troubleshooting network problems, mobile apps, and security issues. I cover all five of the top-level domains, as well as the subdomains listed in the official CompTIA exam objectives, over 12 chapters.

While you don't have to already be an expert in mobile technologies, of course having experience in some of these areas, such as wireless and networking, definitely helps. Good solid background knowledge in basic networking, including protocols and network devices, will give you an advantage in your studies for this exam. Of course, you'll get a good background in all of these technologies in the first five chapters of the book. The remaining chapters discuss how to implement these technologies, integrating them with an existing enterprise network. Topics such as centralized device management, device provisioning, dealing with personal data on mobile devices, and security are prevalent in Chapters 6 through 12.

Passing the certification exam not only places you in a class of professionals recognized for their experience and expertise in this field, but also serves to quantify and validate your knowledge of mobile device technologies. This book is designed to help get you there.

How to Use This Book

This book covers everything you'll need to know for CompTIA's Mobility+ (MB0-001) certification examination. Each chapter covers specific objectives and details for the exam, as defined by CompTIA. I've arrange these objectives in a manner that makes fairly logical sense from a learning perspective, and I think you'll find that arrangement will help you in learning the material.

Each chapter has several components designed to effectively communicate the information you'll need for the exam:

- The **Certification Objectives** covered in the first section of each chapter will identify the major topics within the chapter and help you to map out your study.

- **Tips** are included in each chapter, offering useful information on how concepts you'll study apply in the real world. Often, these Tips will give you a bit more information on a topic covered in the text.

- **Exam Tips** are included to point out an area you need to focus on for the exam. Note that they won't give you any exam answers, but they do help you by letting you know about important topics that you may see on the test.

- Sometimes **Notes** are included in a chapter as well. These are bits of information that are relevant to the discussion and point out extra information.

- Some chapters have **step-by-step exercises** designed to provide a hands-on experience and to reinforce the chapter information. As your devices and circumstances are no doubt different from mine, these may, from time to time, need a little adjustment on your end.

The Examination

As I have several years of experience in taking CompTIA examinations, let me say that I'd love for this to be your one-stop shopping source for exam material to help you study for and pass the exam. Unfortunately, as with any sort of certification exam, there's probably no good single source of information to study for the test. This is because various materials offer different perspectives, depth of information, and explanations that people process and learn from differently. Having said that, I believe that there's more than enough information in this book for you to both pass the exam and be very successful in real life as a CompTIA-certified Mobility+ professional. The other thing you'll need, which in my mind is just as important as good study material, is experience. There's no substitute for practical, hands-on experience, and you should make every effort to get experience on all aspects of the CompTIA Mobility+ exam material that I discuss in this book.

In terms of prerequisites and experience, there's no hard and fast rule that you must have a certain certification, type of training, or years of experience to take this exam. However, CompTIA recommends that you possess their CompTIA Network+ certification or equivalent working knowledge prior to sitting for the test. They also recommend that you have at least 18 months of relevant work experience in administering mobile devices in the enterprise.

The exam consists of 100 questions, covering 5 domains. Each domain covers a specific topic, and is broken down into subdomains, if you will. Because domain objectives can change at CompTIA's discretion, the best source for you to get up-to-date information on the domains and objectives is CompTIA's web site, where the most current exam information will be posted. While I've made every effort to ensure that the most current exam information appears in this book, remember to check CompTIA's site for any exam or objective changes before, and during, your study effort, and definitely before you take the exam.

As you might imagine, the 5 domains are broken out over the 100 questions, but there is not an equal percentage of questions. If you take a look at the most current version of the CompTIA objectives from the site, you'll see that some domain objectives may have more questions than others, based upon how important the topic is to the exam. For example, Objective 1.0, Over-the-Air Technologies, comprises 13 percent of the questions on the exam, per the exam information from CompTIA. This domain is broken down into topics such as networking technologies, cellular technologies, and Wi-Fi. So, out of 100 questions, you should expect to get about 13 of them from this particular domain. Of course, this is just an estimate, as CompTIA can alter the exam and its objectives as it desires. But you can pretty much bet that this is a fair estimate. Other domains appear on the exam at different percentages as well. One test taking strategy that has served countless students is to make sure that you know an area you're comfortable with very well, especially if it has a higher percentage of exam questions, but you should also try to focus most of your studying on the areas you are unsure about, of course. You won't be able to pass the test by just being very good at one particular domain, so you should make every effort to learn the domains that you may not know as well.

Finally, you should know that you have a maximum of 90 minutes to take the exam. The exam has a passing score of 720, on a scale of 100–900. CompTIA, like many certification bodies, uses a type of scale scoring system, so it may be difficult at first to determine exactly how well you did based upon the score you get. However, a score of 720 is roughly equivalent to answering 80 questions out of 100 correctly on the test.

Objective Map

The following map has been constructed to allow you to cross-reference the official exam objectives with the objectives as they are presented and covered in this book. References have been provided for the objective exactly as the exam vendor presents it, the section of the exam guide that covers that objective, and a chapter and page reference.

Exam MB0-001

Official Exam Objective	All-in-One Coverage	Chapter No.	Page No.
1.0 Over-the-Air Technologies			
1.1 Compare and contrast different cellular technologies.	Introduction to Cellular Technologies Signaling Technologies and Standards	4	107, 115
1.2 Given a scenario, configure and implement Wi-Fi client technologies using appropriate options.	Introduction to Wi-Fi Wi-Fi Standards Authentication and Encryption	5	133, 143, 153
1.3 Compare and contrast RF principles and their functionality.	Radio Frequency (RF) Fundamentals Antennas Faraday Cages	3	81, 92, 99
1.4 Interpret site survey to ensure over-the-air communication.	Site Surveys	5	157
2.0 Network Infrastructure			
2.1 Compare and contrast physical and logical infrastructure technologies and protocols.	Network Topologies	2	51
2.2 Explain the technologies used for traversing wireless to wired networks.	Network Devices Networking Technologies	2	39, 62
2.3 Explain the layers of the OSI model.	The OSI Model	1	8
2.4 Explain disaster recovery principles and how it affects mobile devices.	Disaster Recovery Principles and the Mobile Infrastructure	6	198
2.5 Compare and contrast common network ports and protocols for mobile devices.	Ports and Protocols	1	18

Official Exam Objective	All-in-One Coverage	Chapter No.	Page No.
3.0 Mobile Device Management			
3.1 Explain policy required to certify device capabilities.	Policy and the Mobile Infrastructure	6	182
3.2 Compare and contrast mobility solutions to enterprise requirements.	Enterprise Mobile Device Infrastructure Requirements	6	186
3.3 Install and configure mobile solutions based on given requirements.	Install and Configure Requirements-Based Mobile Solutions	7	225
3.4 Implement mobile device on-boarding and off-boarding procedures.	Mobile Device On-Boarding and Off-Boarding	7	236
3.5 Implement mobile device operations and management procedures.	Implementing Mobile Device Operations and Management	7	246
3.6 Execute best practice for mobile device backup, data recovery and data segregation.	Best Practices for Mobile Device Data Protection	6	209
3.7 Use best practices to maintain awareness of new technologies including changes that affect mobile devices.	New Technologies and the Enterprise Mobile Infrastructure	6	213
3.8 Configure and deploy mobile applications and associated technologies	Configuring and Deploying Mobile Applications and Associated Technologies	7	256
4.0 Security			
4.1 Identify various encryption methods for securing mobile environments.	Encryption	9	313
4.2 Configure access control on the mobile device using best practices.	Access Control	9	330

Official Exam Objective	All-in-One Coverage	Chapter No.	Page No.
4.3 Explain monitoring and reporting techniques to address security requirements	Monitoring and Reporting	11	388
4.4 Explain risks, threats, and mitigation strategies affecting the mobile ecosystem.	Mobile Infrastructure Risks	8	275
4.5 Given a scenario, execute appropriate incident response and remediation steps	Incident Response	8	299
5.0 Troubleshooting			
5.1 Given a scenario, implement the following troubleshooting methodology.	Troubleshooting Methodology	10	351
5.2 Given a scenario, troubleshoot common device problems.	Troubleshooting Device Problems	12	419
5.3 Given a scenario, troubleshoot common application problems.	Troubleshooting Common Application Problems	12	433
5.4 Given a scenario, troubleshoot common over-the-air connectivity problems.	Troubleshooting Network and Over-the-Air Connectivity	10	360
5.5 Given a scenario, troubleshoot common security problems.	Common Security Problems	11	394

Basic Networking Concepts

In this chapter, you will

- Learn basic networking concepts
- Discover how the OSI model works
- Explore common protocols and ports that apply to both wired and wireless networks

Have you ever thought about how a house or a building is designed and constructed? There are a lot of steps that go into building a house, from drawing up the blueprints to selecting a site, clearing the land, laying the foundation, putting up walls, wiring and plumbing it, and then putting the finishing touches of paint on it and decorating it. While you might really want to jump in and start doing some things ahead of other not-so-fun steps, you have to build the house in order, or you'll have a shaky foundation with bad plumbing and wiring. Learning about mobile technologies and how they work is a similar process. You have to learn some things before others in order to have a good foundation of knowledge so that you can understand the more complex topics that come later. So this chapter is about learning some of the fundamental (and very important) concepts involved with networking. Mobile devices still rely on the basic networking technologies in order to connect to each other and the Internet, transfer data, and download the cool videos and other content we've come to expect.

CompTIA recommends that you have taken their Network+ certification exam and have at least 18 months experience working with mobile devices, networking, and other related technologies before sitting for the Mobility+ exam. This isn't required, of course, but you still should have some networking experience and solid knowledge of how networks work. That's what the next two chapters cover, before you dive into fun stuff like radio frequency theory, cellular technologies, and wireless protocols. Understand up front that I'm not going to give you the entire Network+ course in only two chapters, but I will cover some of the fundamental (and important) concepts needed to give you a good foundation for the rest of the book. In most cases, some of what these chapters cover will be a refresher to you; in others, it may be material you haven't seen before. The CompTIA Mobility+ exam objectives covered during this chapter are part of 2.2., "Explain the technologies used for traversing wireless to wired networks," and all of 2.3, "Explain the layers of the OSI model," and 2.5, "Compare and contrast common network ports and protocols for mobile devices." These objectives cover the fundamental principles of networking that you'll need to understand the technologies discussed in later chapters.

Introduction to Networking

All networks, whether wired or wireless, have a great many things in common. In this chapter and the next, I'll discuss a lot of those common characteristics that networks have. This chapter covers the basic networking concepts and definitions, the OSI model, and TCP/IP, and I'll spend a lot of time talking about network protocols. The next chapter discusses network design topologies (how networks are physically and logically laid out), as well as the devices, protocols, and technologies that make networks work. First, let's talk about how networks actually work and why we have them, along with some basic components, and some basic terms, concepts, and definitions, to make sure you start off on the right foot.

How Networks Are Built

Why do we have networks? Well, the fundamental reason is to share data and information between computers and other devices. Data isn't transferred between computers through osmosis, so they have to be connected in some way. That's where media comes in. Media is what is used to connect two or more computers together. It could be wired or wireless media. Computers and programs also have to be told the rules for communications and sharing data between them. That's where protocols come in.

Basic Terms and Concepts

Before we get in too far, it's probably helpful to establish common ground with terms and concepts. For those of you who may have had some network experience and knowledge before picking up this book, the next few sections will serve as a good refresher. For those who don't have much in the way of network knowledge, these few sections in this chapter, as well as the next, will give you the right amount of knowledge to meet the objectives for the exam. I'll spend the next few pages talking about some basic network concepts, but keep in mind that some of these topics are covered in greater depth throughout the rest of this chapter, as well as in Chapter 2.

Network Cards

Network cards are what allow a device to connect to the network (of course). They are known inside the profession as Network Interface Cards (NICs), and can come in a wide variety of shapes and sizes. Traditional cards that connect a standard PC to the network may be the kind that plug into a slot inside the PC's chassis, or be in the form of a USB device. NICs are typically used for specific media; that is to say that you'll see wired NICs with an Ethernet interface on them in which to insert a cable with an RJ-45 plug into, and there are, of course, wireless NICs that may have antennas attached to them. NICs that are built into the main boards of mobile devices may be combination boards, meaning that they could have both wired and wireless capabilities; this is especially true in the case of laptops, but not really for tablets or smartphones, which primarily use wireless media.

A network card has a unique address built into it; this is often referred to as "burned" into a chip on the card. It's called a hardware address, a physical address, or, more commonly, a Media Access Control (MAC) address. This MAC address is a 12-digit (48-bit) hexadecimal address, which looks something like this: 5C:26:0A:3C:2C:38. You also may see it separated by dashes occasionally (5C-26-0A-3C-2C-38). The first six digits (5C:26:0A in this case) denote the Organizationally Unique Identifier (OUI) for the manufacturer of the card (in this case, Intel Corp.). The last six digits are the unique identifier for the card itself. All network cards, whether they are wired or wireless, have a MAC address. It physically can't be changed on the card itself, but there are ways to change the MAC address that appears for the card on the network via the operating system or software.

EXAM TIP A MAC address is the hardware address of the NIC. It's "burned" into a chip on the NIC, and therefore is not easily changeable. Remember how a MAC address is constructed for the exam: It's a 12-digit (48-bit) hexadecimal number.

Figure 1-1 shows some examples of various wired and wireless cards that can be found in PCs, laptops, and tablets. From left to right, you see a card for a wired Ethernet

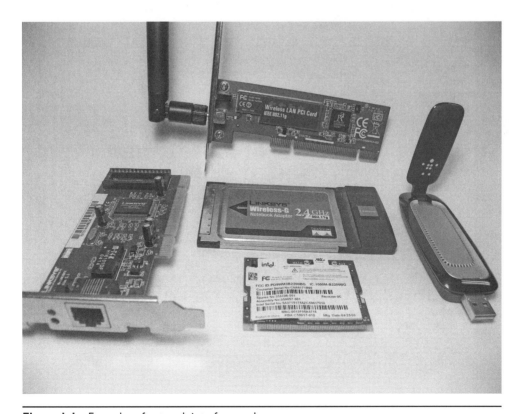

Figure 1-1   Examples of network interface cards

network, a wireless card for a PC (top), an older PCMCIA card (nowadays just called a PC card, middle), an internal wireless card from an older tablet PC (bottom), and a newer USB wireless adapter.

Media

Media, as you probably already know or guessed, is the method over which data is sent from one device to another. The basic types of media are *wired* and *wireless*. Wired media uses copper or fiber cabling connected to devices, while wireless media can use radio, microwaves, or even infrared light to send and receive data. As a network professional, you'd be expected to know the various types of wired and wireless media and their detailed characteristics. However, an in-depth discussion of wired media is really outside the scope of this book because our focus is on mobile technologies, and obviously, wired media and mobility are almost mutually exclusive. I'll discuss particulars of wired media where needed, but I will talk in-depth about wireless media over the next several chapters, particularly Chapters 3, 4, and 5. Figure 1-2 shows an example of wired media in the form of a twisted-pair copper Ethernet cable attached to an RJ-45 connector. For wired networks, this is the most commonly seen cable and connector.

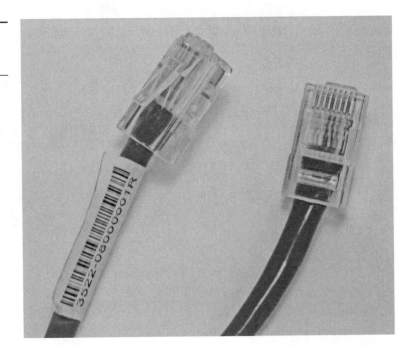

Figure 1-2
An example of wired media

Clients

In the broadest sense of the term, a *client* is a computer or other device on the network that receives some type of network services from a server (a device that offers those services). These services could be file sharing services, such as when a client connects to a shared folder on a server to access a document, for example, or they could be services that give the client network information (such as an IP address) or help the client connect to the Web. In the early days of networking, it was pretty easy to distinguish a client from a server. A client was any computer that requested data from the server and processed that data for a user. Likewise, a server "served" up data for any client that needed it, or provided network services. These days, the distinction is sometimes difficult to make. Often, a device is acting as both a client and a server in various situations—sometimes receiving services, sometimes giving them to the network. Additionally, most of the time, clients and servers aren't just computers anymore. A client can be a laptop, tablet, or even a mobile phone. And likewise, these devices can all be servers in some way.

Servers

As mentioned in the preceding section, a server offers network services. This could be file sharing services, but it could also be network services. There are services that give a client Internet Protocol (IP) addressing information to allow the client to connect to the network, there are services that help authenticate clients to the network, and there are services that provide name resolution services for clients so they can access resources on the Internet. While any device can technically be a server (and sometimes a computer operates as both a client and a server), for large network infrastructures, we tend to think of servers as powerful computers dedicated to that role. Enterprise-level servers are higher-end computers with additional memory, faster Central Processing Units (CPUs), more storage space, and more robust hardware. Servers also usually run enterprise-level operating systems (OS), such as Red Hat Enterprise Linux, Mac OS X Server, or Windows Server 2012, instead of desktop-class operating systems. These OSes are better suited to handle multiple simultaneous requests, are able to contain central databases of user accounts, and can manage networked resources more efficiently. In some cases a server may be dedicated to a specific task, such as providing a web site or application. You also may have servers that provide multiple services, such as network name resolution and authentication services.

Networked Applications

Back in the early days of networking, programs and applications installed on a client were standalone—that is, they were installed on the local machine and ran with no need for any other components on another box. Everything an application needed to run was installed along with it on the computer. Nowadays, that isn't always the case. A lot of applications fall in the category of networked applications. These networked (or in some cases, network-aware) programs require services on the network to function. In some cases, a portion of the application is installed on the user's computer (the client), and some of it is stored on the server. This architecture is called, not coincidentally, *client-server architecture*. Many software programs, such as databases, for example, fit into this category. Other applications use what's called an *N-tiered architecture*. This type of design means that an application and its components could be located on multiple other servers (called *tiers*). For example, a 3-tiered application would have a client component running on the user's computer (maybe the graphical interface portion) and communicate via the network to another component running on a different computer (maybe a logic or processing component), such as a web site, which may link to a back-end database running on a third computer or server.

Network Devices

You typically don't see a large network consisting of just a bunch of computers connected by single cables to a server. We typically see various network devices integrated into the network that provide connection points (at a minimum), network traffic management services, and security services. For the purposes of this discussion, I'm naming only a few of the many devices that you can see on a network. The devices include hubs, switches, bridges, wireless access points, routers, firewalls, and so on. For example, a device that provides just the basic connection services is called a hub, and is really no more than a place to plug in cables that connect to multiple computers on a network. In Chapter 2, I go in-depth and explain the purpose and function of a wide variety of network devices.

One important item of note: Regardless of whether something connected to the network is considered a client, server, or network device, anything that is connected to the network and communicates with it is often referred to generically as a *host*. Generally, a host is anything such as a computer, server, network device, or wireless device that processes data. A *network host* has a network card that is connected and sends and receives network traffic (and may also be called a network node), whereas a *standalone host* may not have a network connection.

Decentralized vs. Centralized Networks

You also may need to know about the concept of centralized networks versus decentralized ones. A decentralized network is one where all connected hosts are considered equal, and have their own security requirements, usernames, passwords, resources, and so on. These may or may not be shared with other hosts. These types of networks are also called *peer-to-peer* networks, and are usually small and informal (meaning no real rules with regards to security or connections). Ad-hoc networks (which are two or more wireless or mobile devices connecting to each other) can be considered as peer-to-peer, or decentralized networks, for example. You learn more about ad-hoc networks in Chapter 5.

Centralized networks, on the other hand, are those that are typically larger, have more clients, and likely have some servers providing specific services to the clients. These networks usually have centralized security and authentication mechanisms implemented that control how users access the network. You'll also usually see more specialized network devices, such as firewalls and Virtual Private Network (VPN) concentrators, as well as different network architectures, such as a Demilitarized Zone (DMZ) and extranets. These specialized devices, and the architectures that support them, will be covered in Chapter 2. There are usually formal rules in place with regards to sharing resources, usually implemented through resource permissions. The distinction of centralized and decentralized networks is important because the focus of the Mobility+ exam and this book is on an enterprise-level network, which is usually a large, centralized one. Figures 1-3 and 1-4 show examples, respectively, of a decentralized, or peer-to-peer network (possibly a small home or business network that also has wireless clients attached to a wireless access point), and a centralized one (in this case a larger business network).

 TIP Understanding the concepts of decentralized and centralized networks will help you better understand network topologies, which are discussed in Chapter 2.

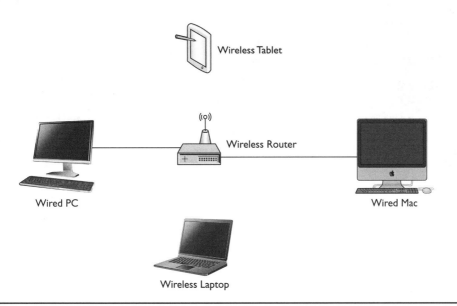

Figure 1-3 A decentralized (peer-to-peer) network

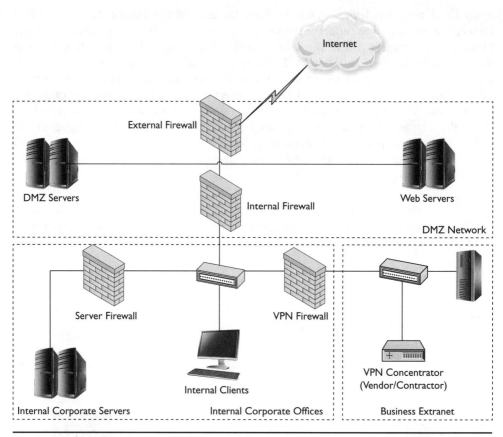

Figure 1-4 Example of a centralized enterprise-level network

The OSI Model

Now that you have a picture of networks in your mind, it's time to take a look, notionally, at how data travels logically over the network. Back when networks were first being built, there were no real standards. Manufacturers and vendors made equipment without regard to how it would work together to send and receive data, and in some cases, without regard to how a user's applications would format, send, and receive data to these network devices and other hosts. The Open Systems Interconnection (OSI) Model was developed to help people understand how networks ought to work. This helped vendors work toward common interoperability standards, develop rules for data communication (called *protocols*), and ensure that data could travel between hosts regardless of device manufacturer or application developer.

Keep in mind that the OSI model is just that—a model. Vendors build devices, develop technologies, and create protocols (or protocol stacks, as they are sometimes called) that conform to the model, more or less. The model is designed to show how networks (and their components) should work, but it's not mandatory. Interoperability among the different standards, protocols, technologies, and devices are what the model is all about. Over the next several sections, I'll discuss the model in depth.

It is worth mentioning that the OSI model is not a protocol stack itself—it's only a model that protocol stacks, such as the TCP/IP suite of protocols, can map to. While the OSI model is just something that shows how things could or should be, a protocol stack is an actual implementation of different protocols that may or may not conform to the OSI model. Think of the OSI model as the map for a road, telling you how it ought to look, while the protocol stack is the actual rules telling you how you must drive on the road. The map can show you the location of the Stop sign, but it doesn't tell you the procedures you must use in stopping for it, looking both ways for oncoming traffic, and then proceeding on your way. That's what protocols do.

TCP/IP is one such protocol stack that very closely maps to the OSI model. There are other protocol stacks that also map to the OSI model (to varying degrees), and some that don't, as well. You'll find in talking to network folks, and reading through hundreds of books on networking, that a lot of people sometimes mix discussions on the OSI model and TCP/IP. Just understand that the protocols being discussed aren't OSI; they are TCP/IP. The objectives for the exam don't specifically call for a discussion on TCP/IP, but you almost can't have a discussion of the OSI model without referring back to TCP/IP for context, and vice versa. So I'll offer a quick overview of the TCP/IP protocol later in the chapter.

How the OSI Model Works

The OSI model can be visualized as being composed of "layers." These layers represent different transformations of data as it logically travels between sending and receiving hosts. The OSI model is a seven-layer model. Like a cake, it has layers stacked from bottom to top. Each layer is responsible for different interactions with data. Each layer performs an action on data designed to get it from the sending to the receiving hosts. In some cases, this might be to prepare the data to go out over the transmission medium, or to change the data format, or even to encrypt the data. As you can see in Figure 1-5, those layers, from bottom to top, numbered one through seven, are: the physical layer, the data link layer, the network layer, the transport layer, the session layer, the presentation layer, and, finally, at the top is the application layer. Each layer has its job to do, has specific network devices that work at that layer, and has protocols at that layer that help it to do its job. Over the next several pages, I'll explore those layers in detail.

 EXAM TIP   Some people use a mnemonic to memorize the order of the OSI model layers. One popular such mnemonic is **A**ll **P**eople **S**eem **T**o **N**eed **D**ata **P**rocessing (from top to bottom), to correspond to the different OSI model layers (application, presentation, session, transport, network, data link, and physical). Use whatever works for you to help you remember them for the exam!

Figure 1-5
The layers of the
OSI model

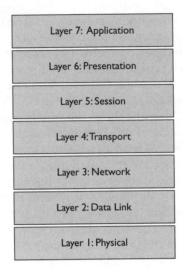

Layer 7: Application

Layer 6: Presentation

Layer 5: Session

Layer 4: Transport

Layer 3: Network

Layer 2: Data Link

Layer 1: Physical

How Data Travels Through the OSI Model

As data is sent from one computer to another, it logically "travels" *down* the OSI layers, and when it is received, it travels back *up* the layers, in reverse order, as shown in Figure 1-6. As it travels down the layers to be sent, data is transformed or modified in some way to enable it to progress to the next layer. Each layer is usually agnostic to what the layer above or below is doing to the data. Data gets sent through each layer as

Figure 1-6
Data traveling
down and up the
layers of the OSI
model

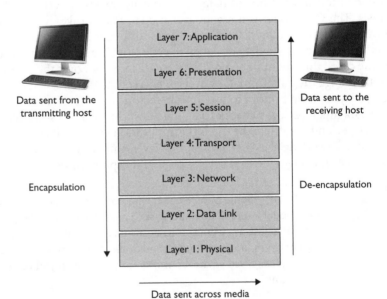

Data sent from the
transmitting host

Data sent to the
receiving host

Layer 7: Application

Layer 6: Presentation

Layer 5: Session

Layer 4: Transport

Layer 3: Network

Layer 2: Data Link

Layer 1: Physical

Encapsulation

De-encapsulation

Data sent across media

a discrete unit. Think of it as a system of envelopes, if you will. When data, in its own sealed envelope, is passed from one upper layer down to another, the layer that gets it doesn't care about what the layer above it has done to the data. It just knows it has an envelope full of data, for the most part. This data unit or envelope can be different for each layer, and is called a Protocol Data Unit (PDU). Each layer may add additional data to the beginning of the higher layer's PDU, such as addressing information, and this is called a *header*. Rarely, layers could also add data to the end of the existing data unit, and this is called a *footer*. When a layer adds its own header (or footer) to the existing data, it creates a new envelope, or PDU, and passes the modified data envelope down to the next layer. This whole process of adding data to the existing PDU is called *encapsulation*. Figure 1-7 illustrates the concept of encapsulation.

The encapsulation process happens as the sending host is transmitting data out to the network; the data is transformed by the different layers, and sent out. On the receiving end, the data goes through the exact opposite process, called de-encapsulation. Each receiving layer (from bottom to top, since it's receiving data now) gets a piece of data in an envelope from the layer below it, strips off that layer's header (and footer, if applicable) information, reads the data, and passes it up to the next layer. This process repeats through all seven layers, for all data. Because network communication happens fairly quickly, this whole process happens lightning fast, to every piece of data that is transferred between hosts.

 TIP The concept of encapsulation is a common one throughout networking and data security. Understand that encapsulation is really the process of putting data in one form inside data of another form. In the case of the OSI model, it's putting data from an upper layer into a PDU for the next layer.

The Transmission Control Protocol/Internet Protocol (TCP/IP) Suite

As stated previously, you can't really have a discussion about the OSI model without also talking about TCP/IP. TCP/IP became the de-facto protocol suite of choice for the Internet after it was adopted by the folks who created the early Internet to assure communications interoperability. TCP/IP replaced, over time, many vendor-proprietary protocols, such as those belonging to Novell, Microsoft, and Apple, not only in Internet communications, but also within local area networks as well. TCP/IP is a protocol suite (or stack, as it is sometimes called) that consists of many different protocols. The

Figure 1-7
Encapsulation of
PDUs

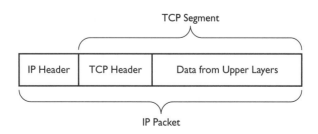

TCP/IP stack is illustrated by a four-layer suite, compared to the seven layers of the OSI model. These layers, while a bit different in name and function, pretty much map directly to similar layers in the OSI model. Figure 1-8 shows the TCP/IP layers, with their respective mappings to the OSI model layers. Note that two layers in TCP/IP map to more than one OSI layer, but their function is virtually identical.

After an explanation of the TCP/IP protocol suite, you'll likely better understand the functions of the OSI model layers that I am about to cover given that I do sprinkle references to TCP/IP in there for context. Look at the next several sections as only an introduction to each layer; I won't discuss all of the particulars associated with each layer, such as devices and protocols, in depth. I will look at protocols later in the chapter, and I'll talk more about which devices work at each layer in Chapter 2.

The Physical Layer

The physical layer of the OSI model is where you see electrical connections and signaling happen, using media. This layer is concerned with electrical impulses that travel across wires, forming patterns of 1s and 0s. Keep in mind that it's not just about wires at this level; wireless media (radio waves, microwaves, infrared light, and so on) also work at this level. The devices at this level (primarily hubs) don't really have or need a clue about the particulars of the data that travel through them. They are just physical connection points that understand and pass electrical energy.

There are no real "protocols" at this level, per se, but there are industry standards that apply at his level—usually for electrical signaling, wiring, cabling, power, and so on. For example, the Telecommunications Industry Association/Electronic Industries Alliance (TIA/EIA) T568A and T56B standards prescribe a wiring sequence for straight-through and crossover cables, usually seen in wired network implementations, that dictates which wires in a copper twisted-pair cable transmit data, and which receive.

Figure 1-8
TCP/IP protocol
suite layers
mapping to the
OSI model layers

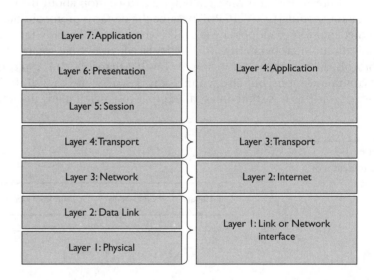

Some of the higher-level standards that I discuss next in the data link layer—Ethernet, for example—do prescribe certain characteristics that physical media and devices must have to meet the standard, however, so it's not unusual to see physical connection and signaling characteristics included in them.

The Data Link Layer

The data link layer is really where you start to see an organization of 1s and 0s that resemble some kind of coherent, organized data. Data at this level is called a *frame*. A frame can have both a header and footer; in fact, it's one of the only PDUs to have both. Frames don't communicate in terms of IP addresses; those are left up to the higher layers. Standards at this layer determine signaling, data organization, media contention and access methods, and so on. Ethernet is a main standard; in fact, it's probably the most common standard at this layer. Ethernet includes the Institute of Electrical and Electronics Engineers (IEEE) 802.3 specifications, and this covers topics such as media sharing, collision control, framing, and so on. Other standards at this level include popular wireless standards, such as Bluetooth (802.15), and Wi-Fi (the IEEE 802.11 standards), both discussed in Chapter 5. At this level, you have two sublayers that also perform different functions, the Logical Link Control sublayer (LLC) and the Media Access Control sublayer (MAC).

Logical Link Control Sublayer

The Logical Link Control sublayer is one of two sublayers of the data link layer. This sublayer helps to ensure that multiple protocols can be used over the same network medium and interface. It allows protocols and their respective data, as well as multiple simultaneous data conversations within a host to compete for and use the network. It can *multiplex* (or combine) multiple data streams to send out over the network card and media during transmission, as well as *demultiplex* (or demux), or separate, the data when it is received by the host.

Media Access Control Sublayer

The Media Access Control (MAC) sublayer of the data link layer is concerned with hardware addressing in order to get data to the correct network node or host. The MAC sublayer also handles media contention, sharing, and access between nodes. The MAC sublayer also handles appending the footer information to the frame, which is a frame check sequence (FCS), used to ensure the integrity of the frame itself.

The Network Layer

The network layer is a critical one in terms of communication between hosts and networks. This layer is concerned with routing between *local* networks and hosts (networks and hosts that are on the same physical and/or logical segment) and *remote networks* and hosts (any network/host that's not on the same physical or logical segment). This layer is also concerned with logical addressing. A logical address is one that is assigned arbitrarily from an administrator, based upon the addressing scheme of the organization.

Logical addresses are usually seen as Internet Protocol (IP) addresses because IP is located at this layer and is also responsible for logical addressing, in addition to routing functions. Keep in mind that a logical address is assigned by an administrator and can be changed easily if desired, whereas the physical or hardware address I discussed earlier is hard-coded into the network card of the host. There are two current versions of IP in use: Version 4 is the most prevalent, with version 6 in the implementation stages for several years now. An IP version 4 address is usually seen as a series of four numbers separated by dots (called the quad dotted decimal format). An example of an IP version 4 address would be something like 192.168.23.42, and represents a specific host on a network. A version 6 address is a much longer address, represented by eight groups of four hexadecimal numbers. IP addressing will be covered in greater detail in Chapter 2.

Devices that work at the network layer include routers and some switches that work at both layers 2 and 3 of the OSI model. Protocols that you will see at this layer include IP, of course, but also Internet Control Message Protocol (ICMP), Internet Group Management Protocol (IGMP), and IP Security protocol (IPsec). ICMP is a maintenance protocol that is basically used to determine if a host is alive on the network and can send and receive traffic. If you've ever used the PING command to see if a host is "up" on the network, then you have used ICMP before.

IGMP is a protocol used in what's called *multicasting*. In a normal host-to-host communication, this is called *unicasting* because one host is communicating with only one other host. This is opposed to *broadcasting*, where one host is trying to communicate with *every* host on the subnet. Multicasting is kind of the happy midway point, where hosts belonging to a multicast group communicate certain traffic only amongst themselves, sending out messages to only a select group of hosts. IPsec is a security protocol used to provide security (through encryption and authentication services) to IP traffic. Regardless of the protocol used at this layer, the PDU for the network layer is called a *packet*. Sometimes networking folks use the term "packet" generically to mean *any* data in general that travels over the network, but in the true definition of the term, it is a protocol data unit found only at this layer of OSI.

The Transport Layer

Understanding the transport layer is critical to understanding other, more advanced concepts in networking and security. The transport layer uses two key TCP/IP protocols, which are important in getting data from the sender to the receiver. The transport layer is concerned with communicating with its corresponding layer on the receiving side of the communications, and works to establish connections between hosts, take care of data flow control, get data to the host in the right order (sequencing), retransmit lost data, and maintain services matched to corresponding ports on the receiving end.

TCP

The Transmission Control Protocol (TCP) is one of the two protocols that reside at the transport layer. Its PDU is called a *segment*. TCP provides transport services for higher-layer application protocols such as the Hypertext Transfer Protocol (HTTP), Simple

Mail Transfer Protocol (SMTP), and many, many other ones. TCP provides a number of services for the application layer protocols that use it.

First, TCP is a connection-based protocol. This means that for every communications session, it attempts to establish a confirmed connection with the receiving host (between the transport layers on both sides of the connection). Second, because data that is sent and received over a TCP/IP-based network may get broken up into chunks and routed over different paths before it gets to its destination, TCP can ensure that the data is reassembled in the correct order when it's received, regardless of the order it is received in, by assigning sequence numbers to each segment. Finally, TCP also takes care of error correction and the retransmission of data in case the receiving host tells the sending one that it did not receive a particular segment.

A TCP segment has several components that you should be aware of. One component is a flag. A flag is simply a one or zero in a particular bit field of the segment. There are eight possible flags (and eight flag bits), although most of the time you only hear about six of them in network studies. If a flag is set to a "1," the flag is turned on, and a "0" indicates it's turned off. TCP uses these flags set in various combinations in each segment to indicate the state of the communications between two hosts. The six most common flags you'll see as a networking professional are SYN (for SYNchronization), ACK (ACKnowledgement), FIN (FINish), PSH (PuSH), URG (URGent), and RST (for ReSeT). Each of these mean different things to a TCP communications session, both when set alone and in combination with other flags.

You can see one prime example of how TCP uses these flags when it is first establishing a communications session with another host. In order to establish the session, TCP performs what is called a "three-way" handshake with a receiving host. This handshake is used to negotiate the connection, establish sequencing numbers that both hosts use to track the segments they've sent and received, and make sure that they are synchronized. Figure 1-9 shows how a three way handshake is performed. First, the transmitting host sends a segment with the SYN flag set. This means that it wants to synchronize the communications session. The receiving host replies back with its own SYN flag set, as well as an ACK (for acknowledgement) flag turned on. To complete the handshake, the first host sends its own acknowledgement, in the form of a reply segment with the ACK flag set. Each host has its own set of sequence numbers that are sent back and forth during the three-way handshake, in order to establish the order it will send its segments and let the other host know what that order is. That way, if any segments arrive to the destination host out of order, that host can determine the order of the segments.

Figure 1-9 The TCP three-way handshake

1. SYN

2. SYN/ACK

3. ACK

Because of all of the control functions, such as connection establishment, error correction, segment retransmission, and flow control, TCP incurs a lot of performance overhead. For the application-layer protocols that use it, it really isn't so bad because these protocols need those particular functions in order to work, and they are designed to take this overhead into account. Other protocols, because they would be impaired by all of the delays and overhead involved with using TCP, use the other protocol found at the transport layer—the User Datagram Protocol.

UDP

User Datagram Protocol (UDP) is almost the polar opposite of TCP. UDP is a simple protocol that's been described as a fire-and-forget type of protocol. Not every upper-layer protocol requires the overhead associated with connection establishment or error correction, for example. Some protocols would be slowed down or otherwise inhibited by all of the overhead that TCP uses for those services, so UDP doesn't care about establishing a connection, error checking, data loss, sequencing, or any of the other overhead stuff its brother protocol at the transport layer is concerned with. UDP does its job and doesn't bother to verify that the traffic made it to the destination. UDP is used for a lot of streaming connections, such as video and audio, and for connecting where the sending application doesn't care whether or not the receiver gets every single bit of data. If there's an issue with the transmission, a higher level protocol or application may tell UDP to resend, but UDP itself isn't concerned about retransmitting. The PDU for UDP is a *datagram*, as opposed to the segment for TCP.

Some popular examples of application layer protocols that use UDP include the Domain Name Service (DNS) (for queries between a client and a DNS server), the Dynamic Host Configuration Protocol (DHCP), and the Simple Network Management Protocol (SNMP). These and other protocols that use UDP are discussed later in this chapter. UDP, like TCP, uses ports to distinguish different types of application-layer traffic it supports. Ports are explained better later on, but understand for now that TCP and UDP are the only TCP/IP protocols that use ports and port numbers, at least as far as the discussion in this book goes.

 EXAM TIP Understanding the transport layer protocols, TCP and UDP, is critical to understanding how data is exchanged between hosts. For the exam, you should know which protocol establishes connections and controls flow and errors (TCP), and which one (UDP) does not.

The Session Layer

Layer 5 of the OSI model is the session layer. The session layer's job is to help establish and maintain communications sessions between two applications on two hosts. When two applications on different network systems communicate back and forth, the session layer manages that connection. Connections to the host machines themselves are managed by lower level protocols, such as the transport layer, but those layers help the session layer protocol get the communications to the application layer level. Remote

Procedure Call (RPC), designed to make client/server communication easier and safer, is a good example of a session layer protocol.

Some applications don't necessarily require a high degree of session layer involvement, such as web browsing, where a simple request is sent and a simple response is received. Those application transactions may be controlled at some other level. Applications that require session layer management include those that may have several constant or multi-part transactions between them, such as databases, for example, where a transaction session must be managed to ensure it completes properly. Note that the PDU for the session layer (as well as both the presentation and application layers above it) is, simply, *data*.

The Presentation Layer

The presentation layer's job is to change the format of data into something that the layer above it and the layer below it can understand, depending upon the direction the data's going at the time. Data coming from the application layer may be in the form of a program-specific file or format, such as an email message, a web request, or even a PDF file. This data must be translated or reformatted into something that the network can use, regardless of its upper-layer proprietary or application-specific format. The presentation layer changes data into a common lower-level format, such as ASCII or EBCDIC (two common data exchange formats), which can be understood outside the application itself, and passes that data down to the session layer. For data coming back up the OSI model, the presentation layer does the opposite, translating data into the format used for the application from these lower-level common formats. Think of the presentation layer as a data formatter or translator.

The Application Layer

Many people believe that the application layer is where applications and programs reside, but this isn't the case. The application layer contains protocols that interface directly with user's programs and the operating system, and is likely the layer that most entry-level network professionals are most familiar with. The application layer is tied heavily to the transport layer, as application layer protocols all use either TCP or UDP as their transport protocols, and in turn use TCP and UDP ports (discussed shortly) to interface with the network.

Application layer protocols provide an interface with user applications and programs, such as web browsers, email clients, databases, and so on, and may run in the form of services on a host. The terms "protocol" and "service" are sometimes used interchangeably, but they aren't exactly the same thing. A service is the hosts' process or program that is actually using a corresponding protocol. For example, the DNS service runs on a host as a client or server service, making queries to resolve IP addresses to common Internet names, or transferring DNS database data, all using the DNS protocol. Other protocols and services work similarly. Popular application layer protocols include HTTP (web services), SMTP and POP3 (email services), and others. The second half of this chapter, explores many of these application-layer protocols in depth.

EXAM TIP Know the different PDUs for the exam for each OSI model layer. The application, presentation, and session layers all use the term *data* to represent their PDUs. The data link layer uses frames, the network layer uses packets, and the transport layer uses segments for TCP and datagrams for UDP. The physical layer does not have PDUs.

Ports and Protocols

The term *port* can get confusing in networking because it's used in different contexts to mean different things. First, a port can be a physical opening in a computer or device that you plug a cable into. You've probably heard of USB ports, serial ports or parallel ports, or RJ-45 ports on a switch. These are all places to plug in physical cables with particular plugs attached to them.

You also may have heard the term "port" used when talking about building software for a particular operating system platform, such as Windows or Linux. Software is said to be "ported" when it has been rewritten or compiled to run on a new or different operating system. For example, a "port" of Microsoft Office has been written to run on Macs, under Mac's OS X family of operating systems.

Finally, the type of port discussed in this chapter deals with TCP and UDP ports that different protocols can use. For our purposes in this chapter, a port is a number assigned to a particular protocol or commonly used by a protocol or service, to identify particular traffic destined for a host. A port number, when combined with an IP address, is called a *socket*.

Let's look at an example of why we need ports. Let's say that you live in an apartment building on 123 Main Street. The mail carrier brings the mail for everyone who lives there in that building, at that address. Now, the carrier doesn't (usually) just dump everyone's mail into one big bucket for everyone to search through for their own mail. The carrier, instead, places it into individual mailboxes, with the apartment number stenciled to the mailbox to identify the resident. That way, people who live there can get their individual pieces of mail from their assigned mailbox. So mail for Mr. Jones, in apartment A, would get addressed to 123 Main Street, Apt. A. Port numbers essentially fill the same function for a host and network traffic being received by the host. Like the apartment building in my analogy, the host has an IP address, which is its "main" address. However, the user on that host may be doing Internet surfing, reading email, and streaming music, all at once on the computer. So, how does the operating system on the host know which application gets what traffic and separate it for the applications? That's where the ports come in. Ports give the network applications on the computer a way to identify which traffic belongs to them, much like the mail for the different residents in the apartment analogy.

How Ports Work

One important thing to keep in mind is that application layer protocols, like HTTP, for example, all use either TCP or UDP as their preferred transport protocol. You also might have some protocols that use both for different purposes. In any event, ports are a function of the transport layer protocol used, either TCP or UDP. So, HTTP, described shortly, which uses TCP as its transport protocol, uses TCP port 80. In some cases, an application layer protocol may use multiple ports (FTP, also discussed shortly, uses both TCP ports 20 and 21, for example). Now, you might be asking yourself questions like "What are all the possible ports you could have?" or "Who gets to say which proto-col uses which port?" Well, let's try to answer those questions now.

First, the port numbers range from 0 through 65,535. These are all the possible port numbers you could have. This range is broken up into a few categories, described shortly. The Internet Assigned Numbers Authority (IANA) is the organization officially responsible for assigning and maintaining port numbers to protocols.

Another thing you should know is that by convention, when we speak of ports and protocols, normally we are speaking from the *destination* port point of view. Both the sending and receiving computers use port numbers to identify traffic, and usually when we tie a port number to a protocol, it's in the context of the computer running a service that is said to be *listening* on a particular port for that protocol. So when we talk about HTTP using TCP port 80, we're talking about the server running the HTTP service that receives and fulfills HTTP requests. The client sending the request uses a different *source* port number for its communications with the HTTP service on the destination server. Source and destination ports are used by both hosts during a particular session to deliver the traffic sent and received between each one to the application that requested it or receives it. Source ports change randomly with every communications session, so it wouldn't really do any good for us to use them as reference points when discussing ports and protocols. That's why we typically are referring to destination ports when discussing protocols. More on source ports later in the text.

Well-Known Ports

The first breakdown of the range of ports is from 0 through 1023. This range is called the *well-known* or assigned ports. They are called so because they are commonly used by specific protocols and services that make the Internet "run." This happened over time, maybe due to popularity of the protocol, tradition, convention, or assignment by IANA. Examples of protocols that use the well-known ports include HTTP, FTP, SSH, and most of the others discussed are coming up in the rest of the chapter.

One thing to keep in mind is that although these ports are commonly used by their respective protocols, there's no real law that says it absolutely must be this way. For example, HTTP uses TCP port 80, but there's no law that says an administrator can't configure that service to run on a different port, say TCP port 3333. There may be some specific reason an administrator might want to do that, say adding another separate web site to the web server when there's already one on it running over port 80. However, because most client software and network-aware programs are built with the well-known ports in mind, changing a port arbitrarily with no good reason may cause some network applications to break because they may be trying to communicate on the wrong port. Usually, if a default port is changed, other configuration changes have to be made with other network programs that use that protocol so that everything still works right. For the most part, people don't change the default ports; they usually just stick to the well-known ones unless there's a good reason for it.

Registered Ports

Registered ports are in the range from 1024–49151. These are formally "registered" by IANA when an entity requests an assigned port to a protocol. A software company may develop a new application that uses one or more TCP or UDP ports, and they apply to IANA to have them formally registered. This puts them on the list of registered ports, so other developers know to try to not use them in their applications. It also helps administrators to be able to more easily identify an application running on a host. As mentioned, there isn't any law requiring an application to use any of these ports or to even register with IANA, unless an organization requires their own developers to do so. It's just the smart thing to do in order to ensure interoperability with other applications and protocols, as well as establish order to the thousands of ports in this range.

Dynamic and Ephemeral Ports

The last group in the range of ports is called the *dynamic range*. This covers port numbers 49152–65535. You'll also hear this range called the private, unregistered, or ephemeral range. These ports are typically unavailable for registration with IANA; they can be used by anyone for any application. Often they may be used as test port numbers for developmental or new protocols, or for in-house use and development. They can also be used for applications that need temporary port numbers only for a single specific communications session between two hosts. This is actually what ephemeral ports are, those that are temporary and may change from communications session to session.

Source ports are typically considered ephemeral ports; each operating system (Windows, Linux, Unix, Mac OS X, BSD, and so on) has its own preferred ephemeral port range from where it assigns random source port numbers for a given communications session. As I mentioned previously, by convention, when we discuss port numbers, we are usually talking about destination port numbers because protocols are typically tied to them. There may be instances when you are looking at logs or a network traffic capture and need to know the source port, but in casual reference, assume that the destination or listening port is the one being referred to.

Figure 1-10 illustrates the discussion and the following example. In the figure, Client A is communicating with Web Server 1, which is running a web site. The server is listening for HTTP requests on TCP port 80, the well-known port for the HTTP protocol. Client A is using a source port number of 56187 for this particular communications session. Any subsequent session that the client has with this server or any other, will likely use another random port number in the dynamic range. For this session, however, both the client and server will use these two port numbers to communicate back and forth. When the client is sending data, it's with a source port of 56187, and a destination port of 80. When the server is sending something back to the client, from its perspective, it's using a source port of 80, and a destination port of 56187.

Similarly, in a separate communication with the same server, Client B is using a source port of 63127, and the destination port of 80. Note that this web server uses port 80 for all of its communications with any client that wants to communicate with its HTTP service, for both its source and destination ports when it is communicating with the client. That part doesn't change. In yet another unrelated communications

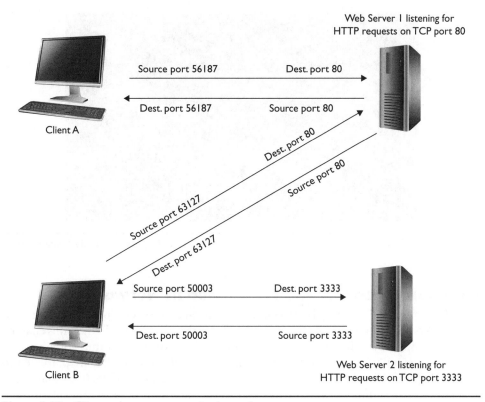

Web Server 1 listening for
HTTP requests on TCP port 80

Source port 56187   Dest. port 80

Dest. port 56187   Source port 80

Client A

Dest. port 80

Source port 80

Source port 63127

Dest. port 63127

Source port 50003   Dest. port 3333

Dest. port 50003   Source port 3333

Client B

Web Server 2 listening for
HTTP requests on TCP port 3333

Figure 1-10   Client and server communications using ports

session, Client B is exchanging data with Web Server 2, which is listening to its HTTP service on TCP port 3333, a non-standard port for HTTP. TCP port 3333 is normally assigned to another protocol, but if that protocol isn't being used on that server, it shouldn't affect anything. The administrator may have a good reason for running HTTP on that particular port, and it may be an internal-only web server that external clients couldn't connect to. The client, however, will have to be configured to communicate with Web Server 2's HTTP service on that port, or the communications won't work. Client B is also using a different source port to talk to Server 2 than it is with Server 1. Note the port ranges that each host is using for its source and destination ports. In Figure 1-11, the output of running the `netstat` command on a Windows 7 host shows the source and destination ports (shown with Local Address and Foreign Address, respectively) that the computer is listening on and has an established connection to.

 EXAM TIP   Make sure you are familiar with the concept of source and destination ports for the exam.

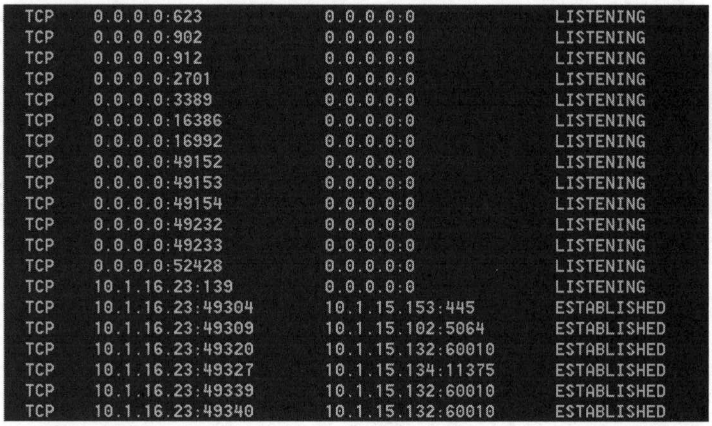

```
TCP    0.0.0.0:623        0.0.0.0:0            LISTENING
TCP    0.0.0.0:902        0.0.0.0:0            LISTENING
TCP    0.0.0.0:912        0.0.0.0:0            LISTENING
TCP    0.0.0.0:2701       0.0.0.0:0            LISTENING
TCP    0.0.0.0:3389       0.0.0.0:0            LISTENING
TCP    0.0.0.0:16386      0.0.0.0:0            LISTENING
TCP    0.0.0.0:16992      0.0.0.0:0            LISTENING
TCP    0.0.0.0:49152      0.0.0.0:0            LISTENING
TCP    0.0.0.0:49153      0.0.0.0:0            LISTENING
TCP    0.0.0.0:49154      0.0.0.0:0            LISTENING
TCP    0.0.0.0:49232      0.0.0.0:0            LISTENING
TCP    0.0.0.0:49233      0.0.0.0:0            LISTENING
TCP    0.0.0.0:52428      0.0.0.0:0            LISTENING
TCP    10.1.16.23:139     0.0.0.0:0            LISTENING
TCP    10.1.16.23:49304   10.1.15.153:445      ESTABLISHED
TCP    10.1.16.23:49309   10.1.15.102:5064     ESTABLISHED
TCP    10.1.16.23:49320   10.1.15.132:60010    ESTABLISHED
TCP    10.1.16.23:49327   10.1.15.134:11375    ESTABLISHED
TCP    10.1.16.23:49339   10.1.15.132:60010    ESTABLISHED
TCP    10.1.16.23:49340   10.1.15.132:60010    ESTABLISHED
```

Figure 1-11   Output of the netstat command on Windows 7 showing source and destination ports

Common Protocols and Ports Used in Networks

The objectives for the CompTIA Mobility+ exam include several protocols and their associated ports that you should know for the exam, so I will discuss them in this chapter. Now that you know what protocols and ports are, let's drill down into some specific ones used generally in network communications, as well as some that are particular to mobile devices. Note that the order of the protocols listed here differs from that of the exam objectives simply because, in some cases, it makes more sense to explain them in a different order, or to group similar protocols together. Also, a couple of protocols have been added here and there that aren't listed in the exam objectives, but you need to be aware of them as a knowledgeable mobile technology professional. After the descriptions of the protocols and ports that follow, there is a summary in Table 1-1 later in the chapter to help you review all of these before the exam.

HTTP

Hypertext Transfer Protocol (HTTP) is the protocol responsible for transferring web content to and from a browser. It's likely the most commonly used protocol on the Internet. HTTP has been the standard protocol for requesting and receiving both static and dynamic web content for many years, and is universally compatible (to one degree or another) across browsers, operating systems, and devices. HTTP uses TCP port 80 by default, and is an unencrypted protocol. For traffic that requires security, most web sites use HTTPS, which is really nothing more than HTTP sent over an encrypted communications session using SSL.

SSL

Secure Sockets Layer (SSL) is a protocol developed to provide for encryption and authentication services for secure communications sessions. Usually you see this implemented most commonly with web traffic that uses a browser client and communicates with a secure web server. HTTP is also the most common protocol used in conjunction with SSL, but technically, you can use a lot of other protocols over SSL that would not otherwise be secure to send their data over this secure session. You also don't necessarily need a browser or web traffic to use SSL. Many client-server applications are SSL-aware and can use it to establish encrypted communications. SSL has been replaced in some implementations with another secure communications protocol, Transport Layer Security (TLS), but SSL still is in wide use. SSL communicates using TCP port 443.

HTTPS

Hypertext Transfer Protocol—Secure (HTTPS) is the version of HTTP that uses SSL to communicate securely between a web client and server. Don't confuse HTTPS with yet another version of secure HTTP, which is called, confusingly enough, Secure-HTTP (S-HTTP). S-HTTP is a very different iteration of HTTP. S-HTTP does not use SSL, and encrypts only a small portion of the web traffic, whereas HTTPS sends HTTP data over an SSL-secured session, effectively encrypting the entire message. HTTPS, in conjunction with SSL, uses public key cryptography in the form of digital certificates to identify and authenticate the server the host is communicating with. It uses a session key to encrypt sensitive data between the client and the server for the duration of the communications session. More about cryptography and how it fits into networks is discussed in Chapter 9. Because HTTPS uses SSL, it is also associated with TCP port 443.

DNS

The Domain Name Service (DNS) is a technology that converts Internet names (such as www.google.com or www.comptia.org) to IP addresses. Most applications and utilities used to connect to resources on the Internet, and even local networks, use IP addresses to find the requested resources. Human beings, however, don't really use IP addresses, and most people probably couldn't remember more than a few at a time. We prefer easy-to-remember names, and can easily remember hundreds of Internet names and Uniform Resource Locators (URLs). URLs are not only easier to remember but are more descriptive of the site you may want to visit. DNS uses both TCP port 53 (for updates between DNS servers) and UDP port 53 (for queries).

DHCP

The Dynamic Host Configuration Protocol (DHCP) has a task that makes administrators' jobs much easier. Instead of having to go host-by-host and manually configure IP addressing information on each device, DHCP will sit as a service on the network (running on a particular device called a DHCP server), and dynamically provide the right network addressing information to each host that comes up on the network and asks for it. When a host boots up, if it is configured to use DHCP, it will broadcast out a particular message using UDP port 67, asking for any DHCP server to respond with a reply. A DHCP server that "hears" the message will reply back to the requesting host on UDP port 68. DHCP can also be used to send additional information to a client, including domain, routing, and DNS server information.

Telnet

Telnet is an older protocol and is used to remotely access and administer a host. The telnet server listens on TCP port 23, waiting for connections. A host with a telnet client on it can be used to connect to the server, gain a command prompt or shell, and perform administrative commands on the host. Telnet is not a secure protocol at all; it does not encrypt any data that is sent over it and offers very weak authentication. It has largely been replaced by more secure protocols, such as Secure Shell (SSH), but you still may see it in use from time to time on older devices. Telnet servers can be anything from a Windows or Linux box to a Cisco switch or router.

SSH

Secure Shell (SSH) is a modern replacement for telnet and some other unsecure remote administration protocols. SSH is usually found by default on Linux and Unix hosts, but can be easily installed on Windows as well. SSH offers secure authentication mechanisms, even those beyond simple username and password authentication, and also encrypts the entire communications session used between two hosts. Like telnet, it can be used to remotely administer a host, but in addition to that function, it can also be used to transfer and copy files from one host to another, as well as offer encryption for non-secure protocols that can be sent over an SSH session. Most SSH implementations also offer a suite of tools to go with it, including utilities to generate keys, copy files from host to host (SCP, or Secure Copy), or set up a more permanent secure file transfer site. SSH uses TCP port 22.

 TIP Secure Shell is not only a protocol, but also includes a set of commands and utilities that can enable you to connect securely to a host in order to administer it, copy files, and run programs. SSH offers both encryption and authentication, and replaces older, non-secure protocols such as telnet.

FTP

File Transfer Protocol (FTP) is a method used to copy files to and from a server running the FTP service. FTP uses two separate TCP ports for this, ports 20 and 21. Port 20 is for data transfer and 21 is used for control (FTP commands). FTP can be used anonymously (with no authentication required), or it can require a username and password combination. Unfortunately, because everything over FTP, such as telnet, is unencrypted, this is not a secure way of passing either data or authentication credentials. There are some other variations of FTP that are more secure, and are covered shortly.

SFTP

SFTP stands for SSH-FTP and is part of the SSH suite of commands and protocols. It is used to securely transfer files from one host to another. You'd think that SFTP is a version of FTP that uses SSH, but, unfortunately, you'd be wrong about that. That particular scenario also exists, though, and is known as *FTP over SSH*, and involves tunneling (similar to encapsulating) plain old FTP through a secure SSH session. This only adds to the confusing list of acronyms and protocol names you already have to know for the exam, but SFTP really has nothing to do with FTP at all. It is a secure protocol that does transfer files between two hosts, but that's pretty much where the similarity ends. You could have a server running SFTP that many SSH clients connect to in order to download or upload files, and they wouldn't be able to use FTP at all because they are different protocols. The same server, separately, could also run the FTP service however, for non-secure transfers, but this could get confusing for the clients! Like SSH, SFTP uses TCP port 22 and encrypts both authentication credentials and data for a secure transfer.

FTPS

To add yet another confusing FTP-related protocol to your growing list of confusing protocols, let's discuss FTPS, which stands for FTP-Secure. FTPS is an enhanced version of FTP that uses either the Secure Sockets Layer (SSL) or Transport Layer Security (TLS) protocols to secure a file transfer session. Which secure protocol it uses (SSL or TLS) depends upon how both the server and the clients are configured; if they can't negotiate on the use of either of these secure protocols, then the server can either refuse the connection or allow normal FTP (unsecured) to happen. FTPS uses TCP port 990.

 EXAM TIP Understand the differences between plain FTP, SFTP, FTP over SSH, and FTPS.

SMTP

Simple Mail Transfer Protocol (SMTP) is the first of several email-related protocols I'll discuss. SMTP has one primary function in life: to send email. It doesn't concern itself with receiving it or storing it; its job is just to send it. SMTP runs as a service on email servers and uses TCP port 25. Like a lot of other older protocols, it was not originally designed to be secure because it does not natively offer encryption of authentication services to protect itself or its data. But, as you'll read in the next few sections, there are ways to secure SMTP and your email. Alternate SMTP, as it has been called, is not really a different form of SMTP; it just uses a different port, TCP 587. This can be used in circumstances when Internet service providers block the standard SMTP port, 25. This is often done because of the history of SMTP abuse by spammers and other malicious users.

POP3

As stated previously, SMTP doesn't have the task of receiving or managing emails; that's the job of other protocols. One of these primary protocols that deals with the receiving aspects of email is the Post Office Protocol (POP), version 3 (POP3). POP3 runs on email clients—i.e., hosts that users receive, read, and send email over. POP3 isn't the only protocol that deals with managing and receiving email; I'll discuss a couple of others next. POP3 supports pretty much all of the typical user-level clients, such as Microsoft Outlook, Thunderbird, Eudora, and so on. Keep in mind that the relationship between POP3 and SMTP is a client-server model, as discussed previously. POP3 can also run on email servers as well to handle other email management tasks. POP3 uses TCP port 110, but you may also see it being run over an SSL or TLS connection and using TCP port 995.

IMAP

Internet Message Access Protocol (IMAP) is another protocol, like POP3, that handles email receipt and management, usually at the client level. The current version is IMAP4. IMAP has several advantages over POP3 as the email client protocol, such as the ability to have multiple clients and connections to the email box on the server (POP3 only allows one connection at a time), the ability to receive new messages automatically while connected (POP3 requires you to periodically reconnect or manually "refresh" to retrieve new messages because it maintains a connection to the server for only enough time to download new messages), and so on. IMAP4 uses TCP port 143. IMAP can also be used over an SSL or TLS connection to secure it because it really has no security mechanisms built-in. When used over SSL, IMAP (sometimes seen as IMAPS) uses TCP port 993. Figure 1-12 shows an example of IMAP configured to receive email over port 993 when connecting to Google's Gmail.

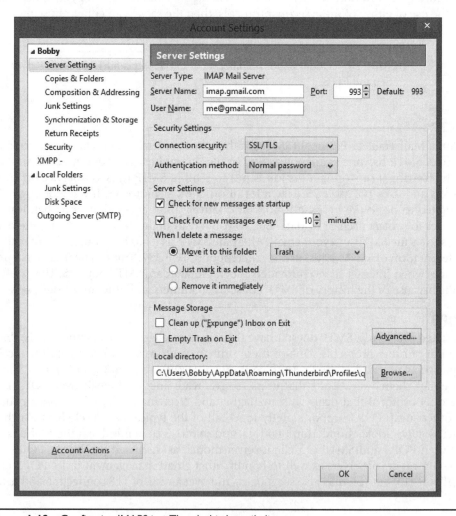

Figure 1-12 Configuring IMAPS in a Thunderbird email client

SSMTP

Secure SMTP (SSMTP—also called SMTPS, standing for SMTP Secure) simply uses SMTP over a Secure Sockets Layer (SSL) connection to secure it because SMTP does not natively offer any security features, such as encryption or authentication. Instead of using the standard SSL TCP port, 443, it uses TCP 465. Configuring SMTPS is quite easy. In fact, Figure 1-13 shows the default configuration for SMTPS in the Thunderbird email client when connecting to Gmail.

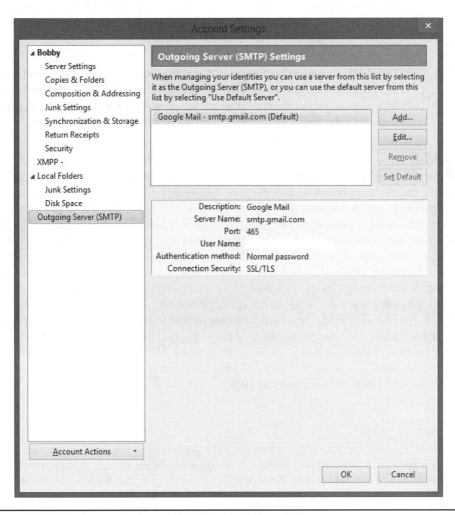

Figure 1-13   Configuring the Secure SMTP Thunderbird email client

MAPI

MAPI stands for Messaging Application Programming Interface and was developed by Microsoft. It's not strictly an email or networking protocol; rather, it's a messaging architecture that is used to connect Microsoft mail clients (such as Outlook) to Microsoft Exchange servers and Active Directory resources. MAPI allows rich integration with non-email content, as well as non-email services (e.g, news feeds) through Microsoft servers, as well as other third-party applications that use the MAPI interface. MAPI does not use a particular port; it depends upon other Microsoft services, such as Remote Procedure Call, Active Directory, and so on to communicate with Microsoft Exchange servers. The servers, on the other hand, typically use standard email protocols, such as SMTP, to manage email outside the organization's network.

 EXAM TIP Understand the group of email protocols—SMTP, POP3, IMAP4, and MAPI, as well as their secure equivalents (SSMTP, IMAPS, and so on)—and how you can secure them. Also watch for port numbers given on the exam that may help distinguish between a protocol and its secure equivalent.

LDAP/AD

Lightweight Directory Access Protocol (LDAP) is a protocol used as part of a distributed network and resource database architecture. A directory services (DS) system is used to store information about resources (users, printers, and other objects in the network) in a database that is centrally controlled and administered, but distributed across several servers for performance and redundancy purposes. The most popular distributed database of this type is probably Active Directory (AD), Microsoft's implementation of a DS architecture. There are, of course, other directory services implementations, such as Novell's eDirectory, OpenLDAP, and many others. LDAP is used to query databases for information about the objects that reside in those networks. An LDAP query, for example, may be used to locate a particular shared folder on the network, or a particular printer that users can send print jobs to. LDAP can also be used to update these same databases as information about the objects changes. LDAP uses TCP and UDP ports 389 for its traffic. Although you may still see some texts that refer to port 636 used by LDAP over SSL, this implementation largely went away with LDAP version 2. LDAP version 3 is the most current implementation.

Airsync

Airsync is the previous name for Microsoft Exchange Active Sync (back in version 1.0), and is a protocol designed to enable centralized mobile device management and synchronization of data. Not to be confused with the popular Android music app of the same name, Airsync originally was designed to be used between Microsoft Exchange mail servers and Microsoft Pocket PC mobile devices. It has since evolved to become a standard protocol supporting devices and infrastructures across multiple platforms and manufacturers. Active Sync is now at version 14.1 and uses both TCP and UDP ports 2175. You may see it show up every now and then as Microsoft Desktop AirSync Protocol.

EXAM TIP Although Airsync is listed in the CompTIA Mobility+ objectives, the protocol is now called Microsoft Exchange Active Sync. You may see either name on the exam, so be aware of both.

APNs

Apple Push Notification service (APNs) is a service that provides updates to Apple iOS and Mac OS devices through a network connection over TCP port 2195. APNs uses "push" notifications, meaning that data such as text messages, notifications, alerts, and so on, are pushed (or sent) to the Apple device without the need to initiate the connection by the end user. These push notifications are limited to a size of 256 kilobytes (KB). APNs also has a feedback service that allows the server sending the push notifications to receive notices of failed notification delivery. This may happen in the event the server is trying to send a notification to an app that isn't installed on the device, or the device has been offline for a while. The APNs feedback service uses TCP port 2196.

RDP

Remote Desktop Protocol (RDP) is a Microsoft proprietary protocol used for connecting to, administering, and using a remote Windows host. RDP's primary function is to easily provide a full, rich graphical desktop experience to end users from the remote host with which they are communicating. RDP uses TCP port 3389. All versions of Windows since Windows XP have included RDP client software, and there are numerous third-party programs as well that can use RDP to remotely connect to Windows systems.

SRP

Server Routing Protocol (SRP) is a proprietary protocol used by BlackBerry Limited (formerly Research in Motion, or RIM) brand products, to manage mobile devices within a BlackBerry enterprise infrastructure. SRP is used to provide enterprise-wide mobile device management (MDM), allowing smartphones and other BlackBerry devices to receive updates, have remote maintenance tasks performed on them, and allow user data to be remotely backed up to a central server. While SRP uses TCP port 4101 to communicate with BlackBerry mobile devices, other ports, such as 3101 and 3200, are also used between the Enterprise Server components.

 NOTE While the exam objectives ask for only specific port numbers associated with protocols, it's a good idea to be aware of others that you may see either on the exam or in the real world.

Jabber

Jabber is the original name given to the Extensible Messaging and Presence Protocol (XMPP), an open-standard protocol used to provide XML-based instant messaging to clients. Jabber.org also manages a worldwide instant messaging infrastructure as well, based upon XMPP. XMPP uses TCP port 5222 for client connections, with TCP 5223 for XMPP over SSL. Note that Jabber, in a different context, is also the name of the Cisco Systems instant messaging product as well, but they are different products.

GCM

I've discussed protocols for Microsoft, Apple, and BlackBerry, so we should give equal time to Google's Android platform protocol, the Google Cloud Messaging (GCM) service. GCM is used to send data from servers to Google apps on mobile devices (or to Chrome-based browser apps on other devices). Like APNs discussed earlier, GCM uses a push-notification model, sending data over an established connection to the device without the need for user intervention. GCM uses TCP ports 5228–5230 for communications between devices and the Android Marketplace. Table 1-1 lists the protocols discussed in this chapter and is a convenient study tool to use right before the exam.

Protocol	Transport Protocol	Port Number	Notes
HTTP	TCP	80	
SSL	TCP	443	
HTTPS	TCP	443	HTTP using SSL
DNS	TCP/UDP	53	TCP for zone transfers, UDP for queries
DHCP	UDP	67, 68	
Telnet	TCP	23	
SSH	TCP	22	
FTP	TCP	20, 21	
SFTP	TCP	22	SSH FTP
FTPS	TCP	990	FTP-Secure (SSL or TLS)
SMTP	TCP	25	
POP3	TCP	110 (995 over SSL)	
MAPI	TCP/UDP	135, 137, 139, 445	Not purely a "protocol"
IMAP	TCP	143 (993 over SSL)	
SSMTP	TCP	465	
Alternate SMTP	TCP	587	
LDAP/AD	TCP	389 (636 over SSL)	
Airsync	TCP	2175	Now called Microsoft Exchange Active Sync
APNs	TCP	2195	
Feedback	TCP	2196	APNs Feedback Service
RDP	TCP	3389	Microsoft proprietary protocol
SRP	TCP	4101	BlackBerry proprietary protocol
Jabber	TCP	5222 (5223 for connections over SSL)	Extensible Messaging and Presence Protocol (XMPP)
GCM	TCP	5228-5230	

Table 1-1 Summary of Common Network Protocols and Ports to Know for the Exam

Chapter Review

This chapter has covered the basics of networking, providing some definitions of terms, including *client, server,* and others. I discussed at length the OSI model—a seven-layer model consisting of the physical, data link, network, transport, session, presentation, and application layers. OSI is not a protocol stack but is used to describe how network communications should take place in order to maintain standards and interoperability. The primary protocol stack discussed in this chapter was the TCP/IP stack, which consists of four layers that map well to the OSI model. These are the application layer (which maps to OSI's application, presentation, and session layers), the transport layer (which maps to OSI's corresponding transport layer), the Internet layer (corresponding to OSI's network layer), and the network interface layer (sometimes also called the link layer), which maps to the bottom two layers of the OSI model, the data link and physical layers.

The OSI model prescribes a method for sending data from one host to another by having each layer encapsulate the data from the preceding layer above it, adding headers to it (and a footer in the case of the data link layer), and creating a new protocol data unit (PDU) for each layer. The PDUs for each layer are data for the application, presentation, and session layers; segments for TCP and datagrams for UDP, both at the transport layer; packets for data at the network layer; and frames for data at the data link layer. The physical layer really has no PDU; it consists of electrical impulses that serve to form 1s and 0s on the media. When the data is processed by the receiving host, this process is reversed (called de-encapsulation), and headers are stripped off as the data is passed up the layers of the OSI model.

The application layer is responsible for interfacing its protocols with user programs and the operating system. Examples of these protocols are HTTP, FTP, SMTP, and DNS. The presentation layer is responsible for changing data formats into something more common, such as ASCII or EBCDIC. The session layer manages sessions between applications on two hosts, while the transport layer manages communications sessions between two hosts. The transport layer is responsible for connection establishment, error correction, sequencing, and flow control. This is accomplished by the TCP protocol, but upper-layer protocols that require none of these functions can use the UDP protocol as their transport layer protocol. The transport layer is where ports come into play; each application layer protocol uses particular TCP and/or UDP ports for communications between hosts. The network layer is responsible for routing between networks and logical addressing (most often seen in the form of IP addresses that can be assigned by an administrator). Network layer protocols include IP, ICMP, IGMP, and IPsec. The data link layer is responsible for framing data and getting it on the media. It handles media contention and access, and deals with hardware or physical addresses instead of logical or IP addresses. The data link layer has two sublayers, the Logical Link Control sublayer (LLC) and the Media Access Control sublayer (MAC). Finally, the physical layer deals with electrical impulses that are sent via media (wired or wireless).

I discussed the definition of ports, and also the port ranges for TCP and UDP port numbers. Ports can fall into several ranges, with 0–1023 called the well-known ports, 1024–49151 the registered ports, and 49152–65535 set aside for the dynamic or ephemeral ports. During communications between two hosts, a source port (usually one that changes often, and is in the ephemeral range) and a destination port (typically in the well-known port range, tied to particular application layer and transport layer protocols) are used to track the communications during the session.

I then covered a fairly lengthy list of popular protocols and their corresponding ports that you may see on a wide variety of networks. Some of the more common protocols on most networks are HTTP, FTP, HTTPS, DNS, and DHCP. Many protocols are natively unsecure, such as HTTP, SMTP, FTP, and so on, and you looked at other ways to secure them, including using them in conjunction with both the SSL and SSH protocols. SMTP, IMAP, and POP3 are the core email protocols, with secure versions for each available. This chapter also covered several protocols that are specific to mobile devices, including Apple Push Notification service (APNs), used by Apple devices; Microsoft's Airsync (now Active Sync for Microsoft Exchange); and SRP, used by BlackBerry. GCM is Google's proprietary protocol, which uses push technology to deliver updates and alerts to its family of devices.

Questions

1. Which layer of the OSI model is tasked with carrying electrical impulses across media?

    A. Network layer

    B. Physical layer

    C. Data link layer

    D. Transport layer

2. You are trying to explain the layers of the OSI model to a colleague. Which two layers deal with logical and physical addressing? (Choose two.)

    A. Network layer

    B. Physical layer

    C. Transport layer

    D. Data link layer

3. You are looking at a network traffic log and see TCP connections that use the SYN, SYN/ACK, and ACK flags set. What type of communication is this?

    A. UDP datagram traffic

    B. File transfer over FTP

    C. TCP three-way handshake

    D. Encryption traffic over SSL

4. Which of the following layers of the OSI model use data as their Protocol Data Units? (Choose two.)

   A. Transport layer

   B. Application layer

   C. Presentation layer

   D. Data link layer

5. Which of the following layers is concerned with changing the format of data into something that the layer above it and the layer below it can understand?

   A. Application layer

   B. Presentation layer

   C. Session layer

   D. Transport layer

6. You are teaching a class on the OSI model, and one of the students has a question about how the TCP/IP protocol stack maps to the OSI model. Which of the following best describes the mapping of the TCP/IP layers to the OSI model layers?

   A. TCP/IP is a four-layer stack that maps well to the OSI seven-layer model.

   B. The top two layers of TCP/IP map to the top three layers of the OSI model.

   C. The bottom two layers of the OSI model map directly to the bottom two layers of the TCP/IP protocol stack.

   D. None of the TCP/IP protocol stack layers map directly to any of the OSI model layers.

7. Which of the following transport layer protocols is not concerned with connection establishment, error correction, or flow control?

   A. Internet Protocol (IP)

   B. File Transfer Protocol (FTP)

   C. Transport Control Protocol (TCP)

   D. User Datagram Protocol (UDP)

8. Which layer in the TCP/IP protocol stack is concerned with routing and logical addressing?

   A. Link layer

   B. Internet layer

   C. Application layer

   D. Transport layer

9. As a junior-level network technician, you are asked to build a simple network, connecting just a few hosts together. None of them are dedicated servers, and they are all equal in the network in terms of function. They also must each have their own security requirements, such as usernames and passwords, that aren't shared between them. What is the best network to design that meets these requirements?

   A. Centralized network

   B. N-tier architecture

   C. Client-server architecture

   D. Peer-to-peer network

10. Your server is pushing app notifications to your iOS devices, but you receive some messages back that indicate that the notifications failed for various reasons. Which protocol would have sent those failure notifications back to your server?

   A. SRP

   B. APNs Failure

   C. APNs Return

   D. APNs Feedback

11. You are reviewing your network logs and see a high amount of traffic using TCP port 3389. What kind of connections could account for this traffic?

   A. Remote connections to Windows hosts

   B. Device updates from a BlackBerry Enterprise Server

   C. Remote connections to Linux hosts

   D. Apple push notifications to iOS devices

12. Which two ports are valid for Simple Mail Transfer Protocol (SMTP) and Alternate SMTP? (Choose two.)

   A. TCP port 25

   B. TCP port 110

   C. UDP port 587

   D. TCP port 587

13. You are trying to minimize the number of protocols used on the network that don't communicate securely. Which one of the following protocols should you review for switching to a secure alternative?

   A. SSH

   B. SSMTP

   C. SSL

   D. FTP

14. Which of the following mobile devices would you expect to use TCP port 2195 for its updates?

   A. BlackBerry Curve

   B. Apple iPhone

   C. Android

   D. Windows Phone

15. Which of the following protocols offers some level of encryption and authentication for network traffic that uses it?

   A. FTP

   B. Telnet

   C. SSH

   D. HTTP

16. All of the following are protocols that are routinely used over SSL to provide enhanced security EXCEPT:

   A. HTTP

   B. SMTP

   C. DNS

   D. IMAP

17. Which of the following protocols provides for automatic assignment of IP addressing information to network hosts?

   A. DHCP

   B. DNS

   C. SSH

   D. SMTP

18. DNS uses which of the following port and protocol combinations? (Choose two.)

   A. UDP port 67

   B. TCP port 53

   C. UDP port 68

   D. UDP port 53

19. You are looking at a network log file that shows communications between a client and a web server. You see that two ports were in use during the communications session, port 50123 and port 80. Which port would be the source and which port the destination when the client is sending a request to the web server? (Choose two.)

    A. Destination port: 80

    B. Source port: 50123

    C. Source port: 80

    D. Destination port: 50123

20. You have configured an FTP server to listen on a non-standard port, TCP port 3135. When you try to test the connection using FTP client software from another host, the connection fails. Assuming there's no other problem with the connection to the server, what should you do to complete the configuration so that clients can connect to the new FTP service on the correct port?

    A. Configure the client software to connect using TCP port 3135

    B. Configure the client software to connect using UDP port 3135

    C. Configure the client software to connect using TCP port 21

    D. Reboot the FTP server

Answers

1. **B.** The physical layer is only concerned with electrical impulses.

2. **A, D.** The network layer handles logical (IP) addressing, and the data link layer handles hardware or physical addressing.

3. **C.** This traffic is characteristic of a TCP three-way handshake.

4. **B, C.** The application, presentation, and session layers all use the term data to represent their PDUs. The data link layer uses frames, the network layer uses packets, and the transport layer uses segments for TCP and datagrams for UDP.

5. **B.** The presentation layer is concerned with changing the format of data into something that the layer above it and the layer below it can understand.

6. **A.** TCP/IP is a four-layer stack that maps well to the OSI seven-layer model.

7. **D.** User Datagram Protocol (UDP).

8. **B.** The Internet layer of the TCP/IP protocol stack (which maps to the OSI's network layer) is responsible for routing and logical addressing.

9. **D.** A peer-to-peer network, which is a decentralized model, meets all of the requirements for the network you must design.

10. **D.** APNs Feedback allows the server sending the push notifications to receive notices of failed notification delivery.

11. **A.** Remote connections to Windows hosts.

12. **A, D.** TCP port 25 and TCP port 587.

13. **D.** FTP, like telnet, is unencrypted, so this is not a secure way of passing either data or authentication credentials.

14. **B.** Apple iPhone. TCP port 2195 is used by the Apple Push Notification service (APNs).

15. **C.** SSH offers secure authentication mechanisms and also encrypts the entire communications session used between two hosts.

16. **C.** DNS is not routinely used over SSL to provide enhanced security.

17. **A.** DHCP provides for automatic assignment of IP addressing information to network hosts.

18. **B, D.** DNS uses both TCP and UDP ports 53.

19. **A, B.** Source port: 50123, Destination port: 80. Web traffic (HTTP) uses port 80 by default, and in this case, the client is sending a request, so it is the source sending to a destination.

20. **A.** Configure the client software to connect using TCP port 3135. If the default port is changed on the listening service, the client must be configured to use the new port as its destination connection port.

Network Infrastructure and Technologies

In this chapter, you will

- Discover common and specialized network devices
- Learn about network topologies and architectures
- Learn about network technologies that control and route traffic, and help connect wireless to wired networks

As I mentioned in Chapter 1, CompTIA recommends that you have taken the Network+ certification exam and have at least 18 months experience working with mobile devices, networking, and other related technologies before sitting for the Mobility+ exam. This isn't mandatory, but it is recommended because you should have some networking experience and solid knowledge of how networks work, both for the exam and to understand and effectively work with mobile technologies in your day-to-day job as a mobile technology professional.

The previous chapter discussed very basic networking terms and concepts, such as client, server, and centralized versus decentralized networking. It also opened up the discussion on network protocols found in typical networks and with mobile devices. This chapter continues to refine that discussion by going into more depth on network architecture and topologies, discussing how networks are built from a device perspective, as well as logical and physical design. The CompTIA Mobility+ exam objectives covered during this chapter are 2.1, "Compare and contrast physical and logical infrastructure technologies and protocols," and 2.2, "Explain the technologies used for traversing wireless to wired networks." Coverage of these two objectives will conclude our discussions on the fundamentals of networking, and you will have the information you need to understand the wireless concepts and technologies discussed in Chapters 3 through 5.

Network Devices

Before we delve into architectures and topologies, it's helpful to discuss some more of the building blocks of networks—network devices. Network devices provide functions on the network that help to route traffic to the correct hosts and networks, manage bandwidth, provide security features, and generally ensure that network traffic gets to where it needs to, based upon the requirements of the application, host, protocol, and data.

This section covers network hardware devices, from the very basic hubs to more advanced devices, such as firewalls and remote connection devices. It discusses their function, what protocols or standards may be used in conjunction with them, and at what layer of the OSI model they operate.

Common Network Devices

There are some devices that are fairly common to most networks, whether they are small peer-to-peer networks or large enterprise ones. I've divided the discussion into common devices and specialized ones because you'll see the common ones on almost every network, while you may or may not see the specialized ones unless the network, or a piece of it, has a need for those devices. The common network devices tend to be connection devices, or devices that in some way segment or separate traffic, or even direct it to where it needs to go. Even these common devices offer functions above and beyond simple connections and traffic control, such as security and bandwidth control, but there are devices that also have those functions specifically designed into them, and are largely used for those specific purposes. Still, you should know about all of the various devices covered in the next few pages.

Hubs

A *hub* is probably the most basic network device. Hubs are typically only connection points; that is, they are merely places to plug cables into that offer connections for multiple devices. While there are "smart" hubs that have some lower-level management and control functions, many "dumb" hubs have no such built-in capability and are nothing more than electrical connections. Hubs don't generally offer any control features or security over the devices or traffic that they pass. They will pass all traffic received to other hosts, regardless of what kind it is or which host it is destined for. In some cases, a hub can act as a signal regenerator or amplifier, simply boosting a signal if needed due to cable length or signaling limitations. Hubs that serve as signal amplifiers are called *repeaters*.

You'll find that in many modern networks, hubs aren't used much anymore, especially in larger enterprise-level networks. This is because modern networks require services that hubs don't offer, such as bandwidth and traffic control, and security. At the physical connection level that computers and other hosts connect to each other or to the larger network, devices called *switches* (discussed a bit later) have become more commonplace. You may still occasionally find older hubs connecting hosts on smaller networks in "mom and pop" type businesses or shops, or sometimes in home networks. You also may see them in temporary "LAN party" types of small networks—ones that are just quickly thrown together temporarily for gaming or file sharing. Even those networks are still more likely to use switches, simply because those devices are more commonly found in stores and aren't very expensive. Hubs don't have any protocols that they use, but do have power, wiring, and signaling standards that they adhere to, and they are found at the physical layer of the OSI model. Figure 2-1 shows a typical hub.

Figure 2-1
A small 5-port
Linksys
networking hub

Bridges

Bridges are the next step up from hubs because they offer some very basic traffic flow control features. If a hub can be considered "dumb," then a bridge is only slightly smarter. While a hub will send (forward) all traffic for all hosts out every port, a bridge can spend a few minutes "learning" the network and forward traffic based on a host's MAC address. It learns the network based upon the traffic that flows through it, and starts to figure out what hosts are attached to it.

Most of the time, bridges are used to connect different Ethernet segments together, and filtering traffic based upon a host's MAC address can save on bandwidth and network congestion. Because a bridge uses the hardware (MAC) addresses to manage traffic, it works at layer 2, the data link layer, of the OSI model. Like hubs, you really don't see many bridges used in modern wired networks anymore, but you do still see them sometimes in wireless ones.

Switches

Switches are more commonly found in networking today at the basic host-level than bridges or hubs, and serve as the first-level connection device between hosts and the network. Unlike hubs or bridges, most modern switches are certainly "smart" devices—meaning that they have logic built into them to perform higher-level functions such as traffic management and even some security functions. Like bridges, switches primarily use the MAC address of connected hosts to determine whether to forward or filter traffic. Unlike bridges, however, those decisions aren't only made between segments; they are also made between individual hosts. Because of the different ways hosts communicate with each other, this brings up some interesting discussions on concepts such as full-duplex, collision domains, and broadcast domains.

Typically, communication between two hosts happens as a result of media access and contention. This means that multiple hosts on the network can't all talk at the same time or it would be much like several people in the same room trying to talk to everyone at once, resulting in a lot of confusion at best, and the inability to distinguish words and distinct conversations at worse. With computers, determining who can "talk" and who must "listen" means that the network must have some method of allowing only one host to talk at a time. Let's discuss some concepts that may help to clear up how computers on a network "talk" to each other.

If two people, say Ben and Sam, are having a normal conversation between them, it goes something like this: Sam will speak, while Ben listens. Then Ben speaks, and Sam listens. This goes on and on, back and forth, until the conversation is done. Because both of these fine gentlemen are nice guys, they are going to politely wait until they hear a pause or some other subtle signal to begin speaking, and not interrupt each other during the conversation. This is called half-duplex (or simplex) communication because either party is listening *or* speaking at a given time—not both at the same time. Humans can't process information very effectively by both speaking and listening at the same time. Imagine if both Ben and Sam were trying to speak *and* listen at the same time. There would be confusion and loss of communication. When a party to a conversation can do both at the same time, this is called full-duplex communication.

Computers can have the same issue on a shared medium, which is what many wired and wireless networks use. Imagine four or five computers connected to a hub. Because all the hub is doing is serving as a physical electrical connection, all the hosts are effectively sharing the same cable. This sort of makes communications on a shared media half-duplex. It also means that, because it's a shared media, what one computer sends out can be received by all computers, even if the message is not intended for them. Of course, only the intended receiver will usually process and respond to the message; the other hosts will typically just drop any traffic not intended for them.

If one computer "talks" at the same time as another, this causes confusion and loss of communication. In network speak, this is a *collision*, and both messages sent out by the hosts are effectively garbled and pretty much fail to go across the network to the receiver. The whole of the shared medium with its hosts attached is called *a collision domain*. The Ethernet standard employs methods to detect and avoid collisions on a network, as do many other networking standards, and this can be somewhat effective on a shared medium. The more hosts that are added to a shared media, however, the less effective those methods really are, and collisions are bound to increase, eventually making the network very noisy, with no clear communications.

However, one good way to make collision domains smaller (cutting the number of hosts using a shared medium down) is to use a bridge. Bridges can reduce collision domains, usually by creating two or more separate network segments, and then only forwarding traffic to the segment that has the intended receiving host on it. A much better way, however, is to eliminate the collision domain altogether, using a switch. How is this possible? Let's find out.

Switches enable full-duplex communications from a host to the switch itself. When the switch receives traffic from one host, it can make smart decisions on which receiving host the traffic is intended for on the local network, based, again, on MAC address. So, effectively, all hosts can then talk at once, and the switch itself quickly sends messages directly to and from the sending and receiving hosts, without all hosts "hearing" one another's messages. The switch basically eliminates shared media issues by creating a dedicated link between the host and the switch, eliminating collision domains. This saves bandwidth, eliminates network congestion, and helps make communications clearer.

From a security perspective, this also helps cut down on the risk of "sniffing" traffic by a malicious person because sniffing has to be done on a shared media. Traffic is only forwarded from the sender to the host, except for broadcast traffic (which is discussed shortly in the discussion of routers), so a network sniffer (a device that can intercept and record network traffic) wouldn't receive as much traffic on a switched network.

Keep in mind that while switches work at the data link layer of the OSI model because they primarily use MAC (hardware) addresses to communicate with hosts, there are advanced switches, particularly those that manage Virtual Local Area Networks (VLANs), that can also function at the network layer (layer 3). VLANs and layer 3 functions of a switch are described later in the chapter when discussing some of the more advanced technologies that help manage, secure, and connect networks together. Figure 2-2 provides an example of a Cisco Catalyst 4507 switch.

Routers

Routers are devices that allow traffic to cross between networks, "routing" it to networks that are connected to it or to other routers so traffic can continue on to other remote networks. Remember that a local network is one that is on the same logical network segment. It also may (or may not) be on the same physical segment as well. If it's a local

Figure 2-2 A Cisco Catalyst 4507 switch (Equipment courtesy of Dynetics, Inc.)

network from a logical perspective, that means that the network has the same logical network ID, and has an IP address range that corresponds to that network ID. Don't worry so much about logical IP addressing at this point; IP addressing is discussed in a little more detail later on in the chapter. A remote network, on the other hand, is not on the same logical network as other hosts; therefore, a router is required to send traffic from one network to another. A router can have multiple network interface cards in it that each connects to a different network; even networks connected to the same router, on different interfaces, would be considered "remote" to each other. A remote network also might be located across several other routers from the local network.

In addition to its management of local to remote network traffic, routers also help reduce *broadcast domains*. Remember from the earlier discussion on collision domains that hosts trying to talk to each other at the same time on shared media may cause a collision, garbling all of the network traffic. A broadcast domain is similar, but is the result of hosts that send broadcast messages—to *all* hosts, not just a particular one— over the network. Switches manage collision domains by forwarding on only those messages that are intended for a certain host (called a unicast). They don't, however, limit broadcasts, because broadcasts are directed at every host on the network, so the switch will allow broadcasts in order to make sure the intended receiver (all hosts, in this case) gets the message. Many different protocols use broadcasts a lot, so a network that has a great deal of broadcasting going on suffers from the same congestion issues as a collision domain would, for similar reasons. Switches separate hosts from each other, eliminating collision domains, and, similarly, routers separate networks from each other, so each network is its own broadcast domain. While collision domains can be pretty much eliminated (or at least reduced to the point where only a single host and its switch interface are on the same collision domain), broadcast domains can't really be eliminated; they can only be reduced by creating more physical or logical separate networks. Routers can help reduce broadcast domains by helping to create and manage these separate networks.

Routers work at the network layer (layer 3) of the OSI model, and use various protocols to control and route traffic. IP is obviously needed because of the way logical addressing is used to identify hosts and their networks, but other protocols, such as Router Information Protocol (RIP) and Open Shortest Path First (OSPF), are used specifically for router-to-router communications. Figure 2-3 shows an example of a rack containing three Cisco routers.

 EXAM TIP You aren't required to know details of specific routing protocols for the exam. They are only mentioned here to help explain the function of routers. You will need to know what a router is, what it does, and at what layer of the OSI model it functions, just as you do with the other network devices described in this chapter.

Figure 2-3 A rack containing Cisco 2800, 2900, and 3900 series routers (Equipment courtesy of Dynetics, Inc.)

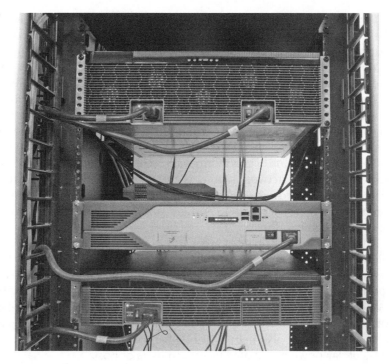

Specialized Network Devices

The next several network devices that will be discussed are those that may or may not be found on networks in either small or large-scale implementations. The hubs, switches, and routers I mentioned are almost always found in networks to some degree, but the devices I'm going to talk about now tend to be more specialized in terms of functions, complexity, and use. These devices include security devices, such as firewalls, Virtual Private Network (VPN) concentrators, and proxies. I'll cover the "textbook" definitions and descriptions of these devices because that's what you'll need for the exam, but, in what I like to call "the real world," you'll also find that there's sometimes no real dividing line between these types of devices. For example, you'll often find multipurpose devices that can be used as firewalls and proxies, and that even have VPN capabilities bundled into them.

Firewall

A firewall is a security device whose purpose is to filter, or inspect, traffic between two hosts or networks. It filters traffic based upon rules that are configured into the device. If traffic matches a rule, then the firewall takes one of several possible actions. It may drop the traffic, effectively blocking it from entering the network; it may reroute the traffic to another device; or it may allow the traffic. It also may alert an administrator to the traffic. A firewall may have multiple network interfaces, each connected to a distinct part of the network. It may "sit" between an external network (such as the Internet) and an internal one, filtering traffic passing between those two networks. Some firewalls may also sit between two internal networks—possibly one for the general population of an organization, and another network that only specific users should access, for example. Multiple firewalls are frequently used in larger networks to partition off protected areas. Firewalls can be either software-based (as a program that runs on top of an existing operating system on a PC or server, for example) or hardware-based, where they are specially configured hardware devices dedicated to only firewall and security functions. Network firewalls, which are really the focus of the discussion here, protect entire network segments and groups of hosts, and are usually enterprise-level dedicated hardware devices. Host-based firewalls usually run as software programs on a PC or a server's OS and are specifically configured to protect that device from undesirable traffic.

A firewall's rules are part of a collection of rules, known as the *ruleset*. Normally, a rule is made up of individual elements that are defined by the administrator. These elements include ports, protocols, IP addresses, times of day, user IDs, domain names, host names, and so on. Combining these elements into a rule helps the administrator specify what kind of traffic is allowed (or not allowed) from a particular source address (or domain) to a particular destination address (or domain). Adding in some of these other elements can also help control whether or not traffic is allowed only at certain times of day or when initiated by certain users or hosts.

One interesting thing about firewall rulesets is that they are usually processed from top to bottom, in order. So if a particular piece of traffic comes into the firewall, it is checked against each rule until a rule "matches" the traffic that is received. The rule will tell the firewall what action to take with the traffic in the event of a match (usually an allow or drop action). If a piece of traffic does not match any of the rules an administrator has defined in the ruleset, the very last rule usually tells the firewall what to do with any unmatching traffic. With most firewalls, this rule is called a *default deny* rule. It is placed at the very bottom of the ruleset, and if the traffic doesn't match any other rule, this last one usually drops any traffic by default, denying any traffic that hasn't been defined. This is to keep any undefined or unwanted traffic from entering the network. Figure 2-4 shows an example of a Cisco Adaptive Security Appliance firewall ruleset.

This discussion of the default deny rule provides a good opportunity to talk about implicit/explicit allow and deny strategies for firewalls. This can be a troublesome concept for networking students. An explicit rule means that the rule is purposely configured (explicitly) for a particular piece of traffic. An explicit allow action for the rule is intended to ensure the traffic is permitted, while an explicit deny is specifically constructed to make sure the traffic never comes through. This is a bit different from implicit rules, which aren't configured with a piece of traffic in mind. Implicit rules may actually not be configured at all; they may be the result of allowing (or denying) any traffic that isn't specifically allowed or denied. Implicit rules can mean that, because a piece of traffic doesn't match any other rule, there's a default action that happens. Understanding the overall security strategy for the network may help you to understand

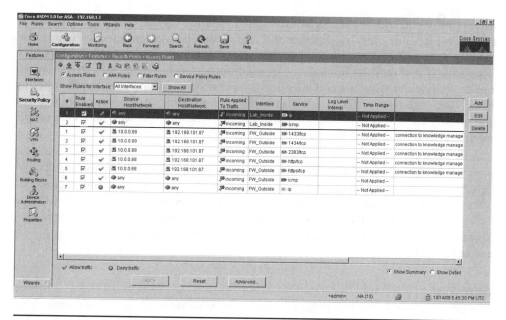

Figure 2-4 An example of a firewall's ruleset.

these types of rules better: Let's say that you've created several rules to block specific ports and protocols to and from particular IP addresses. You may be trying to deny only very specific traffic, while allowing all else to come through. This is a default allow strategy, meaning that unless it's one of those key pieces of traffic you don't want in the network, you're going to allow it. The explicit deny rules are in place to keep the undesired traffic out, but there are implicit rules that allow everything else in, and there also might be a default allow rule at the bottom of the ruleset to accomplish that very thing. It doesn't necessarily have to be there, and by default, if you don't create a rule filtering the traffic, it may be allowed in.

A default deny strategy, on the other hand, has the opposite effect. You would create several rules explicitly allowing only certain types of traffic, under certain conditions. Anything that doesn't have a rule allowing it into the network is denied by default. This is usually the more secure configuration, but it requires some careful planning and occasional adjustment because it can result in a loss of functionality for applications and traffic that weren't included in the ruleset.

Although we tend to think of firewalls as primarily filtering traffic inbound (coming into, or *ingress* traffic) to a network, understand that firewalls work for both inbound and outbound (exiting, or as it's sometimes called, *egress*) traffic. You likely will filter ingress traffic, but you may also have different types of traffic you don't want to leave the network as well. Rules can be set up for either ingress or egress filtering.

Also keep in mind that different firewalls and security devices filter traffic to varying degrees. There are simple firewalls that can only filter based upon a few elements, such as port, protocol, and IP address (source and destination), while other, more advanced ones can filter not only based upon the rule elements I've described here, but also can do deep-traffic inspection and look at the individual pieces of data in traffic to determine if it should be filtered or not. Additionally, other devices, such as routers, can do rudimentary filtering and are often used to do some of the initial traffic filtering before the traffic even gets to a firewall, eliminating some of the obvious stuff that you already know you don't want to allow into the network. This can help balance the workload with the firewall on the network.

 EXAM TIP Watch for questions on the exam that may relate to firewall ruleset troubleshooting. Remember that rules are processed from top to bottom, and when traffic matches a rule, the processing stops. Additional rules are not examined once a match occurs. Rule order is very important; some issues can be caused by a default-deny rule at the top of the ruleset, or by not having one at the bottom.

Gateway

The term *gateway* has a few possible meanings in networking, so there's sometimes some confusion when you hear the term. First and foremost, a gateway is a device (or software program running on a server) that serves as a network translator. A gateway could, for example, translate protocols between two networks that don't use the same protocol suite. Back in the "olden" days of networking, before TCP/IP became pretty much the common protocol suite for network communications, it wasn't unusual to see networks running their own proprietary protocol suite. An Apple network might have run a protocol set known as AppleTalk. Novell networks ran one of several versions of the NetWare network protocol suite, and included protocols such as Internetwork Packet Exchange/Sequenced Packet Exchange (IPX/SPX). While these networks communicated just fine within their own boundaries, talking to another network required a translator, of sorts, that could understand both sets of protocols. If a NetWare network wanted to talk to a TCP/IP network, a gateway served that purpose. This was usually a server that had both protocol stacks installed on it, along with some specialized translation software. Gateways don't only translate network protocols; they can also translate between applications or even between different types of networks, such as between an Ethernet network and a Token Ring network.

You may also hear the term "gateway" when people refer to a *default gateway*, which is popular in networking. A default gateway (also sometimes called the gateway of last resort) is a node, usually a router, that a host sends traffic to when it can't find its destination host on its own logical local network. If a sending host is trying to reach a particular destination host, the first place it tries to find it (usually through broadcasting) is on its own local network. If the destination host doesn't respond, the sending host sends its traffic to the default gateway, in hopes that the gateway (again, the router) will know how to reach it. The router may have the destination network in its routing tables and know how to send to the destination host. If it doesn't, it sends the traffic to its own default gateway (another router), and so on until the traffic reaches the destination. Used in this sense, a gateway is still a translator of sorts, but it translates network routes, so to speak. Most hosts are configured with a default gateway; if they are not, it means that they'll only be able to communicate on their own local networks.

TIP Checking the default gateway is actually a good troubleshooting step to take in the event that a host can't connect to a remote network (such as the Internet). Check to see if a default gateway has been configured for the host. If it hasn't (or if the wrong one is entered), the host won't be able to communicate past its own local network. Entering the IP address of the router may correct the problem.

Proxies

A proxy is a device that makes requests on behalf of other hosts. A proxy is usually a server, but it can also be integrated into a multipurpose security device. Firewalls often have proxy services built into them. You normally see proxies in the form of web proxies, which users contact to make web requests. A proxy server will make a request to the web resource or Internet resource on behalf of the host. This is useful on a network because it can help secure internal hosts. External Internet hosts don't see the request as coming from an individual computer, rather from a network. Another useful function that a proxy can fulfill is preserving bandwidth. It does this by caching frequently requested web resources, such as web pages, and storing them locally so that internal users can access them quickly.

VPN Concentrators

The next device being discussed is a Virtual Private Network (VPN) concentrator. A VPN is a way of connecting remotely to a network as if you were working on the inside of the network. A VPN concentrator allows a remote user to establish a connection from an untrusted network, such as the Internet, authenticate to the internal network, and encrypt traffic from their client through the VPN and into the network. There are two basic types of VPNs, a site-to-site VPN and a client VPN. A site-to-site VPN basically connects remote sites or branch offices, for example, to a larger corporate network. In this scenario, you would have a VPN concentrator at each site. Users at the remote site would go through their VPN concentrator, which would send traffic to the corporate VPN concentrator.

A client VPN is set up for a roaming user who uses VPN client software on his or her desktop or laptop to connect from remote locations to the corporate network. This could be a hotel, or from home—wherever there is an Internet connection. As I mentioned before, a VPN concentrator can also be a feature of a multipurpose security device, such as a firewall. It can also be a standalone box or device, configured on the server. Figure 2-5 illustrates how VPNs work.

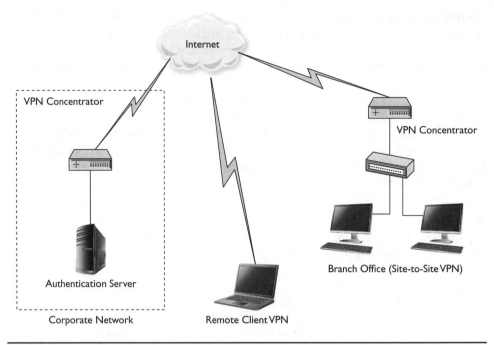

Figure 2-5 Examples of VPNs

Wireless LAN Controller

Although wireless networks are discussed in depth in Chapter 5, I want to mention the network device known as a wireless LAN controller. A wireless LAN controller essentially manages a wireless infrastructure network. It provides a method for wireless clients to connect into a larger wired enterprise network. It can have authentication features, as well as features that ensure that a wireless client meets certain criteria before it is allowed to join the wired network. Wireless LAN controllers can also have firewall and other security functions, and usually connect to some type of authentication server or other security device in the enterprise.

Lightweight and Autonomous Access Points

A lightweight access point (AP) is a device that is part of a larger enterprise-level wireless network. It connects to a wireless LAN controller, via wireless technologies. While the wireless LAN controller is the central manager of the wireless network, the lightweight AP is merely a connection point for wireless clients. It's considered lightweight because it has no real management or control functions. It relies on the wireless LAN controller for these functions, which include security, configuration, and traffic management. Most of the time, the lightweight APs require very little user configuration at the device itself; they are configured via the wireless access controller through a protocol called Lightweight AP Protocol (LWAPP). Lightweight APs typically can't function when disconnected from the wireless LAN controller.

An autonomous access point, on the other hand, is a wireless access point that *can* be managed on its own. Unlike a lightweight access point, it does not require a wireless LAN controller to function. Most access points that you find in homes and small businesses are actually autonomous access points. However, some autonomous APs can be configured to make use of a wireless LAN controller for management and configuration functions. You learn more about APs in Chapter 5 when wireless LAN technologies are discussed.

EXAM TIP   There are several networking devices you'll need to know and understand for the exam. Make sure you're aware of the different roles and function of the specialized network devices: firewalls, gateways, proxies, VPN concentrators, lightweight and autonomous access points, and wireless LAN controllers.

Network Topologies

A network topology is essentially the design of the network. It could be a physical or a logical topology. A physical topology describes how the network is laid out from an actual physical perspective—meaning how the cabling (if it's a wired medium) is run and connected to various devices, in a particular design pattern. A logical topology basically shows how data flows over the network, between devices and over media, regardless of how it's physic ally laid out. You'll find that, for various reasons, a network's logical topology is often somewhat different than its physical one.

EXAM TIP   Watch for topology questions that may distinguish between logical and physical topologies. Remember that a logical topology is how the data and traffic actually flow through a network, regardless of how it's physically laid out. The physical topology is how it is physically connected together. Sometimes these two happen to be laid out the same way, but this is often not the case.

Traditional Topologies

Traditional topologies were originally developed for wired networks, but some of those same concepts can apply to wireless ones as well. Most of the designs for networks revolve around traditional bus, star, ring, and mesh networks discussed in the next few sections. Wireless networks, however, are slightly different. While the physical and logical topologies may resemble some of the traditional ones discussed here, the fact that there is no need for physical connection doesn't change how they may be connected, logically or physically. The real difference is the use of wireless versus wired media.

One concept that is used in both wireless and wired topologies, although a bit differently, is the concept of point-to-point and point-to-multipoint networks. Point-to-point really means, in the strictest sense, one host or station communicating directly with another distinct host. The hosts could be two computers, a computer and a wireless access point, two infrastructure devices, and so on. The connection could be physical, in the case of an ad-hoc wireless network or two computers connected via an Ethernet cable, or it could be logical, meaning that there's no defined physical path between the two hosts, but a logical persistent connection may have been set up between them over multiple possible physical paths. Figure 2-6 shows a simple point-to-point connection.

Point-to-multipoint means that a host can simultaneously communicate with multiple hosts. You typically see point-to-multipoint implementations in both wired and wireless networks. Wireless networks are the subject of Chapter 5, but for now you should know that a point-to-multipoint type of connection allows a host to connect to multiple other nodes on the network simultaneously without having to travel through a central node first. Figure 2-7 shows conceptually how point-to-multipoint works.

Bus

A bus network is the simplest of the network topologies. In its basic form, it essentially connects two hosts together to a shared media. In a wired network, this is typically a single cable. Now, obviously, you don't see it implemented that way much, because most networks have more than two hosts. What you might see are several hosts connected to the same cable, sharing the same medium, possibly though a hub. In the early days of networking, a lot of networks actually had connectors that could connect the host directly to the cable. This would create both a logical and physical bus. When hubs came into use, hosts connected directly into the hubs so you could have several hosts connected in one hub. Physically, this resembled a star, but logically it was still a bus network because the hub only served as an electrical connection point, and you still had multiple hosts sharing one medium. Typically, a bus is a shared medium network, and most of the time will cover one single collision domain. Figure 2-8 shows what a simple bus network might look like.

 TIP In a shared media network, such as Ethernet, collisions occur when two hosts transmit at once. To avoid this, hosts on a shared Ethernet segment will "listen" to see if anyone is transmitting. If they don't detect another transmission, they will go ahead and attempt to transmit. In the event of a collision, the two hosts will stop transmitting and wait a random period of time before they attempt to transmit again, in an effort to avoid another collision.

Figure 2-6
Point-to-point

Figure 2-7

Point-to-multi-point

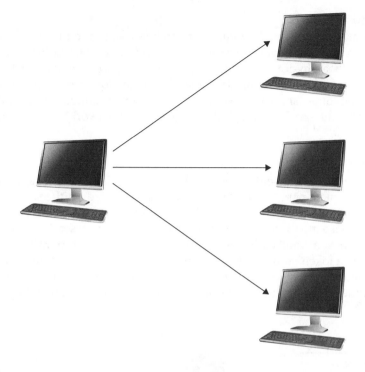

Figure 2-8

A bus network

Star

A star type of network physically resembles a star in terms of its multiple connections into a hub or switch from several hosts. As I mentioned earlier, if the hub is in use, it's typically a physical star and a logical bus configuration. As switches replaced hubs in most networks, this changed the logical configuration as well, simply because the medium was no longer shared and collision domains were almost eliminated. With the switch, you would have both a logical and physical star network. Figure 2-9 shows a typical star network.

Ring

A ring network is seldom seen in local area networks these days. How a ring worked, for the most part, was that each host was connected on the same physical cable, via a special connector. There was typically a host connected on the cable before it, and one connected after it in a series, causing the network to resemble a physical ring. A device known as a Multistation Access Unit, or MAU, was responsible for sending a signal called a token around the ring so that each host in turn would get the token and use it to access the network. If a station had the token at the moment, it could communicate on the network without fear of collisions. Once the host was done communicating, it

Figure 2-9
A simple star
network

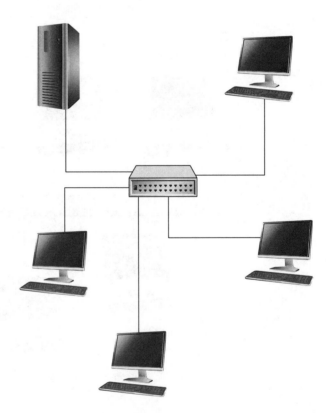

released the token, and the token continued on its way around the ring until another host needed to use it to communicate. As I mentioned, most local area networks don't use these Token Ring configurations any longer. Where you might see a ring is in a larger backbone type of configuration, where there may be large rings of fiber in a wide area network that carry the backbone signal between local network distribution points. When used as a backbone network, you might see a dual ring for redundancy purposes, in case one ring fails at some point in the network. Figure 2-10 shows one way to design a simple ring configuration, but there are other variations that involve multiple MAUs connected to each other and the hosts for redundancy purposes.

TIP A Token Ring network is not considered a shared media network, simply because the token transmitted by the MAU determines which host gets to talk. Only the computer possessing the token at the time can transmit, so there are no collisions on a Token Ring network. This is unlike an Ethernet network, where a host must "listen" to see if anyone is transmitting before they talk, and collisions can occur.

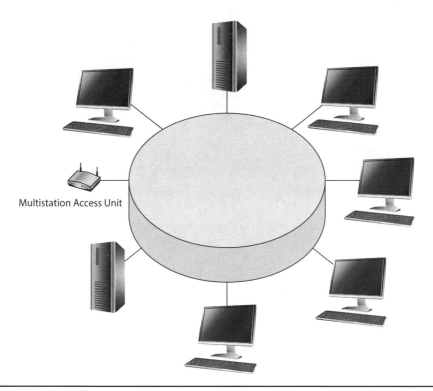

Multistation Access Unit

Figure 2-10 A simple ring topology

Mesh

A mesh network is one that has multiple paths of communication between hosts. These are redundant paths, so that if any one path fails, a host can still communicate with another host on the network through an alternate path. Few wired networks use mesh configurations, however, because of the additional expense, more complex cable runs, and no real need for the level of redundancy you get with mesh networks. Many wireless networks, however, use mesh topologies, allowing them to communicate directly with other wireless devices as well as wireless access points. Figure 2-11 shows a simple mesh network.

Ad-hoc

The term *ad-hoc* is used to describe networks that aren't persistent or that are set up on-the-fly to share files, play games, and so forth. They typically have no centralized management or security involved, as each node in an ad-hoc network is responsible for its own management and security. Ad-hoc networks are typically peer-to-peer in nature and are decentralized. Chapter 5 explores wireless ad-hoc networks in depth and describes how they are set up and used. It also discusses wireless networks that are referred to as "infrastructure mode networks," which are centrally managed and secured.

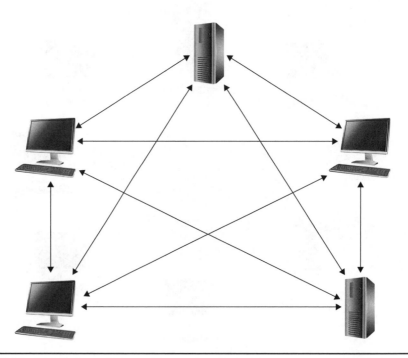

Figure 2-11 A mesh network

EXAM TIP   Make sure you are familiar with the different topologies described previously for the exam, particularly mesh, point-to-point, and point-to-multipoint.

Topology Design

Network topologies are designed based upon several factors. A topology may be based upon the size of the network, the number of hosts on the network, the physical environment, and other considerations. There is likely no one "perfect" topology that fits every situation, so you'll often find several different topologies used in various parts of a large network, and maybe even a couple of different ones in smaller networks. Let's discuss the different factors involved in topology design.

Size

Which topology you use may depend upon size, from a couple of perspectives. First, there's the physical size of the space you have to work with. If your network is confined to a couple of adjoining rooms, you may decide to go with a topology that lends itself to smaller spaces. You may go with a small wireless peer-to-peer type of network that uses a point-to-point type of design, or even a bus design, in the case of a wired network. If it's a larger physical area, you'll definitely have design considerations because of the distance limitations of a cable or wireless signal, and this usually requires additional equipment. More equipment may change the logical and physical topology.

Size can also mean scale. Scale refers to considerations such as the number of network nodes or hosts, services that run over the network, and maybe even the amount of bandwidth used or data sent across the network. Obviously, smaller scale networks, having maybe 10–20 hosts, may be small peer-to-peer networks, or even a star-topology networks that have only a few hosts and switches. Larger networks, however, almost always use more complex topologies such as star or mesh. Ring networks are typically only found in backbone implementations these days, and bus networks are typically only found in very small networks. When designing a topology for a network, scale should be considered, as well as future growth, because redesigning a topology as you are adding more hosts and nodes can be very difficult.

Centralization Model

Chapter 1 discussed decentralized versus centralized network models. Designing a network topology depends to a certain degree on whether the network will be centrally managed or decentralized. This is because the more complex the network, the more complex a topology, and the greater degree of centralized management is necessary. Obviously, in a small decentralized network, such as a peer-to-peer or an ad-hoc network, a topology such as a bus or a small star would be appropriate. However, an enterprise-level network is likely to be complex, spread out to different buildings, and definitely be centrally managed, so a more complex topology will be required. This could be multiple star networks connected together, or some sort of mesh topology, or even a hybrid type of network that uses multiple topologies.

Types of Equipment

The topology used in a network also depends upon the type of equipment available to the network architect. If you only have hubs available, then likely it's going to be some sort of logical bus type of network. Of course this will also limit the scalability of the network, so if the network is expected to sustain larger numbers of hosts or grow significantly, then additional, more complex network equipment will have to be purchased. If the budget allows more sophisticated equipment, such as switches and routers, then obviously either star, mesh, or hybrid topologies will be used. Wireless networks also involve some additional considerations, depending upon whether they are working in infrastructure mode or in ad-hoc mode, and whether they connect to a more complex wired enterprise-level network or not.

Physical Environment

The physical environment has a lot to do with network topology design. The physical environment could include space limitations, obstacles, and walls. There may be multiple floors in the building that limit wireless signals or cause unique considerations in running cable, or multiple buildings in a campus area that may impose distance limitations. You also may have building codes and other factors to contend with. There could be restrictions in a facility on the type of wiring that can be used due to fire code or even electrical codes. There could also be limitations imposed by the owner of the property on businesses that lease space within the facility. In any case, the topology will have to be worked around these physical limitations. In some facilities, there may be limited space for network infrastructure equipment. There may not be a server room to install a great deal of infrastructure equipment in, or the network equipment may be limited to a small closet, which would be prone to excessive heat, and possibly even physical security issues.

One thing to keep in mind about these design limitations is that frequently you'll find yourself using multiple types of topologies in the same network. It's not unusual to see an enterprise-level network that has star, mesh, and even bus topologies in different segments. The central network may use a star topology, while a branch office that's connected via a VPN connection may use a simple bus type of topology. Wireless networks, of course, will typically use a mesh topology.

Network Architectures

Network devices are the building blocks of networks, and topologies are how you connect these devices together to control the logical and physical flow of traffic. The network architecture is how you design and arrange networks that make the most sense from a functional and security point of view. Anyone can throw together a few network devices and computers and make a network, but taking the time and consideration to really look at the network architecture and making decisions about how the network is laid out require some forethought and skill. A typical topology introduces devices such as hubs, switches, and routers into a network to control traffic between hosts and other networks. A network architecture also adds the specialized network devices that I discussed earlier, such as firewalls, VPN concentrators, and so on. This section looks at different network architectures, primarily from a security perspective, but I will also discuss these network architectures from a functional point of view.

Bastion Host

One of the simplest network architectures from a security perspective is one that uses a bastion host. A bastion host is simply a computer, possibly a server, which sits between two networks, separating them. This host usually has multiple network interface cards, with each card connected to a different network. One network may be the Internet, or other public network, while another network may be the private network. It may have firewall software built in, as well as rules configured to filter traffic between the networks. It is also usually "hardened" to withstand any attacks. You'll usually only see a bastion host on very small networks, as its capacity for traffic and functionality may not be very scalable. It also may be part of the larger security architecture. Figure 2-12 depicts a simple bastion host setup.

Intranet

An intranet is the network of computers that belongs to the organization, typically behind a firewall or other security device. It may have internal web servers, file servers, and other network devices and servers that fill requests from internal clients and users. Typically, the intranet is kept separate from external or public networks to minimize unauthorized access and data loss. Only trusted users and clients should be allowed to connect to the intranet from both inside the network and from the outside (typically over a VPN connection).

Internet

Of course, defining the Internet should be an easy thing to do. From our perspective, however, the Internet is the public untrusted network that is external to our private, secure network. Because the Internet is untrusted, there must be a way to filter traffic coming into the private network from the Internet, as well as control traffic going from your private network to the Internet. Security architectures help us to do this and are made up of the specialized security devices discussed earlier, such as firewalls, proxies, VPN concentrators, and so on. A bastion host, as discussed, is one such way to protect a small internal network from the public Internet, but this should only be used for very small networks. Larger networks require larger solutions, such as screening networks, or Demilitarized Zones, which are discussed next.

Network A   Bastion Host   Network B

Figure 2-12   A bastion host separating two networks

Demilitarized Zone

The Demilitarized Zone, or DMZ, as it is usually called, is a special network set up as a buffer between the public or untrusted network (such as the Internet), and the private internal network. A DMZ usually is put in place because there are some assets, such as public web servers, that may be required to be accessible by untrusted clients and users on the Internet. In allowing this access, you also must ensure that these external users can't get to the internal network. So the solution to this is to set up a DMZ that allows some traffic into very specific resources and denies traffic to others. This is accomplished by setting up specific security devices in a certain way to filter and direct traffic. Additionally, networks such as extranets may be set up in a DMZ as well, allowing only trusted external users to access them, and still keeping external users out of the internal private network. While there are several different ways that you can design such a perimeter network, a common solution is shown in Figure 2-13.

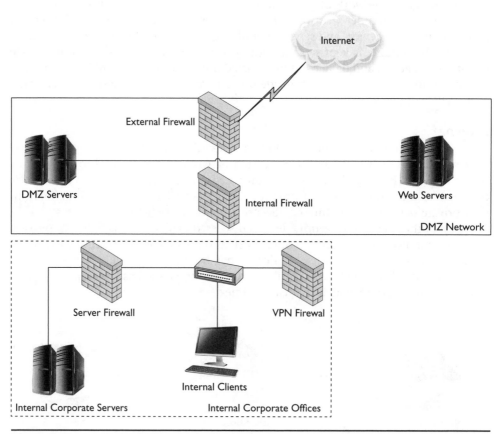

Figure 2-13 A common DMZ architecture

Extranet

An extranet is a specially constructed network that is shared between a private organization and its partners, vendors, or customers. It may have data assets on it that external users, such as business partners, need to access. It is typically physically and logically separated from the internal network. It may be protected by its own security devices, such as firewalls and so on. It may be accessible only through a specific secure connection such as its own VPN connection. Typically, extranets require stronger authentication and encryption mechanisms to protect data from unauthorized access. Figure 2-14 shows how an extranet might be implemented in a corporate network infrastructure.

 EXAM TIP A DMZ is the preferred security architecture for a larger enterprise-level network, and even for some small- to medium-sized networks. Bastion host setups are really only for very small networks, and are really only used for basic traffic filtering and separation.

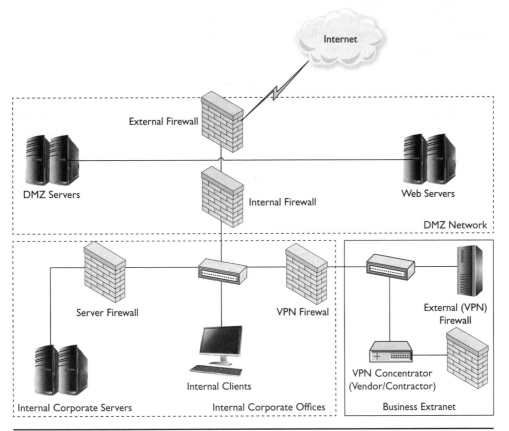

Figure 2-14 An extranet set up off of a corporate network's perimeter

Networking Technologies

Now that I've discussed devices as the building blocks of networks, topologies as the basic design of traffic flow in the network, and architectures used for optimum placement of these devices for security and traffic management purposes, it's time to discuss some of the technologies that you use in networks to help you manage them. Obviously, this book can't cover every single network technology out there, but I'm going to touch on the most important ones, particularly the ones that you are expected to know for the exam. This section covers topics such as IP addressing, subnetting, backhauling, bandwidth management, and quality of service. I am not going to make you an expert on these technologies, but I will cover some of the basic concepts involved.

IP Addressing

As I mentioned both in this chapter and the previous one, computers use logical addressing to communicate with one another on both local and remote networks. Logical addresses are assigned by the administrator based on a predetermined numbering scheme. Contrast this from physical or hardware addresses (also called MAC addresses), which are physically encoded into the network interface card. Logical addresses can be changed at will by the administrator, assuming it doesn't cause any issues on the network.

IP addressing uses the Internet Protocol (IP) versions 4 (IPv4) and 6 (IPv6). For the purposes of this chapter, I'm going to be discussing those IP addresses associated with version 4. IPv4 addresses are 32-bit binary numbers, but to make it easier for humans to read and interpret them, you usually see them in decimal form. They typically come in the form of four numbers, from 0 through 255, separated by a dot. This is commonly called the quad dotted decimal notation, and each "quad" is called an *octet*. An IPv4 address looks something like this: 192.168.13.12.

While it may not be obvious at first, this IP address tells you a couple of things. First, it can tell you the network number or network ID that the host is connected to. It can also tell you the individual number assigned to the host on the network, which is used to uniquely identify the host. Compare this to how telephone numbers work, with an area code, an exchange, and an individual phone number. For example, in the telephone number (628) 555-0129, you know that 628 is the area code and should correspond to a geographical area (in this case, the San Francisco area, starting in February of 2015). 555 is the exchange (in this case a fictional one), and normally is one of several assigned to that area code. The last four digits (0129) are the actual number assigned to the phone line itself, which rings on someone's desk. IP addresses are used similarly, with part of the number being a network number (or network ID), and part of it being the unique host ID assigned to the host or device itself. Unlike a telephone number, however, just by looking at the IP address, you may not be able to discern this information without knowing a little bit about the network itself. That's where something called the subnet mask can help. A subnet mask, used in conjunction with an IP address, identifies which parts of the IP address are the network ID and which is the host address. Subnet masks will be covered in more detail in the next few paragraphs.

When IP addressing was originally created, IP addresses were divided into default classes. There were three primary classes: class A, class B, and class C. These were typically used to divide up networks and hosts. There are also two other classes, classes D and E. Class D addresses are used for multicast groups, and class E addresses are used for experimental purposes. I'm not going to worry so much about these two classes of addresses for my discussion here. Class A, B, and C addresses are assigned different address ranges by default. These ranges make up the first octet, or set of numbers, in the IP address. By knowing the ranges assigned to each class, you can easily determine what class an IP address falls into. The class A range is assigned 0 through127 in the first octet, so a class A address might look something like this: 12.47.121.42. The class B range is assigned numbers from 128 through 191 in the first octet, and class C has the numbers 192 through 223 in the first octet.

Just because the first octet contains the numbers associated with each class doesn't necessarily mean that each class has that number of networks. In fact, with each class, you get a certain maximum number of both networks and hosts. This is different for each class. In class A networks, you have 128 possible networks, which isn't very many. However, you have 16,777,216 theoretical host addresses available per network, which is quite a few if you have an extremely large network. In the class B network, you have 16,384 possible network IDs, with 65,536 possible hosts per network. Why the difference? Well, because each class of network uses a different number of binary digits (bits) to distinguish between network IDs and host IDs. A class A network uses only the first octet, which is eight bits. This gives the class A network fewer network IDs, but many possible host IDs. The class B network, on the other hand, trades host IDs for more network IDs by using more bits beyond the first octet as network bits. The class B network uses a full two octets (16 bits) dedicated to network IDs. And, of course, the class C network trades even more host bits for network bits, resulting in the first three octets (24 bits) dedicated to network IDs, and the last octet dedicated to host IDs. This gives a class C range a whopping 2,097,152 possible networks, with each network getting only a possible 256 hosts each (0–255 in the last octet). Of course, these are theoretical numbers, as some network and host IDs are reserved. For example, in a given host range, zero is usually reserved for the network ID, and 255 as the broadcast address, so a more practical number of hosts might be 254. So how can you determine which are network bits and which are host bits? The answer is the subnet mask.

A subnet mask, when paired with an IP address, is that extra bit of information you need to figure out which are networks and which are hosts. Like the IP address, the subnet mask also comes in a quad dotted decimal format, but the numbers used are from a finite set of numbers. The range for subnet masks is also 0 through 255, like IP addresses, but there are only certain numbers that are used, based upon the way binary digits work. Let's discuss the default subnet masks first, and you'll begin to understand what I mean.

A default subnet mask is one that goes with an address class, and, if there are no changes to the default way the network IP addressing scheme is divided out, the default mask is used and is pretty simple to figure out. The subnet mask tells you which octets cover the network ID range, and which can be used to assign to hosts as addresses. A class A subnet mask, for example, is eight bits in length and effectively tells you that the first octet is the network range, and all of the other octets are possible host IDs. Here's an example that may help: Let's say you have an IP address of 12.63.41.7. The default subnet mask for a class A address is 255.0.0.0. The first octet (255) of the mask tells you that the first octet of the IP address (12) is the network ID, and all other bits in the address (63.41.7) are used for host IDs. This gives you one network with the potential for 16,777,216 hosts. That's quite a bit. Usually, only larger organizations require that many hosts, but the way IP address ranges were assigned in the early days of the Internet wasn't very efficient.

Table 2-1 summarizes the different classes of IP addresses, along with the number of networks, hosts, and bit distribution associated with each. Note that for class D and E address ranges, some items, such as network and host bits, are not defined because those ranges don't work in the same way and are not assigned the same way as class A, B, and C addresses.

Subnetting

We mentioned previously that the IP address tells you two important things: which network ID the host uses, and what the unique host ID is. Of course, just by looking at the address and without more information, you may not be able to tell much, unless it's assigned in the default configuration for that range. To be honest, that doesn't happen much anymore for a few reasons. First, in the early days of the Internet, these address ranges weren't assigned very efficiently to organizations. There weren't as many networks and hosts out there on the Internet as there are today, and no one expected the technological explosion that has become the Internet as we know it. It wasn't unusual to see even smaller organizations get assigned a class A address, even though they may have had only a few hosts on their network. Because of the way the address ranges were assigned on a first-come, first serve basis early on, without much thought for

Class	Network Range (1st Octet) and Start/Stop Addresses	Network Bits	Host Bits	Possible Networks	Possible Hosts	Default Subnet Mask
A	0–127 (0.0.0.0– 127.255.255.255)	8	24	128	16,777,216	255.0.0.0
B	128–191 (128.0.0.0– 191.255.255.255)	16	16	16,384	65,536	255.255.0.0
C	192–223 (192.0.0.0– 223.255.255.255)	24	8	2,097,152	256	255.255.255.0
D (multicast)	224–239 (224.0.0.0– 239.255.255.255)	N/A	N/A	N/A	N/A	N/A
E (experimental /reserved)	240–255 (240.0.0.0– 255.255.255.255)	N/A	N/A	N/A	N/A	N/A

Table 2-1 IP Address Classes and Ranges

growth, they were used very inefficiently. This inefficient method of assigning addresses contributed, in addition to the massive amounts of hosts that are now on today's Internet, to the IPv4 address space being virtually exhausted by February of 2012. While IPv6 was invented in 1995 to help solve this problem, it has been adopted very slowly on the Internet, and still has not been fully implemented as of this writing.

In an effort to more efficiently use the limited IP address space, subnetting came about in order to help better manage the IP address classes. Subnetting essentially allows a network administrator to break the rules of the default address classes by using more or less bits as needed for additional networks or hosts. Subnetting typically means that the default subnet masks will be changed, effectively changing the way network and host IDs are interpreted in an IP address. Here's an example that may help you understand this a bit better: Suppose you have a class A address assigned to your organization. Let's say it's the 13.0.0.0 address range. You know that you now have one network, but over 16 million hosts possible. Well, you actually need a few networks, and not that many hosts. So how can you solve that problem? By subnetting the default network and host bits. Instead of using the 255.0.0.0, the default mask, let's take some bits from the next octet, which are usually reserved for host bits, and use them as network bits. You could subnet it down to a mask that uses 255.192.0.0 (2 extra bits, for a 10-bit mask), which gives you four networks, with 4,194,302 hosts per network. If that's not enough, you could increase the network bits even more, to say the full second octet, so you have a 16-bit mask for the class A address. That gives you 256 networks now, with 65,534 theoretical host IDs.

Of course, you can also do this with class B and C network addresses as well, but the smaller the address space you have to work with, the less effective this is. Think of subnetting a class C network down, which has only 256 possible hosts to work with anyway! This process of moving and borrowing bits also works in reverse (taking away network bits to make them host bits), and this is called *supernetting*. You might do this if, for example, you have several small class C networks that you'd like to make into one larger network ID. The address ranges, of course, would have to be contiguous for this to work.

Although this section isn't meant to make you a subnetting wizard (there are entire chapters of books written for that!), you can probably already see that this technique can help you more efficiently manage your assigned address range. There are other subnet masks you may use as well, but subnetting requires some thoughtful planning about your current network capacity and how it will scale in the future. It also requires a bit more in-depth knowledge about IP addressing and how binary numbers work than covered here. (I've avoided talking about it in this chapter, due to space and scope constraints.)

EXAM TIP　You likely will not have to do any serious subnetting for the exam; that's really more of an advanced topic you'll see on the Network+ exam. Be prepared, however, to visually identify default subnet masks, and possibly incorrect ones.

Special IP Address Ranges

No conversation about the different IP address ranges would be complete without talking about a few of the special ranges that you may encounter when working with networks. The first special range I'll discuss is one of the class A ranges. This is the 127 network, and although some books may not include it in the class A range, doing the math would actually show you that it is, in fact, part of the class A address space. The 127 network is completely reserved for the loopback functionality on a host and will not work as a normal IP address. The loopback basically allows an administrator to attempt to communicate with the host's own TCP/IP stack installed on the computer. This is used to troubleshoot network connectivity issues, and can tell you if your own host's TCP/IP stack is working correctly. You may have seen references to the loopback address of 127.0.0.1 when viewing the network configuration on host. This is the most commonly seen of the loopback addresses, although technically, the entire 127 range can be used for loopback testing.

Another special IP address range is actually three different ranges distributed across class A, B, and C ranges. These are known as the private IP address ranges, and are used by organizations as their internal address space. There are advantages to using these private IP address ranges because they are not routable to the Internet. This means that the host can't use a private IP address to communicate to the Internet without going through a network device that translates a private IP address to a valid public IP address. If an organization has only a small class C address assigned to it, for example, it may use a private IP address range behind its router or firewall that could support hundreds or thousands of hosts, all with a single public small class C address. The private IP address ranges have default subnet masks assigned to them, but they are not necessarily the same ones that the public ones are assigned. They can also be subnetted as the organization desires. The private IP address ranges are 10.0.0.0 through 10.255.255.255 (class A, with a standard 8-bit mask), 172.16.0.0 through 172.31.255.255 (class B, with a 12-bit mask versus the standard 16-bit mask), and 192.168.0.0 through 192.168.255.255 (class C, with a 16-bit mask instead of the standard 24-bit mask). You will often see private IP address ranges used in conjunction with Network Address Translation (NAT), which is covered in the next section.

One other specialized IP address range is the 169.254.0.0 range (with a standard 16-bit class B subnet mask). This range is used by hosts (particularly Microsoft Windows computers) to automatically assign themselves an IP address if one has not been configured for them by an administrator, or if they cannot connect to a DHCP server to automatically get IP addressing information. This process is called Automatic Private IP Addressing (APIPA). Hosts using IP addresses in this range usually default to APIPA if they can't get an IP address through any other means, but still need to be able to talk on a network. Because this address is not routable, they can only communicate with other hosts in the same IP address range. Using APIPA is a good solution if you have a small network that has no dedicated administrator or DHCP server so the clients can automatically configure themselves and communicate with each other.

TIP From a troubleshooting perspective, if you have a host that can't communicate with other hosts on the network, take a look at its IP address configuration. In addition to checking for the correct IP address, subnet mask, and default gateway, another thing to check is the use of an APIPA address by the host. If this is the case, it's possible there's a connection issue between the host and the DHCP server, or an issue with the DHCP server itself.

NAT

Because the days of every single network getting its own large class A or class B address space are over, a method was needed to allow companies to use private IP addresses on the inside, while only requiring a small number of public IP addresses on the outside of the network. Network Address Translation, or NAT, fills this need and helps to conserve IP addresses, as well as helping to secure your internal network. In NAT, your network has only one or two IP addresses that the public Internet can access. The firewall or security device translates a private address space on the internal side of the device to a public one. So your external IP address may look something like this: 162.43.56.12. Your internal IP addresses, however, could be in the private IP address range of 172.16.30.1 through 172.16.50.255. When a client makes requests to the Internet, NAT translates a private IP address to the public one and the request goes out from the public IP address. When the reply comes back, the security device running NAT translates the reply back to the original IP address that requested the data. Because there are probably several requests going on at one time, the security device keeps up with the different requests and their status by using a table in its memory that contains the different IP addresses and requests that are currently going on. The reply coming back from the Internet sees only the public IP address, so the internal IP address space is protected. There are other variations of NAT as well that use pools of public IP addresses and also may translate particular ports for internal IP addresses.

Virtual LANs

Recall from the earlier discussion on switches that they operate at layer 2 of the OSI model. Switches primarily use MAC addresses to manage and communicate with hosts that are attached to them. Also recall that switches help eliminate collisions by providing a single collision domain between the host and the switch. At the layer above, the network layer (layer 3) routers manage logical networks that use IP, and they also reduce broadcast domains. However, routers can be very expensive and are not always required for a smaller- to medium-size network. Over the past several years, newer switches that operate at both layers 2 and 3 of the OSI model have been developed that have characteristics of both switches and routers. One technology that these layer 3 switches, as they're called, implements are virtual LANs (VLANs). The Cisco Catalyst 4507 model that you saw earlier in the chapter is a layer 3 switch.

VLANs are implemented by switches by creating artificial or virtual local area networks that are created in the configuration of the switch itself. Obviously because these Virtual LANs operate at layer 3, switches also have to use IP addresses in addition to MAC addresses. With a typical layer 2 switch, all hosts plugged into the switch are usually on the same LAN. With a layer 3 switch, however, it's possible to separate hosts into different logical or virtual local area networks. Two hosts can be plugged into two switch ports that are side-by-side, yet be on a totally different logical subnetwork. The switch is configured with different virtual LANs and the respective network IDs, and then hosts are assigned, based upon the switch port, to a particular VLAN.

There are several advantages to this setup. First, layer 3 switches are a lot less expensive than most routers. Second, some of the same fundamental functions that a router performs, such as reducing broadcast domains and routing to another network (within the same larger enterprise-level network), can be performed by these layer 3 switches. Those two advantages aside, there are also some good security reasons for implementing VLANs. Segmenting different hosts into different logical LANs can add to the security posture of the network because this allows administrators to separate sensitive traffic from different hosts. The administrator can also isolate certain hosts that have specialized security needs. From a network management and performance perspective, segmenting these hosts could help to conserve bandwidth by eliminating broadcast domains (something that normal layer 2 switches don't do), and would allow the administrator to change IP addressing schemes and configurations fairly easily as the network changes. That may be difficult to do when using hardware routers because you may have to physically add or change network interfaces on the router and change the actual router configuration whenever the IP addressing scheme on the network must change.

One other feature that makes VLANs so attractive in the network is the use of dynamic VLANs. Dynamic VLANs allow a switch port to be configured so that different hosts may be in different VLANs, based upon MAC address, username, and so on. So, if one computer plugs into the port, it may be in VLAN 1. If a different host, say a laptop, plugs into that same switch port, it is considered a member of VLAN 2 based upon its MAC address. Of course, this gives the administrator a great deal of flexibility over managing hosts on the switch, but it requires advanced configuration and planning.

So, there are several advantages to using layer 3 switches and VLANs. Keep in mind that there are still times when a router is required: usually to route traffic outside of the corporate network to the Internet, or to route internal network traffic that doesn't use VLANs. Figure 2-15 sums up how a VLAN might look for a group of hosts attached to a layer 3 switch.

 TIP Layer 3 switches perform routing only between the VLANs created on them. This is called inter-VLAN routing, and must be set up on the switch. They can't route between normal LANs or to remote networks.

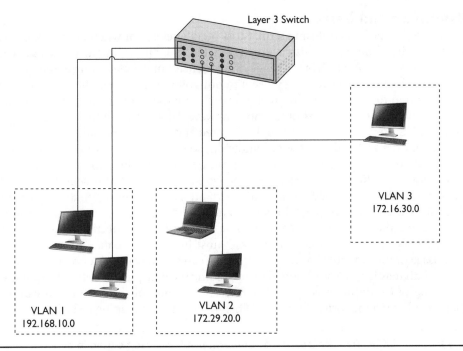

Figure 2-15 A simple VLAN

Network Management

This brings us to the topic of managing network communications traffic and a few of the techniques used to control and manage it between and within networks. There are several ways to manage network traffic, including backhauling, traffic shaping, bandwidth management, and quality of service. I discuss each of these in the text that follows.

Backhauling

Backhauling traffic refers to managing the traffic between major access or distribution points. This typically means managing it over the backbone. The *backbone* is usually the larger capacity pipe that carries traffic between remote networks. Backbones are usually provided by Internet service providers or higher-tier providers, such as communications companies. Backbone traffic is usually carried using some different protocols than what I've discussed so far. Backbone traffic may use wide area network (WAN) protocols that operate at layers 1 and 2 of the OSI model. This type of traffic doesn't really concern itself with higher layer protocols at all.

Bandwidth and Users

Bandwidth is an often confusing term. Most people refer to it when talking about network capacity, although it is also a term associated with radio and other wireless media. In that context, bandwidth is the range of frequencies in the electromagnetic spectrum that a given wireless network may use. I'm not going to explore radio frequency (RF) bandwidth in-depth here; this is much more relevant to our discussion of RF principles covered in Chapter 3. You should know, however, that bandwidth, as a reference to capacity, is a precious commodity on both wired and wireless networks. Several devices and techniques are used to preserve bandwidth so that users may continue to send and receive the data they need without significant performance limitations. There are several factors that affect data capacity in a network: amount of traffic sent over a media, the types of traffic involved (huge multimedia files or streaming media versus small document files, for example), the speed and latency of the different connections in the network, the limitations of the data pipe itself and the connecting equipment, and usage trends on the network. All of these must be carefully balanced to achieve the best usage of the available bandwidth. In some cases, it may not be enough; additional bandwidth may have to be obtained by upgrading the data pipe or network equipment, imposing usage limitations, and other measures. Some of these other measures are discussed here in the form of Quality of Service and traffic shaping techniques.

 EXAM TIP Be aware that the term *bandwidth* can be used in different contexts. For the exam, you may see it used to refer to network capacity, but you will also very likely see it used to refer to RF technologies as well.

QoS

Quality of Service, or QoS, refers to managing traffic such that latency is reduced, retransmissions due to network errors and issues are reduced, and traffic is delivered to the destination network or host in a timely manner. Routers typically perform QoS functions, although other devices that manage backbone traffic can also provide QoS services. QoS services help to manage bandwidth and throughput both within and between networks. Different network issues, such as chatter, routing errors, latency, and network congestion, are factored into QoS services with the goal of reducing transmission and reception issues. QoS features are typically defined in lower-level protocols, although some higher-level applications can be used.

Traffic Shaping and Routing

Traffic shaping is a method of managing traffic to optimize bandwidth and network use. It really involves determining which traffic to send at what point, which means determining traffic priority, allowable delay, and in some cases, traffic routes across remote networks. In traffic shaping, data may be prioritized and routed according to the type of traffic that it is. Traffic shaping can also mean controlling the amount of traffic entering or leaving a network to better manage traffic flow. This may mean limiting the rate of data allowed to flow, or limiting the amount of traffic that flows in a given time period. Traffic may also be routed through other interfaces or devices if a particular route would cause network congestion. Like QoS services, routers and other advanced devices typically performed traffic shaping functions.

Simple Network Management Protocol

Simple Network Management Protocol (SNMP) is a protocol used on TCP/IP networks to manage network devices. SNMP is an application layer protocol that uses UDP port 161 (and both TCP and UDP ports 162) to query and receive data from managed network devices. SNMP uses an agent on each device that communicates with a network management system to give information on the status of the device to include uptime status, connections, networking bandwidth usage, and even some security data. The agent that runs on the device uses a device-unique database, called a Management Information Base (MIB), to collect the data specific to that device that the administrator is interested in receiving. SNMP can be used to manage both wired and wireless devices. Versions of SNMP that you may see in the networking world include older versions 1 and 2, which had very limited security features, and version 3, which is basically the de facto standard for modern networks. Version 3 of SNMP provides device authentication and transmission security for SNMP data.

TIP   SNMP version 3 is the most secure form of the SNMP protocol. It should be used whenever possible, if the device supports it, to help manage network devices.

Power over Ethernet

Power over Ethernet (PoE) is a technology that allows power to be sent over the same cables that provide data to remote network devices, such as wireless access points. This allows a remote device to receive both data and power over the same cable, eliminating the need for additional power cables and connections. PoE uses standard Category 5 (and above) Ethernet cables to deliver power to a device via a standard RJ-45 connection.

EXAM TIP   Remember that PoE requires a minimum of Category 5 Ethernet cable to deliver power and data. Lesser quality cable is not supported.

Chapter Review

This chapter has continued the discussion of network technologies. First, I discussed the different types of devices that serve as building blocks to make up networks. I talked about hubs, which are basically simple connection devices and don't have any traffic management or security functions. Then I discussed bridges, which can segment traffic based upon MAC addresses and joins local area network segments together. Switches, which are more common in modern day networks, are capable of a wide variety of management and security functions. Switches serve to separate collision domains, which can reduce network congestion and help save bandwidth. Switches also use MAC addresses to identify hosts and segment traffic. Switches, like bridges, typically work at layer 2 of the OSI model, but there are layer 3 switches as well. I also discussed another common network device, the router. Routers serve to send traffic to the correct remote networks. Routers separate local area networks and also separate broadcast domains. Routers work at layer 3 of the OSI model and primarily use IP as the protocol to determine the correct network IDs of remote networks.

The chapter also covered specialized network devices. Specialized devices include security devices, such as firewalls, proxies, VPN concentrators, and other types of devices. Firewalls work by filtering traffic based upon rules. Rules can be made up of elements such as IP addresses, ports, protocols, and so on. Firewalls use these rules to determine which traffic should be allowed into a network. Based upon the rules that are configured in the firewall's ruleset, a firewall will allow traffic, reroute traffic, or drop it completely, effectively blocking it. Proxies serve to make requests on behalf of hosts. Proxies are able to hide the host's IP address and the network it's on. Proxies also are able to help conserve bandwidth by caching frequently used web sites. VPN concentrators are devices that are used to authenticate remote hosts or remote sites into the internal corporate network. VPNs can be client-based, or site-to-site, with a remote branch office connecting to the corporate enterprise network as an example.

A wireless LAN controller serves to help authenticate and control traffic from a wireless network going into an enterprise-level wired network. It usually has some basic security and management functions, but may rely on other network devices, such as authentication servers, to provide some functions. Related to wireless LAN controllers are lightweight access points, which basically connect to a wireless LAN controller and provides a connection point for wireless clients. A lightweight access point has no real management or security functions built into it, and requires those functions from the wireless LAN controller itself, so it is typically managed via the wireless LAN controller. You can also have an autonomous access point, which is basically a standalone access point for wireless clients, and does not require a wireless LAN controller to function. All the management functions are performed on the autonomous access point itself, and it can have some basic security functionality built-in as well.

You also learned about network topologies. Topologies are essentially the physical and logical design of a network in terms of how data and traffic flow through the network. The chapter covered several different topologies, including bus, star, ring, mesh, and ad-hoc networks. You learned about the design constraints involved with choosing which topology is best suited for the network. Design constraints include scale, centralization, types of equipment, and physical environment, among other factors.

Network architectures determine the optimum design of the network in terms of device placement, segmentation, security, traffic flow, and other factors. There are several different network architectures you should be aware of for the exam. These include intranet, which is the local internal network, and the Internet, which is the worldwide globally connected network. Extranets are typically networks that are external to the internal network that are specifically used for external clients, customers, and so forth. These extranets may use internal resources but are typically separated in terms of authentication and use.

A very simple network security architecture is that of a bastion host. A *bastion host* is one that has multiple network interfaces, each connected to a separate network. This simple design enables a security buffer between networks such as the Internet, and an internal host. On a larger scale, an important network architecture to consider is the Demilitarized Zone (DMZ). A Demilitarized Zone is a perimeter or buffer network that exists between an internal network and an external untrusted network. A DMZ is typically constructed using external routers and firewalls, which filter traffic and direct it to appropriate resources within the zone. Servers and other resources within a DMZ are typically placed there for external users. A DMZ serves to keep those external users out of the internal network, yet still allow them to use certain corporate resources.

You also learned about several networking technologies that provide higher-level functions for networks and enable data to travel between wired and wireless networks, from local to remote networks, and to the Internet. The first key technology discussed was IP addressing, which is logical addressing, and is used by the administrator to identify networks and hosts. The chapter focused on IP version 4 network addresses and discussed the different classes of IP addresses available. Classes A, B, and C are commonly used to assign to networks and hosts, while classes D and E are used for multicasting and research, respectively. Each class is identified by a specific range of numbers in the first octet. Each class also uses different numbers of bits to specifically identify its range of networks and hosts. Subnet masks are used to further identify the delineations between networks and hosts. Each class has, by default, a certain number of networks and hosts it can use within the class. Subnetting, as I also discussed, allows you to break the rules associated with the different IP address classes and move or borrow bits to gain more network IDs if needed. To conclude the discussion of IP addressing, I also mentioned a few special IP address ranges, including the loopback address range, the private IP address ranges, and the automatic IP addressing range.

Other networking technologies discussed included Network Address Translation, which allows a network to use a private IP address range behind the network device while allowing it to have one or several public IP addresses that are seen from the Internet. This allows an organization to hide its internal address space and still make requests to the Internet. Devices that manage NAT do so through by using a table that maps the private IP addresses to their respective requests and maintains that information for when the replies come back from the Internet. You also learned about virtual LANs, which are implemented using switches that have the ability to work at layer 3 of the OSI model. The switches perform some of the same basic functions that routers can perform, while minimizing the cost and scale incurred by an actual router. Virtual LANs allow an administrator to set up logical LANs through software configuration on the switch and assign different hosts connecting to the various ports on a switch to these different VLANs. Security is one of the benefits associated with virtual LANs because hosts are separated from each other logically, as is their respective traffic. Reduction of broadcast domains is a performance benefit associated with using VLANs.

The chapter ended with a discussion of several aspects of network management that are important to an understanding how networks work. You learned about backhauling traffic, which means managing the traffic at the backbone level between access or distribution points to remote networks. You also learned about the data transfer capacity of bandwidth. There are many technologies that are used to conserve bandwidth, which may involve limiting traffic or restricting certain types of traffic. QoS and traffic shaping are two such technologies. QoS refers to different methods of managing traffic so that latency or delay is reduced, efficient delivery is ensured, and retransmissions due to errors are reduced.

Traffic shaping refers to methods of managing traffic in order to optimize and conserve bandwidth. Traffic shaping methods include determining traffic priority delay and traffic routes across remote networks. In some cases this may mean rerouting traffic, or throttling bandwidth allowed to certain types of traffic.

Two other network management technologies discussed at the conclusion of the chapter included SNMP and PoE. SNMP is a network management protocol that allows administrators to collect data from various network devices, including switches, routers, and servers. SNMP uses an agent on the device to collect the data based upon a device-unique database called an MIB. The data is collected at various time intervals and sent to a centralized network management server. PoE allows both power and data to be sent over a single Ethernet cable to a device such as a wireless access point. This can help eliminate power cables and allow devices to be located in areas where there are no power outlets.

Questions

1. Your supervisor has asked you to select network devices to be used in a new project that involves designing and building a small network for a client. This network requires some very basic management and security functions. Which of the following devices would you choose to connect several hosts together in a small network?

   A. Hub

   B. Switch

   C. Router

   D. Gateway

2. You are trying to explain the different network topologies most commonly used in wired networks. Which of the following topologies is more commonly seen in modern networks as a redundant backbone structure?

   A. Ring

   B. Star

   C. Bus

   D. Mesh

3. Which of the following would result in a logical bus topology? (Choose two.)

   A. Several hosts connected to a single cable

   B. Several hosts connected to a Multistation Access Unit

    C. Several hosts connected via a hub

    D. Several switches connected to a router

4. You need to install a device on the network that allows traffic to flow between the local network and a remote network. Which device would you need to install?

    A. VPN concentrator

    B. Hub

    C. Switch

    D. Router

5. Which of the following devices use the hardware or MAC address to manage and communicate with hosts? (Choose two.)

    A. Router

    B. Bridge

    C. Switch

    D. Hub

6. A technician has recently updated the configuration on the company firewall. After several users complain that they can no longer access any Internet resources, you decide to review the firewall configuration. Which of the following would prevent users from accessing any Internet resources?

    A. A default deny rule at the very top of the ruleset

    B. A default allow rule at the very top of the ruleset

    C. A default allow rule at the very bottom of the ruleset

    D. A default deny rule at the very bottom of the ruleset

7. Which of the following devices could be used to translate different network protocols between two networks?

    A. VPN concentrator

    B. Firewall

    C. Proxy server

    D. Gateway

8. Which the following devices require another device for its configuration management and security functions?

    A. Autonomous Access Point

    B. Wireless LAN controller

    C. Lightweight Access Point

    D. Layer 3 switch

9. Which network topology allows multiple redundant paths of communication between hosts, and is frequently used in wireless networks?

   A. Ring

   B. Star

   C. Bus

   D. Mesh

10. You are designing a network and are trying to determine what some of the limiting factors are that would influence your network topology design. Which of the following are design considerations in choosing a network topology? (Choose two.)

    A. Size

    B. TCP ports

    C. Application layer protocols

    D. Physical environment

11. Which of the following is a simple security architecture that involves a host with two interfaces connecting to different networks and filtering traffic between them?

    A. DMZ

    B. Bastion host

    C. Extranet

    D. Intranet

12. All of the following are reasons to deploy a Demilitarized Zone network EXCEPT:

    A. Separation of internal and external networks

    B. Allowing external users to access certain corporate resources

    C. Preventing external users from accessing restricted internal resources

    D. Allowing external users to access certain restricted internal resources

13. Which of the following IP address classes has a default subnet mask of 16 bits?

    A. Class E

    B. Class C

    C. Class B

    D. Class A

14. You are designing an IP addressing scheme for a medium-sized network that has approximately 30 networks and a few hundred hosts. Which address class would work best for this network?

    A. Class E

    B. Class B

    C. Class C

    D. Class D

15. Your network has been assigned a single class B address of 171.24.0.0. Because this only gives you one network range and many possible host IDs, you decide to subnet it. You would like to have at least 150 to 200 network IDs, with at least 200 host IDs per network. Which of the following bit sizes for a network mask would fit your needs?

    A. 16

    B. 24

    C. 32

    D. 8

16. Which of the following technologies allows you to maintain a larger private IP address space on your internal network, while only using a single public IP address on the external side of your network?

    A. VLAN

    B. NAT

    C. DNS

    D. DHCP

17. Which of the following are advantages to using virtual LANs? (Choose two.)

    A. Combining multiple IP ranges into a single range

    B. Elimination of broadcast domains

    C. Segmentation of sensitive hosts and traffic

    D. Elimination of collision domains

18. Which of the following is a network bandwidth management technique that allows the administrator to prioritize and route important network traffic?

    A. Bridging

    B. Routing

    C. Traffic shaping

    D. NAT

19. All of the following are components of SNMP EXCEPT:

    A. LWAPP

    B. MIB

    C. Agents

    D. Network Management Server

20. Which of the following types of cable does PoE use to deliver power and data to remote network devices?

    A. Serial cable

    B. Ethernet Category 3 or higher

    C. Fiber optic cable

    D. Ethernet Category 5 or higher

Answers

1. **B.** A switch can be used to connect a small network together while still providing some basic security and traffic management functions.

2. **A.** A ring topology is often used in modern networks to provide a redundant backbone infrastructure.

3. **A, C.** Several hosts connected to a single cable, and several hosts connected via a hub would both provide a logical bus topology.

4. **D.** A router allows traffic to flow between a local and a remote network.

5. **B, C.** Both bridges and switches, working at layer 2 of the OSI model, use the hardware (MAC) address to manage and communicate with hosts connected to them.

6. **A.** A default deny rule at the very top of the ruleset would block all traffic coming into the network, including replies to user requests, because all traffic would match it first before any other rules are processed.

7. **D.** A gateway is used to translate different network protocols between networks.

8. **C.** A lightweight access point requires a wireless LAN controller to manage its configuration and security functions.

9. **D.** A mesh topology provides multiple redundant paths between hosts, and can often be found in wireless networks.

10. **A, D.** Both size (numbers of hosts as well as physical space) and physical environment are factors that would influence network topology design.

11. **B.** A bastion host is a computer or device with two interfaces connecting to different networks and filtering traffic between them.

12. **D.** Allowing external users to access certain restricted internal resources is not a valid reason for deploying a DMZ architecture because this scenario actually lessens security. All of the other reasons help to tighten and enforce network security.

13. **C.** A class B IP address has a default subnet mask that uses 16 bits (the first two octets) to delineate network IDs from host IDs.

14. **B.** A class B network, subnetted correctly, would satisfy the requirements for those numbers of networks and hosts. All other classes listed don't support those numbers of hosts or networks.

15. **B.** A 24-bit subnet mask, using the first three octets, would break up the single class B IP address into 254 networks, each having 254 hosts available to it.

16. **B.** Network Address Translation (NAT) allows you to use a private IP address space for the internal network while using a single public IP address for the external side of the network.

17. **B, C.** Virtual LANs allow you to eliminate broadcast domains, something that routers normally do, and also allow you to segment sensitive hosts and traffic into different logical subnets.

18. **C.** Traffic shaping is a bandwidth management technique that can prioritize and route critical traffic along the best path.

19. **A.** The Lightweight Access Point Protocol (LWAPP) is not a component of SNMP.

20. **D.** Power over Ethernet (PoE) uses an Ethernet cable that must be a Category 5 or higher-quality cable to be able to deliver both power and data to remote devices.

Radio Frequency Principles

In this chapter, you will

- Compare and contrast radio frequency (RF) principles and characteristics
- Understand different principles of radio propagation
- Identify different antenna types and characteristics

Now that you've learned the basics of networking and the different components that make up a network, it's time to talk about the essential elements of the primary mobile transmission media—radio waves. While there are a great deal of concepts and terms to understand for the Mobility+ exam, it's important that you grasp the fundamentals of RF, both for a foundational understanding of real-life technologies you'll work with, as well as to help you on the exam. Understanding how radio transmission media works is critical to understanding other concepts, such as wireless networking and cellular technologies, which are discussed in later chapters. This chapter covers CompTIA Mobility+ exam objective 1.3, "Compare and contrast RF principles and their functionality."

Radio Frequency Fundamentals

At its most basic level, radio is really waves of energy that are created when an alternating current on a wire is sent to an antenna, transmitted into the air, and received by a receiver. To continue the discussions from Chapters 1 and 2, radio as a transmission media works at layer 1, the physical layer, of the OSI model. Also remember from discussions in the previous two chapters that, at this layer, you are looking only at signaling, electrical impulses, and so forth. Radio is part of the electromagnetic (EM) spectrum, which comprises a wide range of electromagnetic radiation and energies with different characteristics, including infrared, visible, and ultraviolet light, and microwaves, X-rays, and many others. A simple conceptual diagram of the EM spectrum is shown in Figure 3-1.

As you can see, radio takes up only a portion of the EM spectrum in the lower frequency ranges, but it's very important in that it drives so many of our communications technologies. This chapter is concerned with RF radiation (the radio part of the spectrum), but many of the concepts apply to the entire EM spectrum, as well as other technologies in use, such as cellular, WiMAX, microwave, and so on.

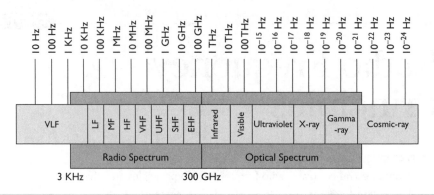

Figure 3-1 The electromagnetic (EM) spectrum

CAUTION Many of the electromagnetic energies in the spectrum are dangerous to humans and animals with prolonged exposure. While RF energies generally are not harmful, you should be aware of some of the hazards of working with any EM radiation in your job. As a Mobility+ certified professional, you may have the occasion one day to work with some of the other energies discussed in this chapter, including microwaves. Be sure to follow proper precautions to protect yourself when doing so.

Radio Frequency Concepts and Characteristics

As mentioned, radio is created by alternating current (AC). Alternating current, by definition, changes direction, back and forth, at very high rates of speed, at different voltages. Information is sent on a wire using this alternating current and then transmitted via an antenna to a receiver. When transmitted, the alternating current model is still used, as the radio waves travel in a similar fashion to AC. In order to understand radio, you must understand some basic characteristics of radio waves, such as frequency, wavelength, amplitude, and phase, which are covered in this chapter.

Frequency

The fundamental RF signal is a *sine wave*, which is shown as an electrical current changing voltage uniformly over a defined time period. All this really means is that, as a wave is changing direction from one side to the other, so to speak, time is passing, of course. If you could visualize this, you'd see something similar to what is illustrated in Figure 3-2. As the wave changes direction, it does this over a measured length of time. A complete change back and forth is called a *cycle*, and by convention, the accepted time period is one second. What we're looking at first is how many times in one second the direction changes, or how many cycles per second. A signal that changes only once over one second is said to have a *frequency* (designated by f) of one cycle per second.

Figure 3-2

Sine wave of an
RF signal

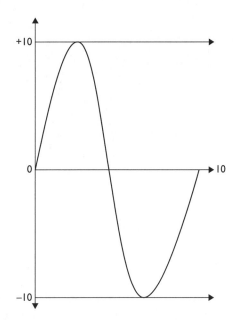

Cycles per second is measured by a unit called a Hertz, named after pioneering physicist Heinrich Hertz, who proved the existence of electromagnetic waves. So a signal that has a frequency of one cycle per second has a frequency of one Hertz, abbreviated as Hz. Obviously, we deal with much higher frequencies in wireless and cellular communications. Typical frequencies used in radio communications are measured in thousands of cycles per second, or kilohertz (abbreviated KHz); millions of cycles per second, or megahertz (MHz); and even billions of cycles per second, or gigahertz (GHz).

NOTE Hertz is used to measure all types of frequencies, not just radio frequencies. It's used to measure acoustic waves and other types of waves as well.

Wavelength

While the frequency of the cycle is measured in terms of so many cycles per second, the actual physical length of a radio wave can be measured as well. This, of course, is called the *wavelength*. The wavelength of a radio wave, designated by the Greek letter lambda (λ), is the distance of one complete cycle from beginning to end. Wavelengths are measured in terms of centimeters, meters, or even kilometers, and are typically expressed in very small numbers—the wavelength of an average Ultra High Frequency (UHF) TV signal can be anywhere from 10 centimeters to 1 meter in length, for example. An example of wavelength is shown in Figure 3-3.

Figure 3-3
Visualization of
wavelength

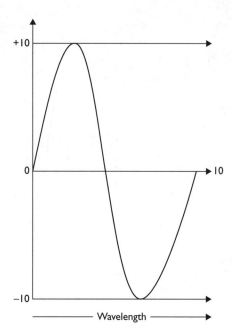

Now, consider that if you increase more waves into a given period of time, you also increase the frequency and the wavelengths are, as a result, shorter. Higher frequencies have shorter wavelengths, and lower frequencies have longer ones. You can get an idea of how this works by looking at Figure 3-4 and comparing it to Figure 3-3. So as you move through the EM spectrum, some energies, like X-rays, are high-frequency and

Figure 3-4
Higher
frequencies mean
shorter
wavelength.

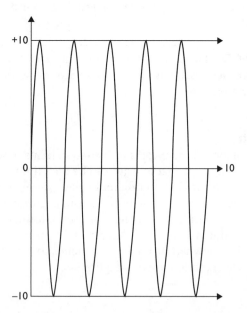

low wavelength, and others, such as radio, are the opposite—lower frequencies and longer wavelengths. Energy with shorter wavelengths (and higher frequencies) tends to be more powerful, but is more limited in distance, while energy with higher wavelengths (and lower frequencies) can travel longer distances, but is less powerful.

Radio frequency ranges in the EM spectrum are from around 3 KHz to about 300 GHz, and are categorized with names such as low, medium, and high frequencies, as well as others. In Table 3-1, you can see the full RF frequency range; along with the names, we call the frequency bands, as well as their wavelengths. Note how the wavelengths *decrease* as the frequencies *increase*.

Amplitude

If wavelength is the actual length or period of a wave measured, then it makes sense that you can also measure a wave's height. This is called the *amplitude* of a wave. As you can see in Figure 3-5, there are a couple of different ways to measure amplitude. First, you have what's called the *peak amplitude*. This is a simple measurement of the height of the wave (in positive numbers) from the zero axis, or starting point, to the highest level (the peak). It's really the absolute value of the height (shown using the mathematical symbol for absolute value, or ||; for example, the absolute value for both –6 and +6 is |6|). The peak-to-peak amplitude, on the other hand, is the measure of the entire range of movement of the wave, from its peak (the highest point) to its trough (its lowest point). In other words, if the peak amplitude covers half of the movement,

Frequency Band	Frequency	Wavelength
Extremely low frequency (ELF)	3–30 Hz	10^5–10^4 kilometers (km)
Super Low Frequency (SLF)	30–300 Hz	10^4–10^3 km
Ultra Low Frequency (ULF)	300–3000 Hz	1000–100 km
Very Low Frequency (VLF)	3–30 KHz	100–10 km
Low Frequency (LF)	30–300 KHz	10–1 km
Medium Frequency (MF)	300 kHz–3 MHz	1 km–100 meters (m)
High Frequency (HF)	3–30 MHz	100–10 m
Very High Frequency (VHF)	30–300 MHz	10–1 m
Ultra High Frequency (UHF)	300 MHz–3 GHz	1 m–10 centimeters (cm)
Super High Frequency (SHF)	3–30 GHz	10 cm–1 cm
Extremely High Frequency (EHF)	30–300 GHz	1 cm–1 millimeter (mm)
Tremendously High Frequency (THF)	300 GHz–3000 GHz	1 mm–0.1 mm

Table 3-1   RF Frequency Range

Figure 3-5
Various
measurements
of amplitude

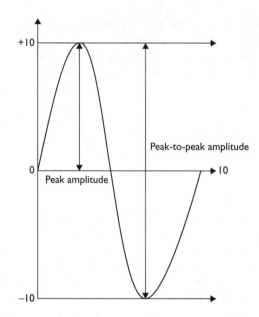

above the zero line, then the peak-to-peak amplitude measures the entire distance the wave travels from 0, in both directions (positive above the line and negative below the line) added together.

Amplitude is related proportionally to both the strength of a signal and its power. Although it's not the only factor that contributes to signal strength, weaker signals have lower amplitude, while stronger, more powerful signals tend to have higher amplitude.

Bandwidth

You've likely heard the term "bandwidth" before. It's one of those terms that, unfortunately, are often misused to mean amount of signal, amount of throughput, and even amount of power. It's none of those! Bandwidth is actually a range of frequencies, (sometimes called a frequency band), and it is measured in Hertz (Hz). A frequency band can be 20 Hz wide, for example, and could reside at any part of the spectrum. It's just the frequency range between given lower and higher frequencies. Let's say that you have a lower frequency of 20 MHz and an upper frequency of 45 MHz. The bandwidth, then, is 25 MHz, which is essentially the difference between the two.

EXAM TIP Remember that bandwidth is the difference between two frequencies, a lower and upper frequency. This is the range between those frequencies.

Modulation

You may have heard of the terms AM and FM (at least if you were alive before the Great Internet Age, anyway) in relation to this old-fashioned thing we called a radio station, and what they stand for. AM is *amplitude modulation*, and FM is *frequency modulation*. Because we now know what amplitude and frequency are, it's the modulation part that's important here, so let's talk about that for a moment.

In order to understand the concept of modulation, let's talk about how data is sent using radio waves. The signal itself is called a *carrier wave*, and, with nothing else done to it, it's just a wave of energy. In order to make the radio wave meaningful as data, you have to change its characteristics. This is really what modulation is—the process of changing or varying properties of a wave or signal. There are actually several ways you can modulate a radio signal to carry data. Some, like AM and FM, are analog, and some are digital. The properties you usually modify using analog methods are either the amplitude or the frequency of the signal.

The carrier wave or signal is changed (modulated) by another signal in such a way that it can represent data. Based upon the method of modulation, the signal can represent 1s and 0s (binary data) or other types of data. Of the two main types of analog modulation, AM and FM, amplitude modulation is the older method and has been around since radio technologies were invented. Amplitude modulation changes the amplitude of each wave. The changes can be made in a discernible pattern, which can represent data. AM is typically used in lower frequency ranges (below 30 MHz) and can be found in commercial radio, aircraft communications, and other shortwave applications.

Frequency modulation is the more modern method of analog modulation, and it changes the frequency of the wave in patterns that, again, can be made to represent data. The two basic types of FM, FM Narrow (FMN) and FM Wide (FMW), can be found in commercial broadcast (FMW) and two-way radio systems (FMN), and are typically found in the range above 30 MHz. Figure 3-6 compares both amplitude and frequency modulation with an unmodulated carrier wave.

Figure 3-6
AM and FM compared to an unchanged carrier signal

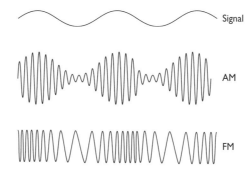

Another analog modulation method worth mentioning here is Single Sideband (SSB). It's primarily used in amateur (ham) radio, as well as marine (commercial fishing/shipping and private boating) applications. Like AM, it is a shortwave radio technology that operates below 30 MHz, but it uses different modulation techniques and equipment that are incompatible with AM.

Since I've talked about analog modulation techniques, it's also worth discussing digital modulation methods as well; you'll be reading about a couple of these coming up in Chapter 5. Digital modulation is a more modern method of signal modulation for several reasons. First, it can make better use of available bandwidth by compressing data. It also can be used to get a better, higher-fidelity signal, which is necessary for data with higher quality-of-service (QoS) requirements. Digital modulation can perform error-checking, and offers the ability to "clean up" a signal on the receiving end–something analog modulation can't do. On the other hand, even a bad analog signal can still produce some useable data or voice, while a bad digital signal may be degraded to the point of total loss of data.

One of the digital modulation techniques of specific interest is the spread spectrum method. Spread spectrum allows a signal to be transmitted over the entire bandwidth of the frequencies within the spectrum. This is accomplished by a variety of methods, including frequency hopping (rapidly changing frequencies within the bandwidth range). In addition to cutting down on interference that results from shifting between frequencies that other transmitters may also use, frequency hopping can lend a limited amount of security to the transmission, as it can be very rapid and hard to predict. Obviously, both transmitter and receiver would have to be configured to change to the same frequencies together. This method, appropriately called *frequency hopping spread spectrum* (FHSS), is used by a wide variety of wireless devices, including cordless phones, baby monitors, and some wireless network technologies that are covered in depth in Chapter 5. Another common method of modulation using spread spectrum is *direct sequence spread spectrum* (DSSS). In DSSS, the data is randomly spread over the entire bandwidth of the frequency range being used, rather than rapidly switching between discrete frequencies the way FHSS does. DSSS is used by the Global Positioning System (GPS), some cellular networks (explained in Chapter 4), and other certain types of wireless networks that are covered in Chapter 5.

Phase

Phase refers to the difference, if any, between two sine waves. Let's suppose that a wave starts at a particular instance, and a second wave, of the same frequency, starts immediately after. There is a delay in both distance and time between the two waves, and they are said to be out of phase with each other. As a result of a phase difference, the waves could overlap with each other and cause distortion in their respective signals. They could also cancel each other out completely. Phase is actually measured as an angle, not a time (frequency) or distance (wavelength) difference. The angle is determined by the difference in wavelengths of the two signals. If the second sine wave starts at a quarter of the length of the first wave, it is 90 degrees out of phase with the first wave, at a half-length, 180 degrees, and so on. Phase difference can be measured from 0 to 360 degrees. Figure 3-7 illustrates the concept of phase.

Figure 3-7

Phase difference
of two RF signals

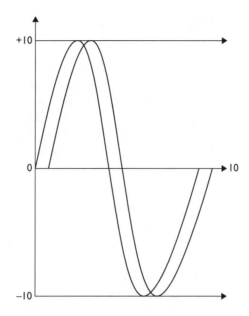

Exercise

1. Look at several devices in your home or work area and determine what radio frequency they use. Usually, this information will be imprinted somewhere on the device. If not, you may have to look in the documentation or manual that came with the device or research the device on the manufacturer's web site. Then research the different wavelengths associated with those frequencies. Create a chart that lists the device, frequency, wavelength, and any other relevant characteristics you find in your research.

2. Using Table 3-1, add another column and research examples of devices that fall into each of these frequency ranges, such as UHF, VHF, and so on. This will help your understanding of what devices are typically found in each range, and help you to become more familiar with the EM spectrum and frequency ranges.

RF Propagation

Radio waves have different characteristics, depending upon the frequency, amplitude, wavelength, and so on. Some radio waves can travel longer distances than others. Some have the ability to penetrate solid objects, while others are bounced off of them. The tendency of radio waves (and other EM energies) to react in these ways is called propagation. Propagation describes how easily radio waves travel through objects, or are absorbed by them, refracted, or reflected off of them. Some transmissions require line-of-sight (LOS), meaning that, regardless of distance, they need have a path fairly free and clear of obstructing objects that may cause issues with propagation. Other factors that influence a radio signal's propagation are attenuation and outside interference.

Absorption

Absorption is what happens when a radio signal is absorbed by a material and cannot fully pass through it. There are many materials that absorb radio waves, to smaller or larger degrees, and can prevent some or even all of a signal from passing through them. This can result in a significant loss of signal. Materials that absorb radio waves include drywall, concrete, fiberglass, carbon fiber products, certain plastics, and even people. Materials that allow, to some degree, radio waves to pass through them include metal and glass.

Refraction

Refraction refers to a characteristic of an RF signal to bend when striking an object. An example of a material that can cause refraction is glass. What happens is the RF signal passes through a material, like glass, and some of it is lost due to this bending or refracting. Some signal may still pass through the material, but it will be reduced and can result in a weaker signal. Refraction is different from *diffraction* in that the RF signal *passes through* the material and is bent, while in diffraction, the RF signal bends *around* the material. Diffraction can also lead to some signal loss.

Reflection

Reflection means that an RF signal "bounces" off of the surface of an object. The surface is usually one that does not lend itself to absorption or refraction. RF signals can bounce off of many objects, and too much bounce can cause loss of signal, poor performance, and a decrease in throughput. Figure 3-8 shows examples of absorption, reflection, and refraction.

 EXAM TIP Understand the differences between absorption, reflection, and refraction for the exam, as these are easily confused terms. Remember that if RF waves are absorbed, they are effectively stopped from propagating. Reflection means that they will bounce off a material, still propagating, but with a reduced signal, and finally, refraction means that they may penetrate a material to a certain degree, but may be split and sent into different directions, also reducing signal.

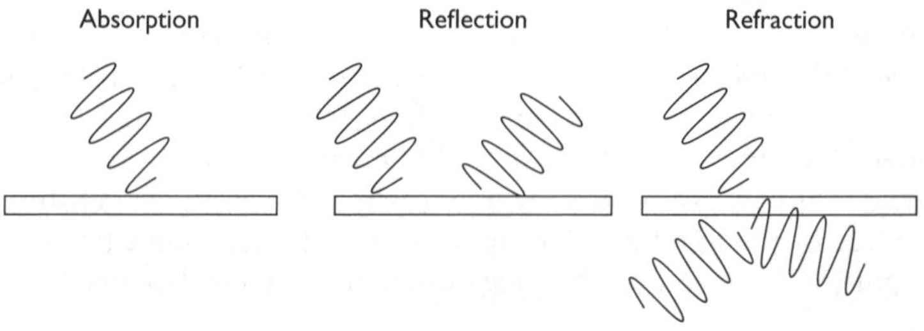

Figure 3-8
Examples of absorption, reflection, and refraction

Absorption   Reflection   Refraction

Attenuation

The weakening of a signal from its source can be affected by many things, including interference, objects that block, reflect, or absorb signals, or the distance the signal travels. Attenuation is the effect of losing a signal over distance. Attenuation, in the context of RF signals, is also referred to as *free space path loss* (FSPL). FSPL is the loss of a signal over distance in a space unobstructed by any object that would otherwise cause refraction, reflection, or absorption. All EM energy attenuates over distance; the farther the energy travels from the source, the weaker it gets. Some energies are able to travel extremely large distances with little or insignificant attenuation, while others attenuate after only a few feet. Sound waves, for example, attenuate such that if you were to stand at one end of a field and yell to a friend on the other end, they may not hear you clearly because the sound simply weakens over the distance. Both wired networks and wireless networks also suffer from attenuation because of the gradual degradation of the signal due to loss of energy over distance, whether it travels over a cable or through the air. Figure 3-9 illustrates the concept of attenuation.

NOTE Attenuation does not take into account any obstacles that may cause absorption, reflection, or refraction of the RF signal, nor does it include any factors such as antenna gain, or limitations of the receiver and transmitter. It's all about signal loss in free (unobstructed) space over distance.

Interestingly enough, in some environments, a device called an *attenuator* is used to intentionally reduce the signal emitted from a radio device, and is connected in-line between the device's antenna and the transmitter. This may be a separate device or built-in as part of the radio equipment. An attenuator may be used in areas where the radio's signal is overpowering other signals, or in areas where the transmit signal is restricted by law or regulation and has to be reduced for compliance with these regulations. Don't confuse this device with an RF filter, which is designed to filter out certain frequency ranges on a radio to prevent interference.

Interference

Interference is the signal distortion effect that outside elements, such as other electromagnetic energies, weather conditions, and so on, have on an RF signal. Sometimes it's called noise, and can come from a wide variety of sources, including other radio transmitters on

Figure 3-9
Attenuation of an
RF signal

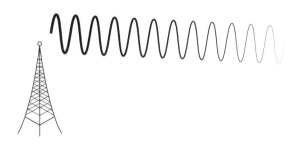

frequencies that are very close together. You could also get noise from heavy machinery, electrical appliances, or anything else close by that generates an electromagnetic field. You may hear the term *signal-to-noise ratio* (SNR) that describes the strength of an RF signal compared to the noise level. I'll also discuss SNR again in Chapter 5 when we discuss wireless technologies.

Antennas

Antennas are used to transmit and receive radio frequency signals. Because of the different characteristics of different RF signals, antennas also can have different characteristics. Some antennas are used to transmit or receive in one direction only, while others transmit and receive in all directions. The particular type of RF technology used, such as microwave, cellular, and radio, also directly affects the type and design of the antenna. Based upon the type of the RF signal they are transmitting or receiving, antennas can have different physical characteristics and appearance.

This section covers the different antennas available to us in radio systems. We'll start out by examining some of the basic of antenna characteristics. Then I'll describe the basic types of antennas you may encounter when designing or building RF systems. There are also several factors that contribute to the selection and use of the various antennas, and I'll cover those as well.

Basic Concepts

Antennas are essentially conductors. On the transmission side, antennas are the part of the radio system that takes the electrical signals produced by the transmitter and turns them into radio waves that are propagated through the air to a receiver. The receiver side also has an antenna that receives the RF signal and converts it back to an electrical signal understood by the receiving equipment. Because of the wide variety of RF signals, the many uses of RF, and the different types of transmitters and receivers, many different types of antennas are available. Factors that influence which antenna you might use for a given setup include frequency range, distance, outdoor or indoor use, antenna height, location, and so on. It's helpful to know some basic characteristics of antennas so you can make the right selection based on your intended use.

Antenna Characteristics

Antennas have several key characteristics that you should be familiar with, not only to help you understand the material for the exam, but also for real-life application. Some antennas are more appropriate for long distances and others for shorter ones. Antennas also are generally used for a particular frequency range and power level. Regardless of type of signal or frequency range, however, there are some technical characteristics of antennas that most of them share, and you will need to know these for the exam. Let's take a look at some of these, including lobes, beamwidth, azimuth and elevation, gain, and polarization in the next few pages.

RF Lobes

RF lobes, or patterns, are the shapes the RF energy takes when emitted from an antenna. We tend to think of radio waves as having the two-dimensional shape and pattern that we see in simple drawings and in ponds when we drop a stone into them, but that isn't necessarily how they would look if we were able to actually see them as they come from the antenna. RF signals are emitted in three dimensions and have shapes peculiar to their specific characteristics, as well the design of the antenna itself. Parts of the physical pattern and shape of the lobe may be useable by the transmitter and receiver, while other parts of it may not be. The "primary" part of the lobe may be what extends out for a distance and is usually the main receivable part of the transmission pattern, while there may be secondary parts that extend sideways or in different directions and may not be useable due to weaker signal strength, for example. Figure 3-10 gives you a bit of a visualization of the concepts of lobes or RF signal patterns.

Beamwidth

Beamwidth refers to the angle of measurement of part of the antenna pattern, or lobe, which is measured as the half-way point of the main lobe, usually at half power. This is sometimes called the Half Power Beamwidth (HPBW), and is the point where the signal power drops to 50 percent (about −3 dB) from the center or peak of the main lobe. As it's an angular measurement, beamwidth is expressed in degrees and can be either a horizontal or vertical measurement. Figure 3-11 shows the portion of a main lobe and its beamwidth, using a parabolic (dish) antenna. Keep in mind that beamwidth measurements and physical pattern, like RF lobes, vary with the physical construction and design of the antenna. One of the things that someone designing a wireless system must do is determine the appropriate beamwidths (both horizontal and vertical) for an antenna to ensure the right RF signal coverage.

RF Power Measurements

Radio signal power is measured in *watts* (abbreviated with W), although we typically also see milliwatts (mW, or 1/1000 of a watt) used to describe smaller power levels from some devices. Watts are a way of measuring the absolute power output. We also,

Figure 3-10
RF signal lobes

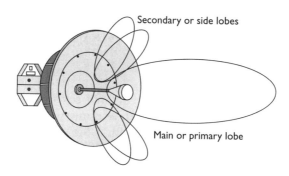

Secondary or side lobes

Main or primary lobe

Figure 3-11
Beamwidth

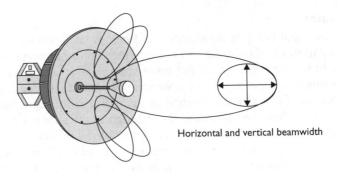

Horizontal and vertical beamwidth

however, use decibels to describe relative changes in power. A decibel is a fundamental unit of measure in radio, abbreviated as dB, which relates to the difference between two signal levels. Decibels can be used to express increases or decreases (changes) in power levels. When used this way, we express it as a dBm, which is a decibel compared to 1 milliwatt (mW) of power, the standard reference power level. There is a mathematical relationship between decibels and milliwatts, based upon logarithmic functions. At the base of the relationship, 0 dBm = 1 mW. As power levels change, the dBm measurement changes as well. While it is a proportional change, it isn't a linear change (a one-for-one change, if you will). You can use the following formula to compute dBm:

$$dBm = 10 \times \log \ (signal \div reference)$$

Because knowing the math isn't important for the exam, I won't go through the steps to compute it, but here's an example: 100 mW is 20 dBm, but doubling the power to 200 mW is only an increase of +3 dB, to 23 dBm. You can see that a mere *increase* of +3 dB *doubles* the signal power. Other changes result in similar proportional changes. So, in the previous discussion about beamwidth, the −3 dB difference accounts for the half-power signal loss that describes the beamwidth. As you'll see in the discussion of antenna characteristics later in the chapter, decibels can be used to help describe antenna power as well.

EXAM TIP You will likely not see any questions on the exam requiring you to know how to calculate dBm, but the relationship between decibels and power is a useful bit of knowledge to have when learning about RF signal power.

Azimuth and Elevation

The *azimuth* of the antenna is the horizontal RF coverage pattern; it can be visualized by thinking of how you would look at a three-dimensional lobe from the top, or looking down on it. *Elevation*, on the other hand, is the vertical RF coverage pattern. It's best visualized by looking at it from a side view, if you will. It may sound confusing at first, but once you grasp the concept that you view elevation from the side (so you can see it from bottom to top), and azimuth from above (so you can see its side-to-side surface), it's really not that difficult. Let's see if I can clarify this using an example. If you were to stand next to a tall building, you'd see its elevation as height, from the bottom of the building to its top floor. Obviously, you couldn't see that if you were hovering in a helicopter directly above the building, but you would, however, see how many city blocks it covered (its azimuth, so to speak). Beamwidths are sometimes referred to as azimuth and elevation beamwidths.

 EXAM TIP   Remember that elevation is viewed from the side, and azimuth is viewed from the top. Don't confuse these for the exam!

Gain

Gain is a measure of how much of the signal's input power is concentrated in a particular direction. It is usually discussed in the context of an omnidirectional antenna, which radiates equally in all directions. Gain is an amplification of the RF signal, and is measured in *decibels isotropic* (dBi). For now, don't worry if that term seems a bit overwhelming; I'll discuss antenna power levels in a few short paragraphs!

There are two types of gain that you need to be concerned with: passive and active. Passive gain is really a function of antenna design and is used to focus the signal by changing the vertical and horizontal beamwidths (signal patterns) to be narrower. An increase in gain results in a narrower beamwidth. As gain decreases, the beamwidth is wider. The exception to this is the omnidirectional antenna, which, because it has a horizontal beamwidth that extends in a full circle (360 degrees), does not decrease its horizontal bandwidth as gain increases, but it still can provide more coverage. Passive gain can be changed by changing antennas, or, in some cases, changing characteristics of an antenna's shape (i.e., by extending a portion of it manually, changing the cone or pole size, and so on). Active gain, on the other hand, is achieved by increasing the signal strength, usually through an amplifier, to the radio transmitter. Remember that gain doesn't necessarily come from changing power levels; it's more of a function of focusing the signal.

Polarization

Polarization of an antenna refers to its horizontal or vertical orientation. The placement of an antenna, horizontal or vertical, can affect how RF waves are both transmitted and received. Normally, the antennas for a given type of radio signal (such as UHF or Wi-Fi, for example) should be similar in design, construction, and use so that the transmitter and receivers are both compatible. If a transmitter antenna is horizontally polarized (oriented), for example, the receiving antenna should also be oriented horizontally. This is so the horizontal and vertical beams of the transmitter can be optimally received. If you've used a computer on a wireless network, for example, with an external antenna that has been moved around or side-to-side, you may have experienced a signal loss due to the antenna orientation changing from vertical to horizontal (or vice versa). This is because a change in polarization of an antenna from what it should be usually decreases the RF signal.

Antenna Power Levels

I briefly discussed RF signal power levels earlier, and how the decibel (dB) is used to reference power of two signal levels. In addition to the radiated RF signal power, decibels can also be used to describe antenna power levels. There are two ways that decibels are used in terms of antenna power. The first applies to *isotropic* antennas, and is abbreviated dBi. Isotropic antennas broadcast signals in all directions equally, in the pattern of a sphere. This is a theoretical scenario because there's no "perfect" isotropic antenna. However, you use this as a starting point. The dBi measurement is a relative power measurement and is used to further define the antenna's dBm power level, taking into account other factors as well, such as power loss from the antenna itself, cables, connectors, and so on. This adjusted power measurement, if you will, is called Equivalent Isotropically Radiated Power (EIRP). Again, I won't go deep into the math for determining the EIRP, but suffice it to say, it can get a little complex. The second way applies to dipole antennas and is less common. Most of the time, you see dBi measurements used in wireless networks. The abbreviation for dipole antenna power measurement is dBd, and can be converted to dBi if needed.

 EXAM TIP While it's useful to know how dBi and dBm relate to power levels, for the exam, you do not need to know the math needed to calculate them.

Types of Antennas

A wide variety of antennas is available, depending upon the particular technology and what the antennas are being used for. The two general categories of antennas discussed are omnidirectional and directional. Most antennas fall into one of these broad categories. It's not unusual in some systems to see multiple types of antennas attached to a tower, to make efficient use of the tower in cases where a system may have multiple transmitters on different frequency ranges. The basic types of antennas covered on the Mobility+ exam are the omnidirectional, semi-directional, bi-directional, Yagi, and the parabolic dish. These will be covered in the next few sections.

Omnidirectional

By the term "omnidirectional," you might be led to believe that this type of antenna can transmit or receive in all directions, and this is true, but only to a certain extent. A "true" omnidirectional would be almost like the "perfect" isotropic antenna discussed previously; that is, it would be able to transmit/receive in all directions, which would, in essence, give you a perfect sphere. This isn't exactly what omnidirectional really means for our purposes. If you could visualize the RF signal pattern coming from an omnidirectional antenna, it would appear to be in the shape of a doughnut (non-filled, of course!) surrounding the top of the antenna, rather than a perfect sphere. This means that it does transmit and receive in all directions, but only in a given plane (usually horizontal). Omnidirectional antennas are used in point-to-multipoint systems, meaning they transmit in all horizontal directions to multiple receivers. An omnidirectional antenna usually comes in the discone, whip, and dipole (think of older TV "rabbit ears" as an example of dipole) varieties. You've likely seen examples of omnidirectional antennas in your home and business wireless network equipment, or attached to cars for mobile radio equipment. Commercial television and radio stations also use omnidirectional antennas in the form of very tall radio transmitting towers. Figure 3-12 shows examples of omnidirectional antennas commonly found in 802.11 wireless networks.

Semi-directional

A semi-directional antenna can transmit and receive signals more focused in a particular direction (or, in some cases, two discrete directions). This may be the case when you are trying to ensure coverage for a particular transmitter and receiver, in a point-to-point setup. Directional antennas come in a wide variety of flavors, including planar, sector, and Yagi (discussed shortly). Planar antennas are flat panels (also sometimes called patch or panel antennas) and can have a horizontal beamwidth as much as 180 degrees, but it is usually less. Sector antennas can cover a particular angle with their RF patterns (think about a slice of pie) and may be mounted in array of several (three or four) antennas on a tower to give a sort of omnidirectional coverage, albeit highly focused in each sector for each antenna in the array. Sector antennas usually have a wide horizontal beamwidth, but a narrow vertical beamwidth.

Figure 3-12
Omnidirectional
antennas used in
802.11 wireless
systems

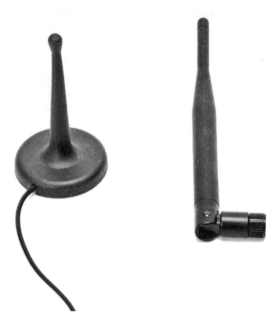

Bi-directional

Bi-directional antennas are a form of directional antennas that transmit or receive in two discrete directions, as opposed to omnidirectional (all directions) or semi-directional (one direction that may be very narrow or very wide). Usually, this is accomplished by actually having two separate antennas constructed together that can send high gain RF signals in two separate (and usually opposite) directions.

Yagi

A Yagi antenna is a directional antenna. If you remember the days of television before cable was common, the old-fashioned TV antennas on the top of a house were typically a type of Yagi antenna. Most of these antennas operated in the UHF and VHF ranges for television reception, although other types of Yagi antennas can operate in other frequency ranges as well. A common Yagi antenna is pictured in Figure 3-13.

TIP You may see Yagi referred to as YAGI in the official exam objectives, but the name comes from one of the co-inventor's last names, Hidetsugu Yagi. Yagi and Shintaro Uda invented the "Yagi-Uda array," or Yagi antenna, in 1926 in Tokyo, Japan. It's not an acronym!

Parabolic Dish

The last type of antenna to discuss is the parabolic dish. It's very likely you've seen variants of this type of antenna before; it is the most commonly used antenna in TV and Internet services from satellite providers. You'll also see it mounted on various kinds of cellular and microwave towers. A parabolic dish is considered a highly directional antenna, meaning that is has very narrow vertical and horizontal beamwidths. It's usually used for longer range communications, up to several miles in most cases, depending upon environmental conditions such as weather and terrain, as well as gain. The RF pattern begins very narrow from the antenna but disperses or spreads very wide over long distances. Figure 3-14 shows a typical parabolic dish.

Figure 3-13
A Yagi antenna

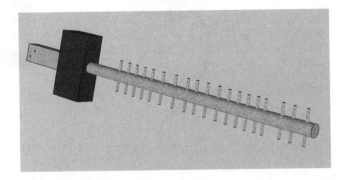

Figure 3-14

A parabolic dish antenna

Faraday Cages

A Faraday cage, named for its inventor, Michael Faraday, is an enclosure created by using nonconductive materials and used to prevent electrical interference. Faraday cages can be very large, surrounding an entire room, or be very small and surround a particular piece of equipment. In addition to preventing electrical interference, Faraday cages can prevent or suppress RF signal leakage. An example of a Faraday cage would be a structure surrounding a secure room that would block incoming or outgoing RF signals, such as cellular phones or other transmitting/receiving devices. Another example, on a smaller scale, might be a box that shields devices from radio frequency interference or from RF leakage. Still another example of Faraday cages may be shielding built into the device itself, in the form of hardened cases, linings, and suppressors, to perform the same functions on a device-by-device basis. If you want to see a common use of a Faraday cage, look no further than your kitchen's microwave oven!

Chapter Review

This chapter covered some fundamental concepts of Radio Frequency (RF) theory. The chapter opened with a discussion of radio's place in the electromagnetic (EM) spectrum, along with other forms of EM energy, such as light, X-rays, and so on. Radio is alternating current that is converted by the transmitter and antenna to energy in the form of waves. A basic radio wave has changes direction over time, and this is called a cycle. Cycles per second are known as Hertz (Hz), and this is the frequency (f) of a signal. Radio frequency ranges in the EM spectrum are from around 3 KHz to about 300 GHz.

Wavelength is the distance of a complete cycle from beginning to end. Higher frequencies have shorter wavelengths, and vice versa. Amplitude is the height of a wave, measured from its start to its peak height, or from its peak to its trough (peak-to-peak amplitude). Amplitude is related to the signal strength. Bandwidth is the difference between a lower and an upper frequency, measured in Hertz. Phase is the angular difference, in degrees, between two waves that begin at different times and distances from each other.

Modulation can be analog or digital, and refers to how we change characteristics of an RF signal to encode data. Popular analog methods of modulation are amplitude modulation (AM) and frequency modulation (FM). Digital methods include spread spectrum methods, which you can see in several Wi-Fi technologies.

The chapter also covered characteristics of RF propagation, which describes the ability of radio waves to travel, often impeded by objects that absorb, refract, or reflect RF signals. Absorption means that the RF signal is absorbed by the material or object it comes in contact with. Refraction allows a signal to pass through an object, but may bend or split the signal and results in signal loss. Reflection causes an RF signal to bounce off of an object. Attenuation is the loss of an RF signal over distance, and propagation can also be affected by interference or noise from the weather, heavy equipment, or even other radio transmitters.

In the discussion of antennas, you first looked at various characteristics of antennas, which often relate to the type of signal, its frequency, and the design and function of the antenna itself. These characteristics included RF lobes, or patterns, of the RF signal transmitted from the antenna; beamwidth, which refers to a measurement in degrees of the half-power point from the main lobe; and azimuth and elevation, which are the horizontal and vertical (respectively) coverage patterns, also measured in degrees. You also learned a bit about the relationship between decibels (dB) and RF signal power, measured in milliwatts (mW). To complete the discussion on antenna characteristics, the chapter covered gain and polarization, and described antenna power levels.

There are several different antenna types, and these can be omnidirectional (transmitting in all directions, but primarily in a 360-degree horizontal plane) or directional. Directional antennas include semi-directional, such as planar, sector, and Yagi, and bi-directional. Yagi antennas are typically seen as the older TV antennas, but are still used today for other purposes as well. Parabolic dish antennas are typically used for a wide variety of applications, such as satellite-based TV and Internet and microwave services.

And, finally, you learned about the Faraday cage, which is an enclosure created by using nonconductive materials to prevent electrical interference and RF leakage.

Questions

1. A frequency of 5 million cycles per second is also expressed as:
   A. 5 dBm
   B. 5 GHz
   C. 5 MHz
   D. 5 dBi

2. A coworker tells you that the company's Wi-Fi network operates at 5 GHz and is in the Very High Frequency (VHF) range. You disagree with the coworker. What frequency range does the 5 GHz wireless network actually operate in?

   A. High Frequency (HF)

   B. Super High Frequency (SHF)

   C. Ultra High Frequency (UHF)

   D. Tremendously High Frequency (THF)

3. If a radio wave has peak amplitude of +3, what would its peak-to-peak amplitude measurement be?

   A. |6|

   B. 0

   C. −3

   D. −6

4. Your company wants you to help design a wireless network that uses digital modulation. You decide to use a modulation method that rapidly changes frequencies within a range. Which modulation method is appropriate in this instance?

   A. Direct Sequence Spread Spectrum (DSSS)

   B. Amplitude Modulation (AM)

   C. Frequency Modulation (FM)

   D. Frequency Hopping Spread Spectrum (FHSS)

5. Which of the following would be the phase difference of a wave that starts at a quarter length of a previous wave?

   A. 4 dB

   B. |4|

   C. 90 degrees

   D. 0.25Hz

6. Your boss insists that doubling the RF signal power from 100 mW to 200 mW requires the decibels to also be doubled. You disagree. What are the decibels required for each power level? (Choose two.)

   A. 100 mW = 20 dBm

   B. 200 mW = −3 dBm

   C. 100 mW = 3 dBm

   D. 200 mW = 23 dBm

7. Which of the following can reduce RF propagation when radio waves partially penetrate an object but are split into multiple paths?

   A. Reflection

   B. Refraction

   C. Diffraction

   D. Absorption

8. What is the signal power difference at the Half Power Beamwidth (HPBW) of an antenna?

   A. 3 dBi

   B. 0.5 dB

   C. –3 dB

   D. –3 dBp

9. You are explaining antenna gain to a coworker. Which of the following are the two types of gain? (Choose two.)

   A. Isotopic

   B. Active

   C. Passive

   D. EIRP

10. You receive complaints from a manager of a small branch office that the wireless network performance has gotten worse over the past few days, but is still working. Which of the following could affect performance of the wireless network?

    A. Antennas repositioned on the wireless router from vertical to horizontal

    B. Power supply unplugged from the wireless router

    C. Active gain set to a higher level on the wireless router

    D. Increase of decibels by +3 dB

11. Which of the following are true statements about azimuth and elevation? (Choose two.)

    A. Azimuth is a vertical RF coverage pattern.

    B. Elevation is viewed from the top.

    C. Elevation is viewed from the side.

    D. Azimuth is a horizontal RF coverage pattern.

12. Which of the following is used to measure isotropic antenna power levels?

    A. dBd

    B. dBi

    C. dBm

    D. dBc

13. You consult for a company that has a sudden Internet outage. They are located in a remote area outside of town that is not serviced by cable or DSL. When you arrive on site, you are told that the Internet connection comes from an antenna on the roof. There are several antennas attached to a mast on the roof. What type of antenna provides Internet service to this company?

    A. Yagi

    B. Parabolic dish

    C. Planar

    D. Dipole

14. How many degrees are covered in an omnidirectional antenna's horizontal plane?

    A. 90

    B. 120

    C. 360

    D. 180

15. Which type of antenna is most likely to be used to receive UHF and VHF television signals?

    A. Planar

    B. Parabolic

    C. Yagi

    D. Sector

16. You need to construct a secure area for electrical equipment to prevent RF interference and leakage. Which of the following would you use?

    A. Faraday cage

    B. Attenuator

    C. RF filter

    D. RF noise generator

17. What is the term for a "perfect" antenna that broadcasts in all directions equally, in the pattern of a sphere?

    A. Semi-directional

    B. Omnidirectional

    C. Dipole

    D. Isotropic

18. What is another term for Free Space Path Loss (FSPL)?

    A. Refraction

    B. Diffraction

    C. Attenuation

    D. Spreading

19. The process of changing or varying properties of a wave or signal is called:

    A. Attenuating

    B. Amplifying

    C. Interference

    D. Modulation

20. What is the term used to describe the proportion of the strength of an RF transmission to the level of interference present?

    A. Attenuation

    B. Decibel

    C. Signal-to-noise ratio

    D. Peak-to-trough ratio

Answers

1. **C.** A frequency of 5 million cycles per second is expressed as 5 MHz, because Hertz (Hz) is the unit of measure for frequency, and 1 MHz is a million Hertz.

2. **B.** The 5 GHz frequency found in some wireless networks operates in the Super High Frequency (SHF) range, which covers 3–30 GHz.

3. **A.** Peak-to-peak amplitude is the absolute value of the amplitude from the peak (highest point above 0) to its trough (lowest point below the zero axis). Because it is a sine wave, both the peak and trough amplitudes will be +3 and –3, respectively, making the absolute value $|6|$.

4. **D.** Frequency Hopping Spread Spectrum (FHSS) is a digital modulation method that rapidly switches, or "hops," frequencies and is used on some types of Wi-Fi networks.

5. **C.** The phase difference between two waves is measured in degrees. At one-quarter the length of the wave, it would be 90 degrees.

6. **A, D.** 100 mW of RF signal power is 20 dBm, and doubling the power to 200 mW only increases the dBm by 3, to 23 dBm.

7. **B.** Refraction occurs when RF signals are bent and split into multiple paths by an object.

8. **C.** A –3 dB loss is equivalent to a half power decrease, while a +3 dB increase effectively doubles the power of an RF signal.

9. **B, C.** Active and passive are two types of antenna gain.

10. **A.** Antennas repositioned on the wireless router from vertical to horizontal could cause a signal to be weaker because of polarization.

11. **C, D.** Elevation is viewed from the side (and azimuth is viewed from the top), and azimuth is a horizontal RF pattern (with elevation being a vertical RF pattern).

12. **B.** Decibels isotropic (dBi) is the measurement of isotropic antenna power.

13. **B.** Because this company is remote and cannot be serviced by cable or DSL providers, it is likely serviced by a satellite Internet provider. In this case, it would be a parabolic dish antenna.

14. **C.** An omnidirectional antenna transmits to 360 degrees (a complete circle) on its horizontal plane.

15. **C.** A Yagi antenna was most commonly used to receive UHF and VHF television signals in the days before cable television.

16. **A.** A Faraday cage can be used to prevent RF interference and leakage.

17. **D.** An isotropic antenna is a "perfect" antenna that broadcasts in all directions equally, in the pattern of a sphere.

18. **C.** Free Space Path Loss is also called attenuation, which is the loss of a signal over distance.

19. **D.** Modulation is the process of changing or varying properties of a wave or signal.

20. **C.** Signal-to-noise ratio (SNR) describes the strength of the RF signal compared to the noise level in the area.

Cellular Technologies

In this chapter, you will

- Explain, compare, and contrast different cellular technologies
- Understand different cellular signaling technologies

Now that I've discussed network and radio frequency technologies, you have the building blocks to go a step further and learn how these standards and technologies are applied to create mobile communications networks. The first application of these basic theories and concepts that I'll talk about is in the realm of cellular technologies. The CompTIA Mobility+ exam objective covered in this chapter is 1.1, "Compare and contrast different cellular technologies."

Introduction to Cellular Technologies

Cellular phones have become quite common in the tech-connected world we live in. Almost everyone has at least one, and they have become part of the "ubiquitous computing" trend we've seen in the past few years, meaning that we expect a constant data connection and accessibility to all of our data and content on every device we own and use. Because "smartphones," as we call them now, do so much more than simply make and receive phone calls—we surf the Web with them, stream music and video, send and receive email, and even perform business functions with them—the technologies and infrastructure that connect them together has to be fast, robust, and secure.

As smartphones and other mobile devices are introduced into the business environment, we also have an increasing need to manage those devices as part of our business infrastructures. This chapter discusses the basics of the transmission and signaling technologies that connect our mobile devices together, and later chapters discuss the security and management aspects of smartphones and other mobile devices in the business enterprise. First, however, let's have a history lesson on how these cellular technologies we so desperately depend on evolved into our high-speed networks of today.

History

In September of 1966, a new, strange, and wonderful television show began, and offered a possible view of the different technologies that we humans might use in our everyday lives in the Future (capitalization intended). It had spaceships and transportation devices that could whisk a person instantaneously from one location to another one far away. It also had very small box-like devices that people entered and viewed data on, and even small handheld radio-like devices that could be carried around with you and used to communicate to other people all over a planet! That show, of course, was *Star Trek*, and it introduced people to a wide array of very cool devices that everyone wanted to have. Some of those cool gadgets introduced on *Star Trek* have actually been invented and have made their way into the mainstream use of us humans still stuck here in the Present Day. Tricorders, as they were called on the show, look strangely like the wireless tablets we now use, and the *Star Trek* communicator looks an awful lot like a smartphone. Sadly, we don't have our own personal spaceships or matter transporters or phasers, but maybe someday!

In any case, in 1966, and in fact, up until 1975, most telephones were still wired. They had no special features, didn't come in dazzling colors, and only made and received phone calls over analog lines. What people referred to as "mobile phones" were really expensive mobile radio sets weighing in at several pounds; they had been developed in the 1940s by AT&T for very large corporations and the military. These devices used standard radio waves and technologies of the day, and usage was typically limited to large trucks and boats—they weren't really "mobile," as we use the term today. Then, in 1973, the first mobile call from a handheld phone was made by Motorola's Dr. Martin Cooper, who called a colleague at Bell Labs. Over the next decade, the technology was developed and finally marketed to the general consumer public.

Consumer adoption of cellular technologies came about over several years and generations of technologies, beginning with the first generation analog technology from 1978 through 1983. Despite limited talk time and battery life, the consumer demand for mobile phones took off fairly strong, spurring further development that led to smaller devices, longer battery life, increased talk time, less time to charge devices, and better quality services. Figure 4-1 shows an example of one of the early clunky Motorola handsets and transceiver sold to the public for use in vehicles. By the 1990s, consumer demand for cellular technologies was well established, and two frontrunners in the primary technologies were also established, Code Division Multiple Access (CDMA) and Global System for Mobile Communication (GSM). The particulars of these two major players in the cellular technology world will be discussed further, but both gave rise to the ubiquitous use and proliferation of cellular phones that we currently have, before eventually evolving into the high-speed mobile broadband that carries most of our data today.

 EXAM TIP You may not get any cellular "history" type of questions on the exam per se, but you will likely get asked about older technologies such as GSM and CDMA.

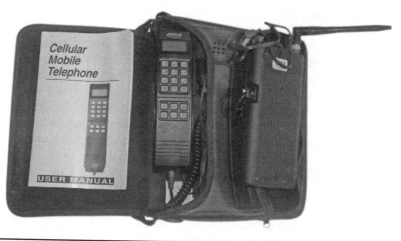

Figure 4-1 An example of an early mobile phone with transceiver

Basic Terms and Concepts

Before you plunge into the alphabet soup of terms associated with cellular technologies, such as CDMA, GSM, and so on, it's probably helpful to go over some basic terms that will help clarify the technologies discussed in this chapter. We've all heard the hype and commercials over the latest and greatest cell technologies. (You've seen the 4G commercial on TV with Justin Beiber and Ozzy Osbourne, right? *How many Gs are there, anyway?* That's a paraphrase, of course. Ozzy states it so much more eloquently in the commercial.) And then there are the "Whose coverage is better?" commercials (amazing the trickery that can be performed with reverse-colored maps). Most new technicians, and certainly the average consumer, don't have a firm grasp on the basic concepts of how cellular really works. Let's try to clear that up with a discussion of some basic terms and concepts.

What's a G, Anyway?

Cellular networks and devices are often referred to by the generation of technology they represent. First generation (or, 1G) is the older analog technologies that first gave birth to cell networks. You likely will not see any 1G devices or networks in the field any longer. 2G (second generation) technologies were the first fully digital-capable networks and devices that were implemented on a widespread basis. Third generation (3G) is what most of us who have had cellular phones for a while now came to know as the standard up until just a few years ago. You also might have seen some carriers advertise for incremental (2.5G, for example) technologies that helped bridge the gap between 2G and 3G. Obviously, with each "G" came better performance, faster data rates, greater coverage, better reliability, and more features. Table 4-1, later in the chapter, lists the major characteristics of the different "G"s that cellular technologies have evolved through.

Frequency Ranges

Now let's talk a moment about frequency ranges. Fresh off of reading Chapter 3, you'll remember (hopefully) how frequencies work and how the electromagnetic (EM) spectrum is divided up into different frequency ranges, based primarily upon the type of electromagnetic energy, or radiation, involved. Earlier cellular technologies use the 850 MHz range, putting it in the lower Ultra High Frequency (UHF) part of the EM spectrum. Later, other technologies used the 900 MHz, 1.8 GHz, and 1.9 GHz ranges, and also UHF frequencies. There were several reasons for using these lower frequency ranges. First, the wavelength is very short, making the devices require less transmit and receive power to operate. Second, these shorter wavelengths are easily reflected off of obstructions and are absorbed by some objects. These frequencies are considered line-of-sight, meaning they require an almost direct, (relatively) unobscured view of the tower or antenna.

 EXAM TIP You also may see 1.8 GHz and 1.9 GHz expressed in terms of MHz for consistency reasons, as 1800 MHz and 1900 MHz, respectively.

As different signaling technologies emerged and evolved, different frequency ranges were used. I'll point out ranges specific to each technology in the sections that follow when I discuss them. As far as frequency usage goes, you'll need to understand *frequency reuse*. It actually has a lot to do with why we have "cells" in the first place, so let's go ahead and talk about cells, towers, and equipment used in cellular technologies.

 TIP The four major frequencies used in cellular, 850 MHz, 900 MHz, 1800 MHz, and 1900 MHz, have been used differently (and at different times) between the United States, Asia, Africa, and Europe. Typical U.S. frequencies are 850 MHz, 900 MHz, and 1900 MHz. Europe has used 450 MHz, 800 MHz, 900 MHz, and 1800 MHz, as well as others.

Cells

Why is it called "cell"? We talk about cell phones and cell technologies all of the time, but most non-techies really don't know what the term "cell" means. So, let's get that out of the way now. "Cells" refer to the areas of coverage from a particular transmitter setup and tower. A cell is a geographical area, and is used because the effective signal range and transmit power of antennas is limited, of course, but also because of the reuse of frequencies in each geographic area. Frequency reuse is important in cellular technologies. Because of the limited availability of frequencies in a band assigned to a carrier versus the enormous number of voice calls that use those frequencies all at the same time, the frequency band can get pretty crowded. Obviously, carriers want their customers to have quality calls that have minimal interference, so frequency reuse methods were devised to carry all of those phone conversations simultaneously and effectively. To prevent frequency interference issues, the frequencies aren't used in adjacent cells; rather, they are used across cells that are non-adjacent so that frequency reuse can be managed better and help to eliminate interference. Frequency use is typically mapped and managed over a seven-cell area. For example, let's say that you have cells labeled as A through G. They use frequencies that we'll arbitrarily call 1 through 7. Now, we can use each of those frequencies within the cells, but we can't reuse them (assign them to other cells) until we are out of range of any cell that already uses them. Doing so could cause interference or coverage issues. So if cell "A" uses frequency 1, we won't allow that frequency to be used again until we get seven cells past that one, say cell "H" several miles away. Of course, cellular devices have to be able to dynamically change frequencies within the band they are using as they travel from cell to cell (this is called *call hand-off*), so most mobile devices are configured to use multiple frequencies within a band.

You'll notice from Figure 4-2 that a cell is hexagonal-shaped. Although various shapes are possible (and have been used), the most common is that of a hexagon. This is considered the ideal standard because of the way towers and transmitting equipment can be positioned within the cells. Because cells can vary in size, additional towers and transmitting/receiving equipment can be placed in different locations within the cell to ensure coverage. Use of a hexagonal cell maximizes coverage and provides for some overlap of coverage areas and frequencies for when a mobile device leaves one coverage area and transitions to another. You may have experienced a temporary loss of coverage where you get no real cell signal in an area. This is referred to as a *dead zone* and is usually the result of a very small area that happens to fall exactly between cell coverage areas. Although this is rare these days, it can still happen despite the increased concentration of cell towers and coverage areas.

EXAM TIP Remember that cells are just geographical coverage areas, typically laid out in the shape of a hexagon. There's no definite standard for how large a cell can be, as it depends on the coverage area terrain, obstacles (such as buildings), and the transmitting power of the equipment in the cell.

Figure 4-2
Conceptual diagram of adjacent "cells" and coverage areas

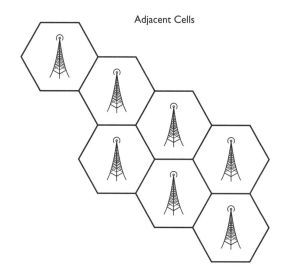

Adjacent Cells

Cellular technologies, of course, use towers with antennas to transmit and receive signals. We discussed different types of antennas in Chapter 3, and cellular towers and antennas work the same as discussed in that chapter. Cell towers can come in different sizes, depending upon the coverage area they are providing. Cell technologies can actually use several different types of antennas, but we usually see the planar variety of antennas (and sometimes some that sort of resemble a Yagi-style of antenna) associated with cell towers. Figures 4-3 and 4-4 show a cell tower and cellular antennas, respectively, similar to what you might see driving around.

Figure 4-3
A rural cell
tower

Figure 4-4
A close-up of cellular antennas mounted on a tower

Cellular Infrastructure

Now that you know what Gs and cells are, let's take a look at some of the generic equipment and infrastructure that cellular technologies use. Obviously, this chapter can't cover every single piece of equipment that cell technologies use, but I'm going to highlight a few important pieces that you may run into as a technology professional who may be working on this equipment. One other item of note is that the signaling technologies and standards discussed later on in the chapter (CDMA, GSM, and so on) may have their own unique pieces of equipment as well, so what I am discussing here is generic to most technologies.

The first piece of equipment, of course, is the mobile device, capable of transmitting and receiving voice and data over a limited distance. The mobile device may be carrier-specific, as well as signaling technology specific. We're well past the days of "devices" meaning only cellular phone handsets. Nowadays, a "device" can mean a smartphone, a tablet with built-in cellular network capabilities, a cellular/broadband USB dongle that is used for laptops, or even a 3G/4G "hotspot" device that allows wireless clients to use cellular broadband to access Internet services. Obviously, the cell tower itself is another piece of equipment. Cell towers can come in all shapes and sizes, and may be designed to cover an area of tens of kilometers, or a small area of just a few kilometers.

Beyond the subscriber-level devices, there are also infrastructure devices that are necessary to make cellular work. Some of these devices sit at the cell level, while others may cover a broader geographic area of several cells. The first basic device I'll discuss is the base station (BS). Each cell has a minimum of one base station, which has its own antenna and coverage area within the cell. A BS is assigned a particular group of frequencies that it uses for devices that connect to it. Other BS in the surrounding area may have different frequencies to prevent interference. The BS is the basic entry point for a mobile device signal into the network.

The base station in turn connects to a "backhaul" network, much like a backbone in a wired network. Several base stations connect via a central backhaul infrastructure to a station known as a Base Station Controller (BSC). The controller connects mobile devices to each other via base stations, as well as to the carrier's network and beyond. The controller can be used to connect base stations from several cells together, help manage frequency reuse across several cells, and handle call handoff as a mobile device transits between base station coverage areas and cells. The BSC also connects to another piece of infrastructure, called a Mobile Switch Center (MSC). The MSC is used to route calls between networks, and provides other services to subscribers as well, such as geo-location services, authentication, and so on. The MSC's routing between networks allows calls to connect to the good old hardline telephone network, the Public Switched Telephone Network (PSTN). Obviously, these are just the key major pieces of a cellular network; there are also dozens of other components that help these devices function and make up the carrier's network, but you should be familiar with the ones we've discussed. Figure 4-5 shows notionally how these pieces of the infrastructure connect together.

In addition to these pieces of equipment, it's probably also worth talking about micro-, pico-, and nanocells. These devices are used to enhance, extend, or add coverage to a particular small area, maybe at the edge of a cell where there is sparse coverage (think about the area on the edge of town, or out on the open plains, for example), or if the coverage area is extremely saturated with devices, closed in, or blocked by heavy obstacles (think airports, parking decks, train stations, football stadiums, and so on) A microcell may have a coverage area of around 100–2000 meters, where a picocell may cover a smaller area of less than 100–200 meters (these numbers are general figures, by the way—there's no formal standard for them). Nanocells may be almost nothing more than signal boosters for a room or small building. Note that these smaller cells can also be set up as only temporary solutions if needed—for example, for an event. A couple of things about these smaller cells: First, there's no set distance that each of these cells cover; this generally depends on the environment it is working in, the equipment itself,

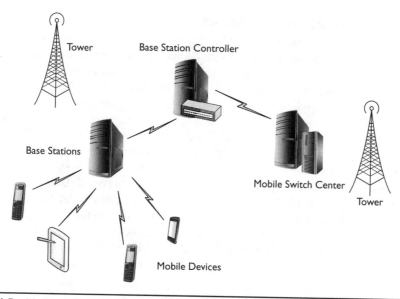

Figure 4-5 Notional diagram of cellular components and infrastructure

and how it's configured. Second, because there's no real standard that defines each of these smaller cells, you will likely hear these terms used interchangeably. For example, AT&T has a product it calls a microcell that has a range of about 40 feet. Go figure.

 EXAM TIP It's likely you won't see much on the exam about the infrastructure devices that make up a cell network because they vary between carriers, but you may see some of the basic terms and concepts tested on the exam, such as cells and the "G"s. Make sure you understand these terms and concepts well.

Signaling Technologies and Standards

There's a veritable alphabet soup of cellular standards and technologies—and to make it even more confusing, different carriers (cellular service providers, such as AT&T, Verizon, Sprint, T-Mobile, and so on) have all, at one time or another, used and promulgated different standards, so, up until very recently no single standard has come out as the "best" one to use. These signaling technologies use various techniques to accommodate the amount of calls assigned to a given frequency band, such as code division, time division, and so on. I'll talk about these techniques, and go in depth on the various signaling technologies used over the generations of cellular up until this point in time.

Code Division Multiple Access

Code Division Multiple Access (CDMA) is one of the older technologies used in cellular communications, and in fact, along with Global System for Mobile Communication (GSM, discussed shortly), evolved into one of the two primary competing technologies that most major carriers used from early on in the evolution of mobile networks. CDMA was developed by a company called Qualcomm and is a spread-spectrum technology. The term *multiple access* in our context means that a technique is used to allow multiple users to access and use a given frequency band simultaneously. There are different ways to do this; some involve altering frequency access times (time division), and some involve frequency or code division. Recall from the discussions on signaling technologies back in Chapter 3 that spread spectrum signaling spreads its data all over a particular frequency range, instead of just one channel. In the case of CDMA, spread spectrum may send multiple calls over a given frequency range of several frequencies at once and is able to identify data from each call by unique codes attached to the data. Early adopters of CDMA included Verizon and Sprint in the United States, although GSM has rapidly become the standard to use in the United States over the past few years.

TIP   You may want to review the discussion on the various spread spectrum technologies from Chapter 3 to help you understand how CDMA works.

CDMA devices have the user, or *subscriber*, information embedded in the phone itself, which makes it difficult to switch devices, even within the same carrier, without some carrier-intervention programming involved. This is unlike GSM, explained shortly, which uses a removable chip containing the subscriber information on it. Because CDMA was rapidly being supplanted by GSM as it became the de facto worldwide standard, it was forced to move through some changes in technologies to make the transition to what we know as 3G networks in order to further transition to fourth-generation technologies. We typically refer to the first iteration of CDMA as "cdmaOne," with "CDMA 2000" as the "transitional" technology used to bridge the gap to 3G. One item of note with CDMA technologies: It was basically implemented almost exclusively in the United States and a very few other countries. This made travel with a CDMA phone to Europe, for example, problematic because European countries did not use CDMA. To get an idea of how CDMA manages multiple calls over the same frequency band, take a look at Figure 4-6. A single call could conceivably spread over several frequencies at once, and when you add other simultaneous calls into the mix, each sharing those same few frequencies within a band, it can get confusing to track them. CDMA assigns a unique "code" to the call to identify it as it rapidly changes frequencies.

Time Division Multiple Access

Time Division Multiple Access (TDMA) can be a bit difficult to understand for a couple of reasons. First, it's unfortunately used in two different contexts. TDMA was, briefly, its own major signaling technology standard used in cellular networks alongside CDMA

Figure 4-6
CDMA managing multiple calls over a frequency band

Frequencies A, B, and C

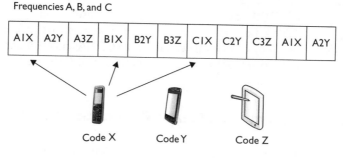

| A1X | A2Y | A3Z | B1X | B2Y | B3Z | C1X | C2Y | C3Z | A1X | A2Y |

Code X   Code Y   Code Z

Mobile Devices 1, 2, and 3

and GSM. This is covered under a standard called IS-136. However, it's also a method for signaling that 2G GSM (and other types of networks, such as an older 2G one called iDEN) is used. That's where the confusion may come in, depending upon whether you're talking about TDMA as a standard or as a frequency access method used by other standards.

Second, instead of using multiple frequencies, and dividing conversations all over them as CDMA does, TDMA basically can take one frequency and divide it up among several phone conversations, with each conversation getting its own "slice" of time allocated to it for the use of the frequency (this is known as time-division access). Now, obviously this happens so quickly that each conversation isn't affected by having to give up the frequency for a bit of time, assuming that the number of conversations and time slice allocations are managed effectively. Figure 4-7 illustrates notionally how TDMA works.

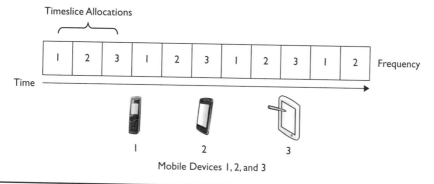

Figure 4-7   Notional view of TDMA

CAUTION Some cellular terms can be used in different contexts, such as TDMA. Remember that TDMA has two meanings: first as a technology used briefly by cellular carriers, alongside CDMA and GSM, and second (and probably more common), as a signaling method that GSM used up until its third generation.

Global System for Mobile Communication (GSM)

GSM is the name for a family of technologies and standards that have slowly evolved over time to become the current fourth generation (4G), and 4G-Long Term Evolution (LTE) implementations, as well as legacy 3G technologies. Although true GSM is pretty much defunct (it's really considered a 2G at best technology), the term is still widely used to describe the family of technologies, and their replacements, that descended from it, such as GPRS, Edge, and others that are covered in the next few sections. You may have heard of the term PCS (Personal Communications Solutions), which was also used to reference the early 2G GSM technologies. Early GSM used TDMA as its signaling method, but the versions that came with 3G and beyond no longer use TDMA. GSM has made the transition to 4G technologies much more uniform and standardized, and emerged as the dominant basic signaling technology used in cellular networks.

GSM was developed in Europe and widely adopted there and elsewhere worldwide, except for primarily the United States. In the United States, CDMA became the dominant standard because of its early adoption by giant carriers such as Verizon and Sprint. As mentioned, this made travel to countries that used GSM problematic for subscribers with CDMA-based devices. The two technologies are incompatible, as are their devices.

Devices that operate under GSM standards typically have two components: the mobile device itself, and the small cardboard-like removable "chip" that fits inside a small slot on the device, the Subscriber Identity Module (SIM). The SIM is sometimes called the SIM *card* (incorrectly, sort of like ATM machine, PIN number, NIC card, and so on, but hey, what can you do?). The two components work together, and the device can't work without the SIM installed in it. The SIM contains information about the subscriber who owns the phone. It is used to authenticate the subscriber to the carrier's network, and can also be used to store some user-related data or content, such as the phone contacts. As opposed to phones that operated on CDMA networks, the use of a SIM provides a bit of freedom of choice and mobility because it's a relatively easy matter to move a SIM from one compatible phone to another to upgrade or change devices and keep the user information intact. Figure 4-8 shows examples of SIMs from an old iPhone 3G (on the left) and one that came from an Android device, a 4G LTE phone.

It's interesting to note that because of worldwide compatibility, as well as the progression to faster and more efficient 4G technologies, most CDMA and TDMA carriers have, over time, switched to the GSM technology standards. Usually this switch has been accomplished using bridging-type of signaling technologies and dual-standard compatible devices, as mentioned earlier. The next few sections detail the evolution of GSM and CDMA through 2G, 2.5G, 3G, and finally, to current 4G standards and

Figure 4-8

L-R: SIMs from an
iPhone 3G and an
Android phone

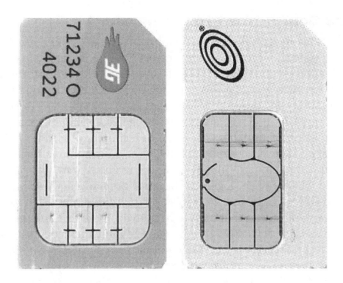

technologies. Although a bit of a history lesson, because these technologies are rapidly becoming obsolete, you may still encounter them in smaller carriers or older implementations. You also may see these terms on the exam!

 CAUTION Although I describe some of the 3G and 4G technologies in this chapter as being from the same GSM family, the truth is that although they may have some characteristics in common with GSM, they are usually significantly different in various ways. I refer to them in such a way because they may have been upgrades from GSM, or have replaced GSM entirely or incrementally. The point is that they are part of the GSM evolutionary path. Some cellular technologists may not refer to them that way in other texts!

CSD

Before I discuss 3G and its transitional technologies, I should briefly talk about Circuit-Switched Data (CSD). CSD was a part of the original GSM technology in the 2G days, when GSM used TDMA as its signaling technology. CSD requires an established circuit (permanent or virtual) to transfer data, so a single data session on a device would require an entire dedicated circuit. This is different than packet switched networks that came later, which don't require a dedicated circuit. Although a much older technology than is used today, it is still possible to run across this technology in isolated areas occasionally. CSD had very low data rates, usually ranging in increments of 2400, 4800, and 9600 bps (bits per second) through 14.4 Kbps (kilobits per second). A later iteration of CSD, High Speed CSD, allowed increased data rates of 38.4–56 Kbps.

Third Generation (3G) Technologies

As carriers planned on the next-generation of cellular, an organization called the 3rd Generation Partnership Project (3GPP) formed in order to facilitate the creation of uniform interoperability standards that could be applied worldwide to the different networks, vendors, and carriers. The goal was to create a set of third-generation standards that ensured interoperability and uniformity, and have them formalized by the International Telecommunications Union (ITU), a globally recognized standards organization. As GSM was already the dominating standard signaling technology for most of the globe, it was pretty much the de facto selection by the 3GPP for transition to a formalized standard. Along with setting the standards, the 3GPP also established a roadmap for transitioning legacy GSM and CDMA technologies to the third generation. These transitional technologies included GPRS, EDGE, and EVDO (all defined and discussed in the upcoming sections). One important note is that although many of these technologies use similar signaling methods as GSM or CDMA, and in fact were implemented along the development lines of those technologies, they are each unique and have their own individual characteristics as well as those in common with GSM and CDMA. Many professionals classify them in the same "families" as GSM or CDMA, while others insist they are separate and different technologies. For our purposes here, I'll discuss them as being part of their respective GSM or CDMA families. Figure 4-9 shows a 3G iPhone as an illustration of how phones changed styles and capabilities from 2G to 3G technologies.

Figure 4-9
An iPhone 3G

GPRS

General Packet Radio Service (GPRS) was considered an "incremental G" designed to carry GSM toward the third generation. Strictly speaking, it's not GSM, but a logical replacement for 2G GSM. It was first introduced as an installed technology in 1999 and used TDMA as its access method. This "2.5G" technology introduced the addition of packet-switching technologies to complement the existing circuit-switching infrastructure used by GSM. While 2G GSM is strictly CSD, GPRS is strictly packet-switching. Packet-switching, rather than requiring an entire dedicated circuit, can send discrete packets over different paths. Information in the packet allows it to be reassembled, in order, with other packets upon arriving at its destination. This was the first cellular technology that allowed TCP/IP protocols and data (meaning Internet access) over mobile devices. This new and improved version of 2G brought the use of Multimedia Messaging Services (MMS) to phones. GPRS was also significant in that it allowed the use of the new (at the time) Wireless Application Protocol (WAP) stack used to provide web-enabled services and content to mobile phones.

EDGE

EDGE stands for Enhanced Data rates for GSM Evolution, and became popular in the United States after it was introduced by Cingular (subsequently bought out by AT&T). It's another one of those incremental generation technologies (considered 2.75G). Like GSM, EDGE used TDMA as its access method, and offered data rates of up to 236 Kbps.

UMTS

Universal Mobile Telephone System (UMTS) came out in 2002, and was the first real third-generation technology offered. It's not really compatible with the original GSM (although, again, it's considered to be in the GSM family by some folks because it was designed to replace GSM technologies versus CDMA ones), but there are devices that were made for downward compatibility reasons that could switch between older 2G+ technologies and 3G. To confuse matters somewhat, with the entry of UMTS and 3G into the picture, GSM switched from using TDMA signaling to one known as Wideband CDMA (WCDMA), which has almost nothing to do with the older CDMA signaling method I've discussed previously! UMTS supported downloads data rates of 384 Kbps to the device, and upload rates of 128 Kbps. Unlike GSM and GPRS, which only supported CSD or packet switching methods, respectively, UMTS could support both.

EVDO

EVDO (Evolution-Data Optimized, or, you may also see from some sources, Evolution-Data Only) is a CDMA 2000 technology designed to bridge the gap between older CDMA (cdmaOne) and 3G. It's considered in the same class as GPRS and EDGE because it's a transitional technology. One issue that affected EVDO was that you typically could not make voice calls and exchange data at the same time. This was not an issue with GSM-related 3G technologies because part of GSM's specifications dictates simultaneous voice and data capability in the GSM evolutionary path. Fortunately, this limitation has been overcome on the CDMA path by transitioning to the newer 3G and 4G technologies. Because it was a transitional technology, it used both CDMA and TDMA access methods, and is a packet-switched technology.

HSPA and HSPA+

HSPA stands for High Speed Packet Access and is a current 3G technology still in wide use as of this writing. It grew from combining two other protocols, High Speed Downlink Packet Access (HSDPA) and High Speed Uplink Packet Access (HSUPA). It uses WCDMA and is a huge improvement over existing 3G technologies. It's primarily used for devices that don't support 4G technologies, or for 4G phones that are forced to use some existing 3G networks occasionally. HSPA's data rates range from around 14 megabits per second (Mbps) download and 5–6 Mbps upload. HSPA is another "descendant" of the GSM family, and was a logical successor to UMTS. HSPA+ technologies (also known as Evolved HSPA) are a further evolution released in 2010, and offer higher speeds and greater bandwidth, in the neighborhood of 168 Mbps download and 22 Mbps upload. HSPA+ is considered a "quasi" 4G technology as well.

Fourth Generation (4G) Technologies

So far, our discussion on signaling technologies up until this point has been more or less a history lesson because today we've already seen the beginning of the end for most of the previously discussed major technologies. GSM was the clear winner in the battle to be the dominant standard as a result of sheer numbers of subscribers worldwide, its more defined standards, and general better performance with each successive generation or even incremental improvement. Most of the technologies discussed here are descendants or extensions of, or replacements for, the GSM family, used to bridge the technologies from 2G through 3G, and finally to the point where GSM and CDMA become a blur and a new generation of cellular technologies are born. That's where the fourth generation (4G) and its technologies come in. The current and emerging 4G technologies include LTE, LTE-Advanced, and WiMAX, which are explained in the next few sections. Figure 4-10 shows a newer 4G Android phone connected to a 4G LTE network.

LTE

Long-Term Evolution (LTE) is one of the latest technology upgrades in the GSM family. It has been in the works since early 2006 and was introduced into networks in 2009. With the introduction of LTE as the first of the fourth-generation technologies, the long-overdue transition from legacy CDMA and GSM technologies is finally happening. It's likely, however, that some of those legacy technologies will still linger for a while until the carriers finally turn off or upgrade the networks that support backwards compatibility with older devices. Although technically LTE does not meet the specifications the 3GPP folks set for true fourth-generation technologies, LTE, along with WiMAX and HSPA+, are sometimes marketed as such because they are vast improvements to original 3G.

As with any new technology, LTE has had some issues in compatibility, with some phones (and carriers) only supporting it for data, while still using 3G networks for voice calls. Additionally, LTE and other 4G technologies require more power, especially if a phone is trying to locate a 4G network that may be out of range, so high usage may drain batteries more quickly in older phones. Newer devices with better 3G/4G integration likely won't have that issue. Some device manufacturers (Apple comes to mind) were a bit slower in producing 4G-compatible phones than other vendors, so product or brand loyalty may prevent some users from jumping onto the 4G bullet train as fast as others.

Figure 4-10 An Android phone connected to a 4G LTE network

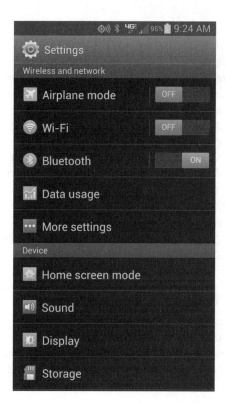

LTE requires a completely new infrastructure than the older 3G network. Because it's primarily IP-based, it focuses more on designs that are reminiscent of high-speed data networks versus older voice ones. LTE offers data rates of 300 Mbps, with upload rates of 75 Mbps, according to the 3GPP specifications. Of course, each carrier markets its own "version" of LTE as being faster and more robust, and covering more geography than the others.

 CAUTION Remember that LTE, HSPA+ and WiMAX aren't considered "true" 4G technologies, as they don't technically meet the ITU standard for 4G. However, they are usually marketed by the carriers as 4G.

LTE-Advanced

LTE-Advanced is the name given to a set of standards approved by the ITU in 2009, and the 3GPP in 2011, which improve upon the LTE specifications significantly. Dubbed a "True 4G" technology by the ITU to differentiate it from other pseudo-4G technologies such as LTE, LTE-Advanced is being introduced into networks worldwide at the time of this writing. LTE-Advanced has a theoretical upper data rate of 1 Gbps (gigabits per second).

WiMAX

WiMAX (an acronym for Worldwide Interoperability for Microwave Access) is the name given to technologies that create a very large wireless network, called a WMAN (wireless metropolitan area network), and usually covers an area the size of the average-sized city. WiMAX is another pseudo-4G broadband technology that uses microwave transmissions to provide wireless connectivity to subscriber stations, typically for those that are mobile or in areas where there is a lack of other services (cable, DSL, and so on). Like other wireless technologies I discuss later in Chapter 5, WiMAX has its own professional trade association comprised of organizations and commercial vendors whose goal is to promulgate the use of the technology. This trade association is known as the WiMAX Forum, created in 2001, and it also helps to develop standards that include interoperability and security. The Institute of Electrical and Electronics Engineers (IEEE) produces a recognized standard for WiMAX technologies, the IEEE 802.16 family of standards, which was most recently updated in 2011. Sprint has been the carrier of note that primarily has used WiMAX, but Sprint is slowly replacing it with other 4G technologies that are faster. WiMAX has a data rate of about 30–40 Mbps. Now that you've looked at all of the "G"s (and minded your Ps and Qs while doing so), Table 4-1 summarizes the major points about each technology.

 EXAM TIP Know the different characteristics of the cellular technology generations (1G–4G) and their associated frequency ranges and technologies. Table 4-1 provides a great last-minute memory review to help you right before you take the exam!

Generation	Technologies	Data Rates
1G (analog)	CDMA/GSM/TDMA (early versions)	28–56 Kbps (kilobits per second)
2G (also known as PCS)	GSM/CDMA	2400 bps to 14.4 Kbps (using CSD) 38.4–56 Kbps (HSCSD)
2.5G	GPRS	56–115 Kbps
2.75G	EDGE	236 Kbps
3G (transitional from CDMA)	EVDO	600 Kbps to 1.4 Mbps (megabits per second)
3G	UMTS	384 Kbps (downlink), 128 Kbps (uplink)
3G (and transitional to 4G)	HSPA, HSPA+	HSPA: 14 Mbps (downlink), 5–6 Mbps (uplink) HSPA+: 168 Mbps (downlink), 22 Mbps (uplink)
4G	LTE, LTE-Advanced, WiMAX	LTE: 300 Mbps (downlink), 75 Mbps (uplink) LTE-Advanced: 1 Gbps (gigabits per second) (theoretical data rate) WiMAX: 30–40 Mbps

Table 4-1 Cellular Technologies Through the Generations

Roaming and Switching Between Network Types

Switching means that a subscriber device has moved into another cell coverage area. Usually this happens when a user moves too far away from the base station or tower in a cell, and the device has to switch to another base station or tower with a stronger signal. This usually happens between cells of the same carrier. The MSC and the mobile unit will detect if transmit/receive power levels get below a certain point, indicating the device is getting out of range of a BS, and switch the device to another BS.

If, on the other hand, the device moves between carriers' cells, perhaps because the user's carrier does not have coverage in a particular area, the device will connect to another carrier's network, assuming it can find one and the technologies are compatible. Most of the time, the user doesn't even know this has happened, except for maybe an icon or alert that pops up on the device itself. This is known as roaming (between carriers), and in some cases, the user may incur usage charges or fees for this. This is especially true when traveling overseas. Most phones have the capability to configure roaming settings, such as use of data while roaming, and so on. Figure 4-11 shows an iPhone 4S's cellular settings screen, where users can turn off cellular data to save on roaming costs, connect to wireless networks, or even turn off faster networks (such as 3G) to fall back to earlier technologies in case the faster networks aren't available. This prevents the device from searching for a network that is out of range, thus saving battery power. Figure 4-12 shows similar settings for an Android smartphone connected to a 4G LTE network on Verizon.

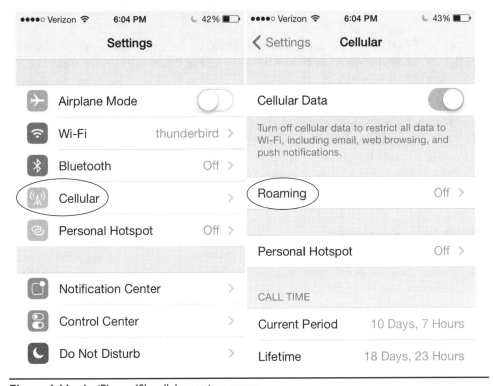

Figure 4-11 An iPhone 4S's cellular settings screen

Figure 4-12
Android's data
network and
roaming settings

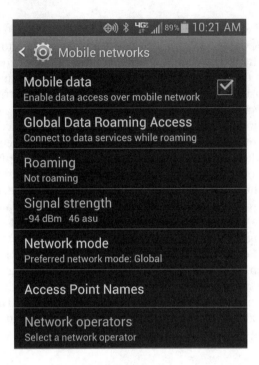

EXAM TIP   Roaming is between cellular carriers, whereas switching is usually between the cells of the same carrier. Roaming may incur fees, whereas switching does not.

Beyond the mere switching of base stations and cells due to signal strength, some switching may occur because the network type may change or not be supported in an area. For example, if a phone is using 4G, and travels to an area where only 3G services are available, it will switch to the supported service once it starts to lose connectivity with the 4G service. In this case, the user may still be connected to her own carrier, but her service may not perform as well as she is used to.

Chapter Review

This chapter has provided an introduction to the mobile cellular and broadband technologies that carry almost all of the data used by mobile phones, tablets, and sometimes even laptops across wide areas. "G" refers to the generation of mobile technologies, including 1G, 2G, 3G, and the latest fourth generation (4G) technologies. These generations have seen several different technologies used, but the most dominant has been CDMA and GSM. CDMA was adopted early on in the United States, while GSM became the de facto standard over the rest of the world. Because GSM and its direct successors eventually won out as a worldwide, formalized standard, CDMA carriers have upgraded and added bridging technologies to slowly move their networks to later generation technologies.

"Cells" are geographic coverage areas, usually architected in the shape of a hexagon. The hexagon shape provides for maximum coverage, while at the same time helping to manage frequency reuse. Frequencies must be reused over a group of cells to prevent interference but cells manage a limited number of frequencies within a given band. Cellular infrastructure is made up of the mobile device, tower, and several infrastructure nodes that assist in frequency handling, call hand-over, backhaul, and connection to the public switched telephone network. The nodes, generically, are the Base Station, Base Station Controller, and the Mobile Switch Center. Additionally, there are smaller devices that can assist in extending a cell's range or providing temporary service to remote areas, or areas where there is high device congestion, such as airports and train stations. These are known as microcells, picocells, and nanocells.

CDMA, one of the primary technologies that developed widespread cellular use, is a spread spectrum technology that slices and spreads multiple calls over a given frequency range, and is able to identify data from each call by unique codes attached to the data. CDMA devices have the user, or subscriber, information embedded in the phone itself, which makes it difficult to switch devices. Most CDMA technologies have evolved to transition to current day 3G and 4G technologies.

TDMA is a method of signaling (and also, briefly, its own carrier technology) that allows several calls to use one frequency, and allocates each call a defined slice of time in which to use the frequency. Circuit-Switched Data requires an established circuit (permanent or virtual) to transfer data, so a single data session on a device would require an entire dedicated circuit. GSM used TDMA as its signaling technology of choice until it transitioned to 3G. GSM is the name for 2G technologies that competed with CDMA and was adopted in most of the world. It was given a formalized set of specifications, and most 3G and now 4G technologies were either extensions or direct replacements of the original GSM. Third-generation technologies that came from the GSM family included transitional technologies such as GPRS and EDGE, while CDMA used EVDO as its transitional technology to 3G. UMTS (replacing GSM) was the first real 3G technology. UMTS moved away from TDMA as its access method in favor of Wideband CDMA (WCDMA), and also changed from circuit-switched to packet-switched networks.

HSPA is a current 3G technology still in wide use, and uses two other protocols, HSDPA and HSUPA. It uses WCDMA, and is a huge improvement over existing 3G technologies. HSPA+ is an improvement to HSPA and is considered an almost-4G technology. LTE is the first of the 4G technologies, although it doesn't meet all of the specs for ITU's 4G standard. Along with WiMAX and HSPA+, it's more of a high-end 3G technology, but it's usually marketed as 4G. LTE-Advanced is considered a true 4G technology by the ITU and 3GPP (standards organizations that dictate specifications and interoperability for cellular technologies). WiMAX is a microwave technology that provides for last-mile subscriber access or backhaul services to broad coverage areas known as wireless MANs, usually the size of a city. WiMAX can be used to service remote areas that don't have cable or DSL services.

Switching is what occurs when a device has to transition to another cell and moves to another tower or BS within the same carrier's network. Roaming means that a device has to leave a cell or coverage area serviced by its carrier and join another carrier's network. This may incur usage charges or fees.

Questions

1. You've been asked to explain to a customer what the "G" in 4G means. You tell the customer which of the following statements regarding cellular generation technologies?

   A. Fourth-generation technologies are the oldest and slowest technologies.

   B. Cellular technologies have evolved over four major generations.

   C. 4G is GSM, while 3G is CDMA.

   D. 4G is the latest phone from Apple.

2. You're helping an engineer design a cell coverage area. What is the optimal design for a cell coverage area?

   A. An octagonal design

   B. A hexagonal design

   C. A rectangular design

   D. A circular design

3. You are explaining how early cellular technologies worked to a friend, and you want to explain circuit-switched data versus packet-switched data techniques. Which of the following statements is true regarding those technologies?

   A. Packet switching must send the entire message in one large chunk of data.

   B. Circuit-switching does not require a dedicated circuit.

   C. Packet-switching does not require a dedicated circuit.

   D. Circuit-switching sends out data in discrete packets that can later be reassembled.

4. Which organization is dedicated to developing and promulgating interoperability standards amongst cellular technologies?

   A. 3GPP

   B. WiMAX Forum

   C. Wi-Fi Alliance

   D. ITC

5. Which of the following cellular technologies was adopted early on in Europe and soon became the dominant cellular standard in the world?

   A. GPRS

   B. LTE

    C. GSM

    D. CDMA

6. You are teaching a class on cellular signaling technologies, and one of students has a question about CDMA access. Which of the following best describes CDMA's frequency access method?

    A. CDMA assigns each call to a particular frequency in a band and identifies the frequency using a unique code.

    B. CDMA allocates only one call per frequency band.

    C. CDMA uses discrete time slices to allocate the same frequency to several calls on a rotating basis, allowing them to each use the frequency for very small lengths of time.

    D. CDMA is a spread spectrum access method that can spread several calls across a frequency band using unique codes to identify the calls.

7. What would a user have to do in order to change devices on a GSM network?

    A. Remove the Subscriber Identity Module from the old device and insert it into the new device.

    B. Contact the carrier and arrange for a new device to be programmed and shipped to the user.

    C. This cannot be done on a GSM network without purchasing a new device and registering as a new user.

    D. Contact the carrier and arrange for a new Subscriber Identity Module to be programmed and shipped to the user.

8. Which technologies can use either circuit-switching or packet-switching methods? (Choose two.)

    A. UMTS

    B. EVDO

    C. GSM

    D. GPRS

9. As a junior-level cellular technician, you've been asked by your supervisor to explain the differences between CDMA and GSM. She wants to test your knowledge of the two technologies and asks you to tell her about 3G transitional technologies for CDMA. Which of the following is applicable to her question?

    A. GPRS

    B. EVDO

    C. EDGE

    D. UMTS

10. What is the term used to formally transition a device from one cell coverage area to another?

   A. Onboarding

   B. Device rehoming

   C. Offboarding

   D. Call hand-off

11. Which cellular infrastructure node is used to route calls between networks?

   A. BS

   B. MSC

   C. BSC

   D. PSTN

12. You've been asked to provide a solution for a congested airport that has hundreds of users with mobile devices passing through it every day. Which extension of a cell network can provide service to a small area that may be remote or saturated with mobile devices, such as an airport?

   A. Kilocell

   B. Minicell

   C. Microcell

   D. Megacell

13. You hear a coworker talking about PCS (Personal Communications Services), and you believe they are incorrect about what PCS is. Which of the following is true regarding PCS?

   A. PCS refers to second-generation GSM services.

   B. PCS is a trademarked name for Sprint's 4G LTE services.

   C. PCS refers to third-generation CDMA technologies.

   D. PCS refers to push-to-talk services offered by some cellular carriers.

14. Which of the following technologies was the first to offer TCP/IP data services?

   A. LTE

   B. EVDO

   C. EDGE

   D. GPRS

15. Which technology supports a data rate of 168 Mbps down and 22 Mbps up?

   A. HSPA

   B. HSPA+

   C. HSDPA

   D. HSUPA

16. Which of the following fourth-generation technologies has an upper theoretical data rate of 1 Gbps?

    A. LTE-Advanced

    B. LTE

    C. WiMAX

    D. HSPA+

17. Which of following technologies are covered by IEEE standard 802.16?

    A. CDMA

    B. LTE

    C. WiMAX

    D. HSPA+

18. You are explaining fourth-generation technologies to a customer. All of the following are characteristics of 4G you can explain to the customer EXCEPT:

    A. Gives the user the ability to stream various media to the device

    B. Provides higher data rates

    C. Requires more power

    D. Uses circuit-switched data methods

19. Which of the following switching methods can UMTS use? (Choose two.)

    A. Message-switched data

    B. Circuit-switched data

    C. Packet-switched data

    D. Call-switched data

20. What is the term used when a device travels out of range of its home carrier coverage and has to use the services of another carrier altogether?

    A. Switching

    B. Roaming

    C. Call hand-off

    D. Dead zone

Answers

1. **B.** Cellular technologies have evolved over four major generations: 1G, 2G, 3G, and 4G.

2. **B.** A hexagonal design is the optimum design for a cell coverage area.

3. **C.** Packet-switching does not require a dedicated circuit and sends data out in discrete packets that can be reassembled after they reach their destination.

4. **A.** The 3rd Generation Partnership Project (3GPP) was formed in order to facilitate the creation of uniform interoperability standards that could be applied worldwide to the different networks, vendors, and carriers.

5. **C.** The Global System for Mobile Communication (GSM) was developed and adopted early in the cellular evolution and became the dominant cellular technology from which most modern standards have evolved.

6. **D.** CDMA is a spread spectrum access method that can spread several calls across a frequency band using unique codes to identify the calls.

7. **A.** The user should only have to remove the Subscriber Identity Module (SIM) from the old device and insert it into the new device.

8. **A, B.** Both UMTS and EVDO can use either circuit-switched or packet-switched methods.

9. **B.** EVDO is a CDMA2000 technology designed to bridge the gap between older CDMA (cdmaOne) and 3G.

10. **D.** Call hand-off is the term used to formally transition a device from one cell coverage area to another.

11. **B.** The Mobile Switch Center (MSC) is used to route calls between networks, and provides other services to subscribers as well, such as geo-location and authentication.

12. **C.** A microcell may have a coverage area of around 100–2000 meters, and may be used to enhance, extend, or add coverage to a remote area or an area that is extremely saturated with devices, closed in, or blocked by heavy obstacles.

13. **A.** PCS refers to second-generation GSM services.

14. **D.** GPRS was the first cellular technology that allowed TCP/IP protocols and data over mobile devices.

15. **B.** HSPA+ supports data rates of 168 Mbps down and 22 Mbps up.

16. **A.** LTE-Advanced has an upper theoretical data rate of 1 Gbps.

17. **C.** WiMAX is covered by IEEE standard 802.16.

18. **D.** 4G technologies do not use circuit-switched data (CSD) methods.

19. **B, C.** UMTS can use both circuit-switched data and packet-switched data methods.

20. **B.** Traveling between coverage areas offered by the user's home carrier and having to connect to another carrier's network is called roaming, and user may incur charges or fees to use that carrier's services.

Wi-Fi Client Technologies

In this chapter, you will

- Configure and implement client Wi-Fi technologies
- Conduct site surveys and interpret site survey information

The next type of mobile transmission technology I'll discuss is the type you normally see nowadays in homes and businesses that transfer the bulk of our data to the Internet. Wireless networking technologies have been around for a number of years and have become much more commonplace in most homes and offices than older cabled networks. Wireless networks enable us to become truly mobile, especially with devices such as laptops and tablets that we use to extensively process, transmit, and receive data. Rather than being bound to a heavy old desktop, we are free to roam throughout the house and the office, or even sit out on the back deck and surf the web, send email, stream music, and (occasionally) even get work done. This chapter covers several different aspects of wireless networks, including wireless standards and organizations, and the different technologies used in wireless networks. You'll take a look at different wireless security technologies and learn about the importance of conducting site surveys. You'll also learn how to interpret the data that you get from those surveys in order to help you plan, implement, and maintain wireless networks. In terms of the CompTIA Mobility+ exam, this chapter focuses on two critical objectives: 1.2, "Given a scenario, configure and implement WiFi client technologies using appropriate options," and 1.4, "Interpret site survey to ensure over the air communication."

Introduction to Wi-Fi

The term "Wi-Fi" is almost universal in describing wireless networks. In particular, we use this phrase to describe 802.11 wireless networks, which have become commonplace in our homes, offices, hotels, and even restaurants and cafés. Wi-Fi as we know it started out several years ago with the approval of particular frequency bands for use in consumer wireless networks by the Federal Communications Commission (FCC), and the subsequent standards issued by the Institute of Electrical and Electronics Engineers (IEEE) . The next few sections discuss a little bit of the history of Wi-Fi, and some of the relevant concepts and terms that you'll need to be familiar with for both the exam and real life.

History

Wireless networking got its start into the mainstream of consumer and business data implementation several years ago with the introduction of the 802.11 wireless standards. Although we refer to the family of wireless standards as the 802.11 standards collectively, the original 802.11 standard was never widely implemented, and for some good reasons. It did not offer higher speeds or data throughput, and devices and equipment used to implement the standard were not widely available. In fact, at the time the original 802.11 wireless standard was approved, most people were only beginning to hear of the World Wide Web and the Internet, and a great many people probably did not even have computers in their homes. Most of the few who did probably used dial-up Internet access or even more expensive Integrated Digital Services Network (ISDN) connections. There were not a lot of companies offering any type of serious Internet service provider (ISP) services, let alone high-speed broadband access. So even though the first wireless standard was approved for use in the consumer market, the consumer market did not really have the infrastructure or demand for it.

That all changed when computers became more common in homes, and everyone began to surf the Web, send and receive email, and discover what the Internet was all about. When that happened, most people wanted more bandwidth and faster connections, so cable and phone companies began to offer higher speed connections in the form of cable modems and Digital Subscriber Line (DSL) connections. In addition, satellite-based Internet services were offered to those who were out of reach of cable company services and beyond the distance required for DSL connections.

As higher bandwidth and speed was introduced into homes and businesses, and became more affordable, people added more devices to their homes. This usually meant someone had to string cable all over the house or the office, and that someone had to be knowledgeable in doing so. The average person didn't necessarily have the ability or the knowledge to create a wired network in their home that serviced more than two or three devices. So, as the number of computers and Internet-connected devices in homes increased, the need for wireless technologies became apparent. Wireless device manufacturers, of course, saw this need and began to produce devices that used newer wireless technologies, offered faster connections, and supported multiple devices. The explosive growth of technology was self-perpetuating because the more wireless capability that was introduced to the consumer, the more devices were produced that required this technology, and the more devices the consumer purchased. In addition to desktop computers, laptops, cell phones, and tablets, even television sets now have wireless capability.

We now live in such a wirelessly connected mobile world that we'd likely be lost without the wireless technologies we've become accustomed to. Because we are used to having information at our fingertips, in near real time, we demand higher bandwidth, faster speeds, and the ability to access information from virtually anywhere in the world. This obviously creates a long-term market for technology professionals who know how to support mobile devices and wireless infrastructures. That's where you come in!

Basic Terms and Concepts

To understand the fundamentals of wireless networking, it's important to be familiar with several terms and concepts associated with wireless networks. Knowledge and understanding of these concepts and terms are required by the CompTIA Mobility+ exam in the exam objectives, and you should take the time and effort to understand these concepts and terms thoroughly before sitting for the exam.

Clients

A *wireless client* is a device that connects to a wireless network. A client can be a laptop, desktop, tablet, smartphone, network server, or just about any other device that has a wireless network card. Try not to confuse the client (device) with the user (person using the device) who is trying to connect to and use the wireless network.

 TIP You'll also see the term *station* (or STA) used in more technical documentation to refer to a wireless client.

Wireless clients typically have a wireless network card capable of using specific wireless technologies to connect to a wireless network. They may come preconfigured to connect using specific information for the network, or they may be configured on the fly. Most wireless devices, especially consumer ones, come with standard operating systems such as Windows, Linux derivatives, or Apple. Some commercial and industry devices come with embedded (chip-based) OSes for specialized use, and have limited configuration options to connect to wireless networks. Clients can have wireless network cards that fit into a slot on the motherboard (as in the case of desktop clients), use older Personal Computer Memory Card International Association (PCMCIA) card slots, connect via USB, or are even integrated into the motherboard. Some examples of common wireless clients are shown in Figure 5-1, as well as a variety of both legacy and newer wireless adapters for clients in Figures 5-2 and 5-3.

Figure 5-1   Common wireless clients

Figure 5-2
Legacy wireless
cards

Figure 5-3
Modern USB
wireless adapters

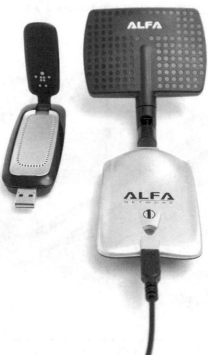

Access Points

Wireless access points are network devices that clients connect to in order to access other devices, other networks, and the Internet. An access point can be a small home-use device, a simple SOHO (Small Office Home Office) device, or an enterprise-level device with several features that control access to multiple networks, including wired networks. These devices can offer security features, accounting features, and control access to both networks and data by specific users. Some examples of common access points are shown in Figure 5-4.

Portable Hotspots

A portable hotspot is typically a small device that has access to cellular technologies such as 3G and 4G, and provides access to these networks for Wi-Fi devices. Most of these devices can be purchased from wireless providers such as Verizon, Sprint, AT&T, T-Mobile, or other carriers, and are usually specific to their type of broadband network. These devices can provide wireless access for up to five or ten devices at a time. They're basically wireless routers that route traffic between Wi-Fi devices and broadband technologies. Depending upon the carrier, some cellular phones, as well as some tablets, can act as portable hotspots. Many of these devices can be purchased as dedicated hotspots.

The popular term used for these portable hotspots, as well as devices like cellular phones that also provide hotspot service, is MiFi, which stands for My Wi-Fi. Novatel Wireless actually owns the trademark to this name in the United States, but it has become common to refer to most of these devices as MiFi-capable. Figure 5-5 shows a screenshot of an Android phone acting as a portable hotspot.

Figure 5-4   Different types of access points

Figure 5-5
An Android
phone acting as a
portable hotspot

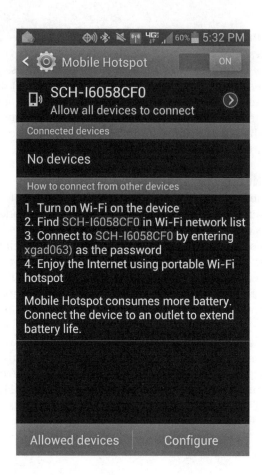

Ad-hoc and Infrastructure Modes

Wireless networks operate in two distinct modes of operation. Whichever mode is used depends upon whether the wireless network is centrally managed or is a simple peer-to-peer network connected to on-the-fly between two or three wireless devices. The first type of network is the *ad-hoc* network. This type of network usually consists of just a few devices that are connected together for the purposes of sharing files or gaming or Internet connection sharing. These networks are typically characterized by low security settings, and are also relatively close to each other. For these networks to work, each client device has to be configured with identical configuration settings, including network name; security settings, such as authentication and encryption; and any passwords or shared keys used to connect to each other. Ad-hoc networks are usually found in places where one user wants to connect to another user's device for a short period of time to share files or play games with the other user and so forth.

The other type of network is called *infrastructure mode* and is usually characterized by several client devices that connect to a central access point, usually a wireless access point (WAP) or, in larger organizations, a wireless LAN controller. The WAP enables clients to connect to one another and to another network, such as a wired network, or even the Internet. Most of the wireless networks found in homes, small offices, and larger corporate or enterprise-level networks, are infrastructure mode networks. Infrastructure mode typically means that the wireless access point is configured with particular settings, such as the network name, security settings, and so on, and the clients that connect to this wireless access point must have the same settings in order to connect to the WAP. Figures 5-6 and 5-7 illustrate both ad-hoc and infrastructure modes.

 EXAM TIP   Understand the difference between ad-hoc and infrastructure mode networks! You may see some questions that test your ability to distinguish between these network types.

IBSS, SSID, BSSID, and ESSID

If you've heard of these terms before, chances are you have been somewhat confused over their meaning. A lot of people, wireless technicians and "experts" included, often (incorrectly) use these terms interchangeably, but the truth is, they don't exactly mean the same thing. Honestly, I've been guilty of that as well, but the CompTIA Mobility+ exam may not be too forgiving if you confuse the terms on the test. Let's see if I can clarify the meaning of these terms for you.

SSID

SSID stands for Service Set Identifier: It basically refers to the unique network name for the wireless network a client is trying to connect to. A wireless client can detect networks on a range of channels and frequencies, and lists those wireless network names

Figure 5-6
Ad-hoc mode

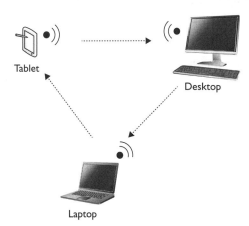

Tablet

Desktop

Laptop

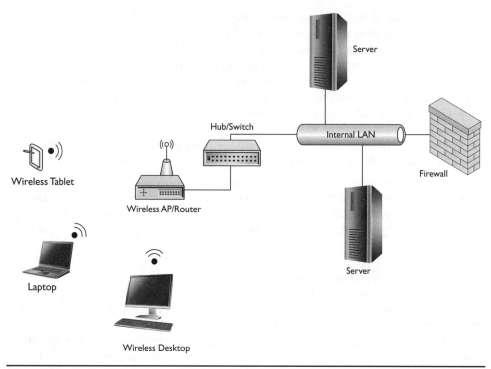

Figure 5-7   Infrastructure mode

it has detected for the user to choose from and connect to. Typically, this requires the wireless access point to "broadcast" the SSID of the network just for that purpose.

Many access points, when they are first powered up, broadcast a default SSID, which can (and should) be changed when the device is configured by the end user or network administrator. Sometimes this doesn't get changed, so it's not unusual for a wireless client to detect different wireless networks, with many of them using the default SSID of Linksys, or Netgear, or some of the other popular manufacturer default SSIDs. This can be confusing for the end user who is trying to connect to the wireless network because he or she may not know which one they really want to connect to just by seeing the SSID. In addition to confusing the end user, it also may give some insight to any would-be hackers detecting these networks; they may assume that because the administrator didn't bother to change the default SSIDs, other default (and unsecure) configuration settings may also still be in place on the WAP.

SSIDs should be changed from their device defaults to a network name that is in line with your organization's naming conventions. Keep in mind that an SSID can be composed of both uppercase and lowercase letters, numerals, and special characters. It can be up to 32 characters in length, and is case sensitive. It may be a name that delineates its purpose or targets a particular client audience, such as *marketing* or *accounting*. It also may be used to separate groups of users and their allowed access. For instance, a wireless SSID of *public* or *guest* would likely be set up as a separate wireless network for a group of users who are allowed to access the network for basic functions, such as email or web surfing, but not allowed to access more sensitive networks or data. On an average street almost anywhere in the United States, you'd probably see many wireless networks with names such as "Anderson" or "RogersFamily" or something similar. While it may provide convenience for guests of these residents to help them connect to the right network, it also gives the bad people an idea of who owns which wireless network and where it's located. Figure 5-8 shows some configuration options in a Linksys wireless router that can be applied to the SSID settings.

Some organizations and individuals intentionally rename the default SSID on their devices to something obscure that doesn't immediately give away information about the network. An SSID such as 0Xb8ZXD really doesn't tell you about the network itself, but it keeps the average person from guessing information from the network name, and possibly to whom it belongs. This technique is called "security through obscurity" and relies on hiding or obscuring information to protect it. The theory here is that if a bad guy doesn't know it's there, it's protected. This typically could not be further from the truth. It really doesn't do anything to stop would-be hackers or malicious users because there are other ways of detecting and intruding upon wireless networks than just detecting the SSID. In fact, one recommendation by security specialists and laymen during the early days of wireless was to turn SSID broadcasting off on the wireless access point. This sounded like a good idea at first, because supposedly it kept unwanted clients from connecting to the wireless network. What it actually did

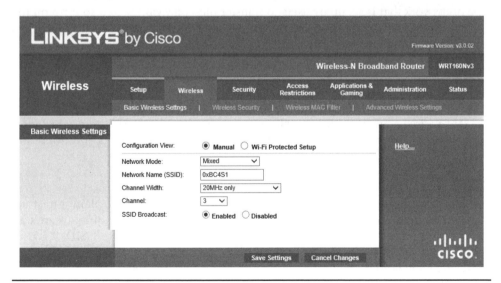

Figure 5-8 SSID configuration settings on a Linksys wireless "N" access point

was confuse legitimate users from easily connecting to the network, unless their client device was preconfigured, or they knew the SSID in advance and manually entered it into their configuration settings. *SSID hiding* or *cloaking*, as it is often called, also sometimes allowed clever hackers to set up rogue access points on the network that did broadcast SSIDs, tricking users into connecting to them. SSID hiding is another example of security through obscurity, and given that wireless networks can be easily detected and attacked by means other than just a broadcasting SSID, it's really not very effective.

The bottom line here is that the SSID is just a network name, and while it should be changed from the defaults on the wireless access points, obscuring it or stopping its broadcast by those devices really doesn't do much for you in terms of security. It may indeed stop people who aren't very knowledgeable about wireless networking from connecting, but beyond that, it should not be used as the sole security measure you rely on to secure your network. I'll cover some better security methods that are much more effective later on in this chapter, and later in the book when we discuss security in depth.

 CAUTION   Don't use SSID cloaking or obfuscation as your sole method of network security as there are other ways to detect and hack into a wireless network even if the SSID is not broadcasting! Use other methods, such as WPA/WPA2 technologies to secure your network.

IBSS

IBSS stands for Independent Basic Service Set, and describes the network in use in an ad-hoc mode network. This network, as described earlier, is basically a device-to-device type of network that is set up by users to transfer files, play games, and so forth on the fly. It's also used when there's no real need for an intermediary device, such as a wireless router, to control the connections.

ESSID

ESSID stands for Extended Service Set ID, and is the network name of a larger infrastructure mode network that has multiple access points, and at some point, connects to a larger wired network. This is usually a corporate network infrastructure, which provides wireless coverage for a large organization, with multiple devices, locations, and user and device characteristics.

BSSID

The BSSID, which stands for Basic Service Set ID, describes the basic plain infrastructure-type of network, typically associated with one wireless access point. There's an important point to consider, however. When referencing a network by BSSID, it is usually the hardware or MAC address of the wireless access point. So, where an SSID and ESSID are referenced as network names, the BSSID is usually the hardware address of a given WAP.

EXAM TIP Understand the differences between SSID, IBSS, ESSID, and BSSID!

Wi-Fi Standards

There are several organizations that control existing Wi-Fi standards. In the United States, the first you should be aware of is the Federal Communications Commission (FCC). This government agency is a regulatory authority for the use of radio frequencies and technologies. Of course, other countries have their own equivalent agencies of the FCC. The FCC allows private and commercial wireless networks to operate on four separate frequency bands. These are the 2.4 GHz Industrial, Scientific, and Medical (ISM) band, the 3.6 GHz band, the 5 GHz Unlicensed National Information Infrastructure (U-NII) band, and the 60 GHz band. Most of the wireless specifications that we are concerned with in this chapter operate in the 2.4 and 5 GHz bands, but the other two do exist and are worth mentioning.

Another organization relevant to this discussion is the Institute of Electrical and Electronics Engineers (abbreviated as IEEE, and typically spoken as "I-triple E"). Their 802 committee is responsible for developing the LAN, MAN, and wireless LAN (WLAN) standards that we use in both wired and wireless networking. The 802.11 standards they developed, which are discussed below, are the ones that dictate signaling, power, security technologies, and other factors used to develop and field a standards-compliant wireless network that is interoperable, regardless of manufacturer and equipment.

Yet another relevant organization that doesn't necessarily produce standards, but does heavily influence them and bring technologies to market, is the Wi-Fi Alliance, a group of manufacturers and other interested parties that formed a trade association of sorts to jointly produce interoperable equipment and technologies. Members of the alliance currently include the big name organizations, such as Cisco, Microsoft, Nokia, Samsung, Dell, Apple, Sony, and many others. The Wi-Fi Alliance promotes wireless technologies with consumers, other organizations, and the government. They also sponsor a Wi-Fi Certification program that certifies technologies and devices as being interoperable and compliant with standards, enabling a manufacturer's devices to be branded with the Wi-Fi Alliance logo if it is submitted for certification and meets requirements. In addition to promoting wireless interoperability, they also serve to advise organizations, like the IEEE, on potential new technology standards.

802.11 Standards

The IEEE 802.11 standards are a set of physical layer signaling and technology standards for implementing WLANs in the 2.4, 3.6, 5, and 60 GHz frequency bands, as allowed by the FCC. While we collectively refer to all of the 802.11 standards under one umbrella, there are several standards, including the a, b, g, i, k, n, and many others. These are referred to as amendments to the original standard and cover changes in technology, security, infrastructure, and so on. The original 802.11 (with no suffix) specification was released in 1997 and set the initial standard for wireless networks. The data rates of the standard were extremely slow, which is why it was not immediately implemented. It specified two net bit rates of 1 or 2 megabits per second (Mbps), which is very slow by today's standards. It was never really widely implemented, and was quickly supplanted by faster standards, the 802.11a and b standards. Since 1997, the original standard has been updated several times, with the latest one being the 802.11-2012 standard. This update also includes all the previous amendments and updates from 1997 to present. The 802.11 standard operated in the 2.4 GHz ISM band and originally provided for both *frequency hopping spread spectrum*, or FHSS, and *direct sequence spread spectrum* (DSSS) technologies. A third physical signaling technology, called *orthogonal frequency division multiplexing* (OFDM), was introduced with later amendments and is used, along with its variants, in newer technologies. The amendments that followed (for example, a, b, g, n, and now ac) all primarily use one of these three physical signaling technologies (with some small variation). The particulars of the relevant amendments you need to cover for the exam are discussed in the sections that follow.

 EXAM TIP You likely won't see anything on the exam relating to the 3.6 or 60 GHz frequency ranges, but it's good to know they exist as part of the 802.11 standards.

802.11a

The 802.11a standard (sometimes called the "a" standard) was one of the first wireless standards to come out, actually about the same time as the wireless "b" standard (802.11b) in 1999. The 802.11a networks offer a throughput of 54 megabits per second (Mbps) and operate in the 5 GHz range. One advantage of this technology when it came out was that there were very few devices that operated in this frequency range, as opposed to the 2.4 GHz frequency range that other wireless standards operated and still operate in, where devices such as cordless phones and microwave ovens can cause interference. Using this frequency range meant less interference from other wireless devices. 802.11a uses a technology called *orthogonal frequency division multiplexing*, or OFDM. OFDM, rather than using just one frequency, uses several adjacent ones to carry data. This helps with issues that may affect any particular frequency, such as interference.

Although superior in many ways to the "b" standard, the "a" standard was not widely adopted when it was first introduced. One possible reason was the proliferation of "b" devices on the market. Another possible reason that led to the slower adoption of the 802.11a standard when it was first introduced was the fact that while this range falls within the unregulated bands in the United States, in Europe the 5 GHz range was still subject to heavier regulation, so 802.11a networks were not widely implemented in Europe. This has since changed, and since 2003, 802.11a wireless has become more widely implemented worldwide. Although the higher data rate and a relative lack of interfering devices in the same frequency range should have hastened consumer adoption of the standard, the wireless "b" standard, which also came out about the same time in 1999, was more widely used.

802.11b

The 802.11b standard was released at almost the same time as the "a" standard, but was marketed more heavily to consumers, so it was more readily adopted and accepted in the consumer market space. This was due primarily to the high proliferation and availability of equipment (access points and network cards) that used wireless "b" technology. 802.11b operates in the unlicensed 2.4 GHz Industrial, Scientific, and Medical (ISM) range, and has a throughput of about 11 Mbps. 802.11b uses a physical layer signaling technology (sometimes abbreviated as PHY) known as *direct sequence spread spectrum* (DSSS), which is included in the original 802.11 specification. DSSS basically means that the wireless signal is spread over the full bandwidth of a frequency or channel.

Although now considered to be very insecure, 802.11b networks introduced the first wireless security standard, known as Wired Equivalent Privacy, or WEP. This was the first attempt at providing some sort of security in wireless networks. I'll cover WEP a bit more later in this chapter.

802.11b networks are not often seen any longer, due to the heavy proliferation of more improved wireless technologies and standards, including the 802.11g and "n" standards. You may typically see these types of networks in homes or offices that are required to support legacy equipment, including older versions of Windows or older hardware that cannot support the newer technologies. While some newer technologies are backwards compatible with the "b" standard networks, this compatibility is being phased out as newer equipment is manufactured and sold.

802.11g

The 802.11g standard had become the most commonly used of the 802.11 standards, but has rapidly been replaced by the newer 802.11n standard. It was introduced in 2003, and was implemented when the bulk of the widespread wireless adoption took place. It is fully backwards compatible with the 802.11b standard, so as consumers upgraded and added to older "b" standard networks, they were able to maintain legacy equipment and networks. 802.11g networks operate in the same 2.4 GHz ISM band that "b" networks operate in, and use OFDM physical signaling technology. Because it is backwards compatible with 802.11b, it can actually support the DSSS technologies used by those legacy networks, but incurs performance degradation on the 802.11g devices when backwards compatibility is used. This is called *mixed mode*, and can be changed on the wireless access point to reflect mixed "b" and "g" modes, or set to allow for wireless "g" only. 802.11g supports data rates of up to 54 Mbps, with legacy equipment running at lower supported rates. Some newer devices have dropped support for the legacy 802.11b networks, as this equipment is phased out and no longer made or sold by the mainstream retailers. Figure 5-9 shows an example of mixed-mode settings on a WAP.

802.11n

The industry has developed a habit over the years of moving faster in implementation than the formal adoption of the standards by the IEEE, as manufacturers produce "draft" standards devices with the understanding that these devices will be fully

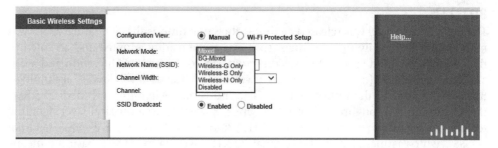

Figure 5-9 Mixed-mode settings on a WAP

(for the most part) compatible with the standards once they are formally approved and released. Such was the case for the 802.11n standard, which was approved as final in 2009. The standard was in draft for many years prior to its official release, but there were devices that were produced and sold as "compatible" with the draft standard long before it was approved. This was the case for a great many pre-release 802.11n standard devices, or "draft-N" devices. These devices were marketed and sold as "draft-N compatible." When the standard was finally released, any devices that weren't fully compatible usually required a minor firmware upgrade.

The 802.11n standard increased data rates significantly, from the 54 Mbps data rates found in 802.11g networks to a theoretical 600 Mbps. This technology uses something called *multiple-input multiple-output* (MIMO) streams, and supports up to four separate streams of 40 MHz each, effectively doubling the channel width of previous technologies from 20 MHz. MIMO uses multiple antennas to separate the spatial streams, and uses a signaling technology called *spatial division multiplexing*, or SDM. There are also significant improvements in security and other enhancements to the 802.11n standard. The "n" standard is backwards compatible with both 802.11b/g technologies. One important feature of 802.11n is that it can operate in both the 2.4 GHz and 5.0 GHz frequency ranges. This allows the user to switch between frequency bands in the event of excessive interference from other devices in the 2.4 GHz range, assuming all of their client devices support it.

802.11n is the current standard for higher-speed wireless access in homes and offices, but is about to be quickly replaced by the newer 802.11ac standards within WLAN environments.

802.11ac

The latest and greatest standard concerning wireless networking is the 802.11ac standard, the draft of which was completed by the IEEE during the writing of this chapter and was approved by the IEEE in January 2014. This standard proposed a much faster speed and increased throughput beyond what we experienced with the 802.11n standard, and was designed for high bandwidth application (think HD video streaming) and high-end enterprise use. It basically widens the RF bandwidth to 160 MHz, and extends and improves the MIMO technology introduced in 802.11n, allowing up to eight separate multiuser MIMO spatial streams. The specification allows for speeds up to 1 Gbps (for multi-station WLANs), and 500 Mbps for single station links. It won't necessarily replace 802.11n because it operates only in the 5 GHz range, and there will still be plenty of devices operating in the ISM 2.4 GHz range for the foreseeable future. It will, however, complement the "n" standard and provide for highly reliable bandwidth for those consumers who choose to upgrade their networks to the new standard.

As with wireless "n" implementation, there were already wireless "ac" draft-compatible chipsets and devices produced and on the market before the standard was approved, accelerating the implementation. These devices are supposedly compatible with the initial approved draft, primarily because there have been no controversies over the technical specifications between vendors and the folks drafting the standards. As the standard became final, these pre-release devices likely only require minor firmware updates to them to ensure they are fully compatible. This fairly close adherence to the proposed standards by early implementations usually happens because the Wi-Fi Alliance members closely monitor and participate in the IEEE working group discussions, providing input and technical assistance to the group as needed, to ensure a quality standard is produced.

Table 5-1 provides a summary of all of the different 802.11 standards discussed so far. It provides a handy study reference to review before the exam to refresh your memory on the characteristics of each of the 802.11 amendments discussed in this chapter. Keep in mind, however, that this chapter has covered only a few of the 802.11 amendments that have been approved and published, particularly the ones you will need to know for the exam. Several others exist (including a few more that I will discuss

802.11 Standard	Data Rates	PHY Signaling Technology	Band	Frequency Range
802.11	1 to 2 Mbps	FHSS/DSSS	ISM, 2.4 GHz	2.4 GHz
802.11a	6–54 Mbps	OFDM	UNII, 5 GHz	5.150–5.250 GHz UNII-1 5.250–5.350 GHz UNII-2 5.725–5.825 GHz UNII-3
802.11b	5.5 and 11 Mbps	DSSS	ISM, 2.4 GHz	2.4–2.4835 GHz
802.11g	Up to 54 Mbps (when used in mixed mode, data rates match legacy devices)	OFDM (but backwards compatible with 802.11b using DSSS and HR/DSSS)	ISM, 2.4 GHz	2.4–2.4835 GHz
802.11n	Up to 600 Mbps (when used in mixed mode, data rates match legacy devices)	HT-OFDM	ISM, 2.4 GHz UNII, 5 GHz	Same as 802.11a/b/g
802.11ac	Up to 1 Gbps	HT-OFDM	UNII, 5 GHz	Same as 802.11a/n

Table 5-1 Summary of Wi-Fi Standards

later in the chapter, such as the 802.11i security amendment), which relate to security, radio frequency management, roaming, and quality of service, to name a few. Some of those are beyond the scope of the exam and this book, however.

Wireless Channels and Frequencies

Before leaving our discussion on 802.11 wireless technologies, let's discuss the frequency ranges of the channels these technologies use. Channels and frequency ranges are broken up based upon the region in which they're used. There are some channels used in Europe while others are used in North America, Japan, and other countries. Some countries even restrict which channels can be used indoors and which can be used outdoors. For example, there are 14 available channels in the 2.4 GHz ISM band, which is used by DSSS and HR/DSSS PHY technologies. In the Americas, only channels 1 through 11 are used, and in Europe 1 through 13. Japan uses all 14 channels; in fact, Japan is the only country authorized to use channel 14, according to the 802.11-2012 standard.

Each of these channels is separated by 5 MHz, and DSSS channels are 22 MHz wide. There are three adjacent non-overlapping channels—channels 1, 6, and 11. So theoretically, you could have three separate wireless access points working in the same area without overlapping channel interference. As you will see in our discussion on site surveys, this is important because you want to minimize Wi-Fi interference from other devices as much as possible. Table 5-2 lists the channels and frequencies for the 2.4 GHz ISM band to demonstrate how the different frequencies are allocated to the channels. Figure 5-10 shows how to change the channels on the Linksys-N wireless AP.

Frequency (GHz)	Channel	Notes
2.412	1	Used by all countries
2.417	2	Used by all countries
2.422	3	Used by all countries
2.427	4	Used by all countries
2.432	5	Used by all countries
2.437	6	Used by all countries
2.442	7	Used by all countries
2.447	8	Used by all countries
2.452	9	Used by all countries
2.457	10	Used by all countries
2.462	11	Used by all countries
2.467	12	Used by Europe, Israel, Japan, and other countries
2.472	13	UsedUsed by Europe, Israel, Japan, and other countries
2.484	14	Only used by Japan per IEEE 802.11-2012 standard

Table 5-2 2.4 GHz ISM Band Frequencies and Channels

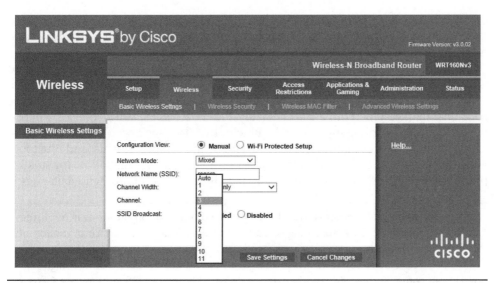

Figure 5-10   Changing the channels on a Linksys wireless AP

Exercise

If you have administrative access to a wireless AP, and have the appropriate permission, review the configuration settings and compare them with the discussions in this chapter. Make note of the SSID, whether it is set to broadcast, the type of authentication and encryption the AP uses, and the channel it is using. What other settings are relevant to our discussion on 802.11 wireless networks?

Bluetooth

The next wireless technology to discuss has become quite commonplace in the personal device market. I am, of course, referring to the Bluetooth technology that many people use to connect small devices together, such as headsets to cell phones and other portable media devices. Bluetooth technologies are used to support a wide array of data requirements, including exchanging contact information, streaming music, using hands-free cell phone devices, and so on. Let's dig into Bluetooth technologies and standards.

802.15

The 802.15 standard is yet another IEEE standard that applies to wireless networking. This standard covers devices that use Bluetooth technologies to connect to exchange data. Bluetooth is an open standard for short-range radio frequency (RF) communication, primarily used in the consumer and, to some extent, the commercial market space. Bluetooth is used mainly to establish ad-hoc wireless personal area networks (WPANs) between devices such as cell phones, laptops, automobiles, medical devices, printers, keyboards, mice, and headsets. Bluetooth allows easy file sharing, elimination of messy cables in small spaces, and even Internet connectivity sharing between devices. Bluetooth operates in the ISM band (2.4000 GHz to 2.4835 GHz range), and uses FHSS signaling technology, similar to some of the IEEE 802.11 standards of technologies. A couple of Bluetooth devices are shown in Figure 5-11.

There are several Bluetooth versions out in the market, as this technology changes and is updated rapidly. The older versions, including 1.2 (2003) and 2.0 + Enhanced Data Rate (EDR, which came out in late 2004), are considered legacy, but are still out in the consumer market. Bluetooth 2.1 + EDR (July 2007) devices are still also heavily in use, but are quickly becoming supplanted by Bluetooth 3.0 + High Speed (HS, appearing in 2009), which provides significant data rate and security improvements, and Bluetooth 4.0 (June 2010), which includes Low Energy (LE) technologies that allow for smaller, power-saving devices. Table 5-3 summarizes the Bluetooth standards, and data rates.

NOTE The effective transmitting range of various specifications is not listed in Table 5-3 because ranges are not tied specifically to a particular version of Bluetooth; rather, they are a product of the power levels set on the device. Bluetooth devices are also categorized in classes: A Class 1 device (high power) can communicate with another device up to 100 meters (328 feet) away, a Class 2 device (medium power) has a range of up to 10 meters (33 feet), and a Class 3 device (low power) has a range of about 1 meter (3 feet).

Figure 5-11
Common consumer Bluetooth devices

	Bluetooth Specification	Data Rate
Table 5-3 Bluetooth Specifications, Data Rates, and Ranges	1.x	Up to 1 Mbps (Basic Rate)
	2.0	Up to 3 Mbps (Enhanced Data Rate)
	3.0	Up to 24 Mbps (High Speed)
	4.0	Up to 24 Mbps (High Speed)

An IEEE standard, Bluetooth also has its industry advocates and trade alliances. The primary one is the Bluetooth Special Interest Group (SIG), originally made up of some of the bigger names in the industry, such as Ericsson, IBM, Intel, Nokia, and Toshiba. It has since grown to include other industry and trade giants, and promotes Bluetooth technologies pretty much the same way the Wi-Fi Alliance does for 802.11 technologies.

PANs

A personal area network, or PAN, is created when two or more Bluetooth devices (or even those that use older infrared technologies) are connected in exchange data. A PAN could consist of a cellular phone and Bluetooth headset, or two cellular phones connected together, or even to laptops or tablets that are connected in exchange information. A Bluetooth PAN can also be called a *piconet*. A Bluetooth PAN has a typical range of no more than about 30 feet, given current Bluetooth capabilities. As such, a PAN will consist of devices that are very close to each other.

PANs aren't just limited to using Bluetooth technologies, however. Other PANs include those connected by infrared, using the wireless Infrared Data Association (IrDA) standard, such as wireless remotes, printers, wireless mice, digitizers, and other serial devices. These devices have been standard for several years now, but are being replaced by those that use other technologies, such as Bluetooth and 802.11 wireless. Still another technology in use to create personal area networks includes one affectionately referred to as ZigBee, which is covered under the IEEE 802.15.4 standard. Developed by the ZigBee Alliance, this technology uses very low-power devices to produce short-range mesh networks, usually in an ad-hoc structure. While consumer adoption of devices that use ZigBee is still very limited, wireless industrial control systems make use of this technology extensively. Figure 5-12 shows an example of a ZigBee USB wireless adapter.

 EXAM TIP You likely won't see ZigBee mentioned on the Mobility+ exam, but you may see it in your work with wireless networks, so it's helpful to be familiar with it.

Another type of PAN worth mentioning uses a technology called Near Field Communications, or NFC. While this technology probably won't be tested on your exam, it's a very useful one to know about, as it is becoming increasingly more prevalent in our mobile connected world. NFC uses chips embedded in mobile devices that create electromagnetic fields when these devices are close to each other. The typical range for

Figure 5-12
A ZigBee USB
adapter

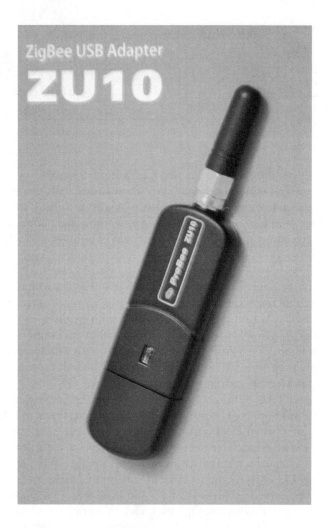

NFC communications is anywhere from a very few centimeters to only a few inches. The devices must be very close to or touching each other, and can be used for data exchange for information such as contact information, small files, and even payment transactions through stored credit card information in mobile applications. This is a very new technology and is just starting to see widespread adoption in newer mobile devices, as well as the infrastructures and applications that support them. Because it's a new technology, certain issues may not be totally worked out just yet—security, for example. Again, while this technology probably won't be tested on your exam, someone sitting for the exam in a few years could see questions on it.

One more wireless technology that I want to briefly mention here is WiMAX. Worldwide Interoperability for Microwave Access, or WiMAX, is based on the IEEE 802.16 standard, and is used for middle-haul or back-haul communications across larger areas, such as metropolitan areas or areas where customers can't get DSL, cable modem, or other broadband access from a typical provider. WiMAX is based upon microwave technologies and enables wireless clients (with a special wireless network card or adapter) to access high-speed mobile broadband from greater distances, without the aid of a portable hotspot connecting them to 3G or 4G cellular technologies. WiMAX isn't one of the wireless LAN or PAN standards covered in-depth in this chapter, but Chapter 4 does give it more detailed attention.

Authentication and Encryption

Unlike wired networks, wireless networks are not physically bounded on either end of the media. In wired networks, there is some degree of protection offered by physical access to the cable, as it usually runs through a wall or ceiling or from a secure network closet to a locked office, for example. Wireless networks do not have this advantage, and even wired networks require significantly more protection than merely limiting physical endpoint access. Because wireless networks send all their data via radio waves through the open air, these radio waves can be intercepted. If the data is not otherwise protected, it is subject to interception and disclosure. There are specific technologies that have been developed to protect both wired and wireless networks because beyond the physical media security requirements, the two types of medium are very similar in most other security requirements. Before we get into the specific wireless security technologies, let's discuss a few concepts and related definitions. Two critical areas of mobile security in general, and 802.11 wireless technologies in particular, are authentication and encryption.

Authentication

Authentication is the overall process involved with identifying a user or device, and then confirming (or authenticating) their identity. It's not enough to simply assert that you are someone in particular; it must be independently confirmed through some secure method or authentication technology. Authentication technologies usually involve some method for passing encrypted credentials across the network to an authentication device or server, which has a database of user credentials. Either (or both) devices and users can be authenticated, depending upon the type of network or the device they are attempting to connect to. Authentication in the context of wireless networks usually involves one of two methods. This first, and least secure, involves supplying a pre-shared key (or passphrase) to a device (typically a client), which is encrypted. The encrypted hash is usually sent to the authenticating device (a wireless access point, for example), which verifies the encrypted hash as having been generated only from the pre-shared key. The second involves the use of public/private key pairs, using an existing Public Key Infrastructure (PKI).

802.11 wireless networks use several different types of authentication, depending upon which technologies are used. For example, open authentication is the very basic type of authentication used by wireless networks, but it's really no authentication all, as any client can connect to any access point as long as they have the password or key. Shared key authentication, on the other hand, is used by WEP. PSK-Personal and PSK-Enterprise are used by WPA. I'll go over these different authentication methods in a little more detail in the upcoming sections. Keep in mind that authentication is *not* the same thing as encryption, which is discussed next, although the technologies discussed in the coming sections typically have the ability to take care of both at the same time.

Encryption

Encryption ensures that the data transmitted and received between devices can't be easily intercepted and read. The data is sufficiently scrambled using one of several methods, and only the devices on the network possessing the correct key can decrypt the data. There are several encryption methods specific to wireless technologies that I'll discuss shortly, but all encryption methods have several things in common. An encryption method requires two important things, the algorithm and key. The algorithm is a mathematical formula, if you will, that determines the precise method for scrambling plaintext data (human readable data) into ciphertext. Encryption algorithms are extremely complex and involve a wide variety of mathematical functions, including very large prime numbers, elliptic curves, and other complex functions. Encryption algorithms are typically publicly known and have been tested thoroughly. The key, on the other hand, is the piece that is kept secret, or at least protected. In the context of wireless technologies, this key is often called a pre-shared key, or passphrase, or a password. In the case of multifactor authentication, the key is a part of a public/private key pair, generated by a Public Key Infrastructure and assigned to the device or user. Together, the key and algorithm provide a secure way to send data from one user or device to another user or device, with confidence that the data sent will not be intercepted or decrypted. While I could fill an entire book with a discussion on cryptography, an in-depth discussion on encryption would be outside the scope of both the exam and this text.

WEP

Wired Equivalent Privacy, or WEP, was the first attempt toward securing wireless networks. WEP was introduced in 802.11b networks in order to provide the same security assurances that a wired network would provide. Obviously, even a wired network is not secure by itself; it requires security measures that include authentication and encryption. WEP uses shared key authentication and employs the same key for authentication and for data encryption. Beyond the use of shared key authentication, WEP is not very secure for a wide variety of reasons. The first involves the key size.

WEP uses small key sizes, which come in two variations: 64-bit or 128-bit. The 802.11b standard requires only a 64-bit key. These keys are static, which means they do not change once they are entered into the device. All devices must have the same key configured on them in order to communicate. If the key is changed, it has to be changed on all of the devices that communicate over the network. As part of its key, WEP uses a 24-bit initialization vector (IV), which serves as a random "seed" to generate the authentication and encryption key. This 64-bit key actually consists of a 40-bit key and a 24-bit IV. Similarly, 128-bit keys actually use a 104- bit key and a 24-bit IV. When WEP was first developed, it was thought to be relatively secure, but rapid increases in CPU power and enhancements in technology enabled people to crack WEP very quickly once they determined how WEP worked and what vulnerabilities it had.

The second issue with WEP is that it uses a poor implementation of the RC4 streaming protocol to encrypt data. While RC4 by itself is not necessarily unsecure, the way WEP utilizes it is not very effective. Additionally, with the small key size, static key, and the small initialization vector that repeats often, poor implementation of RC4 means that WEP is a poor choice for modern wireless security.

WEP is a very weak authentication method and can be easily cracked. Searching for the term "cracking WEP" in Google or YouTube will yield some interesting results on how easily and quickly this can be done. Because of this, the use of WEP is discouraged. However, there are still some legacy devices, particularly the older 802.11b devices, that cannot use stronger, more modern authentication and encryption methods. For these devices, other compensating methods may be used, such as host-based data encryption. In any event, the industry quickly realized it needed a better solution and sought to develop a more secure method of authentication and encryption for wireless networks. That's where WPA comes in.

WPA

Wi-Fi protected access, or WPA, was designed to make up for some of the faults that were inherent in WEP. WPA was developed primarily by the Wi-Fi Alliance and other interested industry parties as a stop-gap method to secure wireless networks while the IEEE worked on a standardized improvement to security. WPA uses passphrases to allow users to create longer, stronger pre-shared keys. A WPA passphrase can be from 8 to 63 ASCII characters (which are case-sensitive), or 64 hexadecimal characters. The actual passphrase is not the key; rather, it is used by the system to generate the 256-bit pre-shared key. This is unlike WEP, which allows only 6 or 10 character passwords (depending upon the key size of 64 or 128 bits, respectively).

WPA uses the Temporal Key Integrity Protocol (TKIP) to dynamically generate 128-bit keys on a per packet basis. In other words, every packet that is transmitted has a new 128-bit key. TKIP was originally included in a later version of WEP, but was not widely implemented. As implemented in WPA, TKIP uses a 48-bit initialization vector and an improved implementation of the RC4 stream cipher for backwards compatibility with WEP.

WPA was released in 2003, and was used for some time as a temporary measure while awaiting the passage of the 802.11i standard by the IEEE. The IEEE finally ratified the 802.11i standard in 2004, which is a formalized version of WPA and is called WPA2. WPA2 offers TKIP for backwards compatibility with WPA devices, but prefers AES (Advanced Encryption Standard) encryption as its default.

The primary differences between WPA and WPA2 are the encryption methods used, but otherwise, either can be used in the home or SOHO environment. WPA2 uses AES to implement the Counter Mode with Cipher Block Chaining Message Authentication Code Protocol (CCMP) required as part of the 802.11i standard. WPA2 also uses TKIP for backwards compatibility with WPA and legacy devices that don't support AES.

Note that when passphrases are used, this is called WPA2-Personal authentication. The other form of WPA is called WPA2-Enterprise, and allows authentication of both the user sitting at the keyboard and the device or client they are communicating from. This type of authentication is normally seen in larger, enterprise-level networks and requires a specialized infrastructure set up to accommodate it. This is also known as IEEE 802.1X authentication, and will be discussed in the next section. A summary of WEP and WPA settings for a WAP is illustrated in Figure 5-13.

802.1X

802.1X is a security protocol that is not necessarily tied to wireless networking. It was originally designed for Ethernet wired networks, but fit in nicely as a solution for enterprise-level wireless authentication and access control. 802.1X is a port-based access control method. It works differently than 802.11 authentication, and has the ability to use a wide variety of authentication and encryption methods. One key point regarding 802.1X is that it offers the capability to authenticate both users and devices to the enterprise network. In other words, not only can the client device authenticate to the access point, but the user at the keyboard of the client device can be authenticated as well to the enterprise network. 802.1X provides additional protections beyond pass-phrase authentication between the client and access point. It also provides for mutual authentication between devices. This means that both the client and access point have to be authenticated to each other, instead of only the client device authenticating to the WAP.

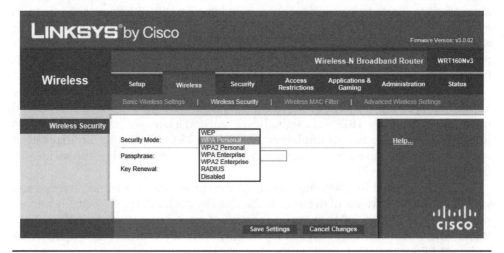

Figure 5-13   Security settings for a WAP

Wireless Security Technology	Protocols	Notes
WEP	RC4 streaming protocol	24-bit IVs, static keys that repeat. Easily broken.
WPA	TKIP	48-bit IVs, dynamic keys.
WPA2	AES-CCMP/TKIP	802.11i standard, backwards compatible with WPA.
802.1X	EAP	Port-based authentication framework; used for WPA/WPA2-Enterprise.

Table 5-4 Summary of Wireless Security Technologies

802.1X uses several different terms than what we've discussed with wireless security, and you need to be aware of these for the exam. The first term is *supplicant*, which is really another name for a wireless client device. An *authenticator* is simply the wireless access point or wireless LAN controller itself. The third term is not one that we find associated with 802.11 wireless security terminologies, and is the *authentication server*, which uses a remote access authentication protocol such as Remote Authentication Dial-In User Service (RADIUS), Terminal Access Controller Access-Control System (TACACS), or its Cisco-developed successor, TACACS+ to provide authentication and accounting services using an enterprise-level user database, such as Microsoft's Active Directory.

While 802.1X is really a port-based protocol, it also serves as an authentication framework that allows other authentication protocols to be used in conjunction with it. The most common one is the Extensible Authentication Protocol (EAP). EAP is the primary authentication protocol used with 802.1X networks. There are several variations of EAP, including EAP-TLS, Protected EAP (PEAP), EAP-MD5, EAP MS-CHAPv2, and others. Some of these are older and have been deprecated to some extent. Because EAP is extensible, it allows a wide variety of authentication methods, including username and password combinations or certificate-based (PKI) authentication.

Table 5-4 provides a summary of wireless security technologies. Review it carefully right before you sit for the exam.

Site Surveys

A *site survey* is an effort to understand the environment in which wireless networks operate for the purposes of determining capacity, coverage, and growth of the network, while also determining any factors that may interfere with the effective operation of the network, such as radio interference and other environmental factors. A site survey is typically performed before a wireless network is implemented, but should also be performed periodically to monitor the growth and performance of the wireless network, and to ensure that the installation stays optimized, to the extent possible, for its efficient use. Several elements must be considered when undertaking a site survey. These include location, environment and physical considerations, interference from other devices, numbers of users and devices, bandwidth, and several other factors. This section looks at these considerations and factors and discusses ways to perform site surveys as effectively as possible.

NOTE We are primarily discussing wireless 802.11 networks, but these same principles can apply to cellular, Bluetooth, ZigBee, and other technologies.

Preparation

Preparation is the key to a successful site survey. Before you go to the site and start measuring coverage and signal strength, you'll need to gather a lot of information, and this is likely the most important step. This information includes the business model of the organization, the details of the site and environment, security documentation, how it will integrate into an existing network, and so forth. Let's discuss these next.

One important consideration that has to be examined before a wireless network is implemented is the reason why the customer wants the network in the first place. That reason usually depends on the type of business or organization. For example, a business that is devoted to manufacturing will have different needs than one specializing in storing inventory in a warehouse. A coffee shop or bookstore will have different needs than a medical clinic. In some cases, security will be an extremely important factor, and in other cases public wireless access must be designed into the network. In other cases, line-of-business applications must be considered due to bandwidth requirements and throughput. The business model of the organization, and the business's reasons for implementing the wireless network, must also be considered.

Perhaps the organization is expanding its operations into a new facility where employees require high mobility solutions. Or maybe they have an existing wired network, and this solution must integrate into it. In any case, before performing a site survey, you will need several pieces of information. You will be able to gather this information in several different ways. You may have to interview the customers and also request documentation artifacts from them, such as blueprints or floor plans, business plans, security plans and procedures, and so on. You also may have to look at public sources of information, such as weather patterns and utilities usage. In addition to experts on wireless technologies, you also may have to include people knowledgeable with building codes and construction, electricity and electrical systems, security, and network systems on the survey team. All of these will help you plan out the wireless network and answer many questions that you will have during the course of the site survey. Let's examine a few particulars involved with wireless networking site surveys.

EXAM TIP Preparation for site surveys is one of the key areas you will be tested on in the exam. Be sure you know the steps that go into good preparation for a wireless site survey.

Capacity

One of the first major considerations in planning a cellular or wireless network is capacity. *Capacity* refers to the ability of the wireless network to handle increasing numbers of users and devices as well as the workload put on the wireless network itself. Capacity planning is extremely important for wireless networks due to usage and growth. When you are looking at capacity, you are thinking of numbers to a certain extent. You will want to know how many users will be on the wireless networks. Each user may have a laptop, tablet, and cellular phone that use the wireless network. You may want to know what kind of technologies each device incorporates, such as 802.11a/b/g or n. This would affect the devices' speed and capacity for data. How many devices will each user have? Is the number of users and devices expected to grow over the next year or two? While capacity does include the numbers of users and the devices they use, it also includes a discussion on the use of those devices and the applications that run on them. You will want to know what the primary applications that people use on their devices will be. Will they simply be emailing and web surfing? Are there any business-unique applications that must send data across the network, such as inventory or job scheduling information? How much bandwidth does this data consume? Will the users be transferring or sharing large files? What are the performance requirements of these applications? These are questions that you must get answered, as completely as possible, to aid in the planning portion of the network.

Coverage

Coverage refers to several different factors in a wireless network. It can refer to the distances involved between the access points and the clients; it can also refer to the signal strength and quality between those clients and access points, as well as transmit and receive power, and bandwidth. Coverage typically addresses specific areas within the facility or organization. For example, in a warehouse, wireless devices such as barcode readers may require coverage from one end of the warehouse to another. Wireless laptops and tablets in the office areas require coverage as well, but this could be from a different access point that operates on an entirely different network segment. Things that can affect coverage include distance between devices, of course, but also objects such as walls, machinery, furniture, and other items that may block RF signals. Placement of wireless access points affects coverage, due to placement location and distance limitations. Coverage is also affected by electrical interference and other environmental factors. The amount of power that an access point and device transmit with can affect coverage area as well. One additional item that can affect coverage is saturation of users and devices within the coverage area. This means that if the capacity of a given coverage area is exceeded, users may experience slow response times and inadequate signal strength.

Coverage is a difficult factor to nail down and optimize because there's so many things that can influence it. Access point placement should be carefully considered to optimize coverage, as well as provide security for these devices. For example, an access point shouldn't necessarily be placed near an outer wall, as this affects both coverage (distance to the opposite side of the facility) and security (eavesdroppers may be able to pick up the wireless signal outside of the facility). So wireless access points may need to be placed more centrally within the facility at designated areas and within certain distances of each other. By and large, they should be placed away from potential interference sources such as break rooms (potential interference from microwave ovens), electrical devices and outlets, machinery and manufacturing equipment, and other high interference areas. They should also be placed so they are not easily accessible by unauthorized personnel. This may include locked wiring closets, server rooms, or even locked cabinets within rooms. To ensure adequate coverage within all areas of the facility, you might consider multiple access points, especially in larger facilities.

You also should consider power levels on the access points. Higher power levels may mean greater coverage, but this has to be balanced with security as well because higher power outputs mean that the wireless signal may be more easily intercepted. You'll also have to consider legal requirements with power output, as transmission power output levels may be regulated in your area.

To check and verify RF coverage, you should test RF reception in several different areas of the facility. For example, you may want to check coverage in different office areas, the warehouse, production areas, within secure areas, and even outside the building to see how much wireless signal leakage there is beyond the walls of the facility. If you find that the coverage is too little in some areas, obviously you'll need to add access points or increase the transmission power on the existing access points. These areas where coverage is limited or nonexistent are referred to as dead spots. If you find that you can easily pick up the wireless signals from the outer edge of the parking lot, you may want to look at relocating wireless access points or reducing the power output on them. When you test coverage in these various areas, try to test the RF transmission and reception under heavy load conditions. In other words, use the typical applications that the organization and its users may use, especially ones that require large amounts of bandwidth, such as file sharing and streaming. You should also test the load in different locations from multiple devices and users. You want to test things such as signal strength and receive signal strength. You'll definitely want to record each of these items as well as other performance factors within each coverage area you test.

Signal Strength

Signal strength refers to the strength of the transmission power from the transmitting device. This is one of the key factors in determining RF coverage over a given area. There are various factors that affect signal strength, especially with the distances involved with wireless networking. Different manufacturers set upper and lower limits on signal strength for their devices, so each can have different capabilities. Transmit power, however, is one consistent way of discussing signal strength with wireless devices. On some devices, the transmit power can be set in its configuration utility, but this can be limited due to restrictions placed on the device by the manufacturer, or even restrictions imposed by local laws. Other devices simply don't allow the transmit power to be changed. Another factor affecting signal strength is the type of antenna used because some antennas allow better signal strength than others. Signal strength is typically measured by the distance from the transmitter, and can be viewed in most devices in terms of dB-millivolts or dB-microvolts per meter.

Receive Signal Strength

While this may seem to be a repetitive term, *receive signal strength* refers to the amount of power received from a wireless transmission, and it is used by wireless LAN client devices. The receive signal strength is influenced by the power of the transmitter, obviously, but also by radio frequency noise in the area. This RF noise could be from other transmitters, electrical interference, blockage by solid objects, and even the weather. There is a measurement called signal-to-noise ratio (or SNR) that describes the difference between the received signal and the noise level. This is measured in dB, and can be calculated by subtracting the received signal (in dBm) from the noise (also in dBm). For example, if the received signal is –70 dBm, and the noise is –90 dBm, then the SNR value is 20 dB. Typical SNR ratios for wireless networks are from around 20 to 25 dB.

Some wireless devices have what's called a *received signal strength indicator* (RSSI) that shows a value for the receive signal strength. Unfortunately, sometimes this is an arbitrary value assigned by the device itself, and there's no really comparing it to the SNR value. It really depends upon the software that comes with the device itself as to whether it shows the actual SNR ratio.

 CAUTION   The RSSI is an arbitrary value assigned to the device itself and can fluctuate based upon several factors, such as transmit or receive power, RF propagation, and even weather conditions. Even two identical devices side-by-side can have different RSSI values. Keep this in mind when using them to determine device placement and power level adjustment.

Interference

Interference (also sometimes referred to as noise) can come from a wide variety of sources. Typically, we look at interference as coming from two different sources, non-wireless and wireless interference. Non-wireless interference can come from devices that may use frequencies near those of the wireless network. These may include cordless telephones, Bluetooth devices, industrial equipment such as manufacturing machinery, radar systems, and yes, even microwave ovens. There are several popular anecdotal stories about wireless technicians trying to find the source of wireless interference that always occurs around lunchtime, with the culprit usually being a microwave oven in the company break room. Machinery or powered equipment can also generate non-wireless interference. Several methods are used to find sources of interference, including spectrum analyzers, which are discussed in the next section.

Wireless interference can come from other 802.11 wireless networks and devices that operate in either the ISM or UNII frequency bands. One such type of interference is called *co-channel interference,* and this usually results from other devices that occupy and use the same channel. The other type of interference is called *adjacent channel interference.* This refers to other devices on overlapping channels. Again, using spectrum analyzer software or hardware can help determine the source of RF interference.

Spectrum Analysis

Spectrum analysis is a term that is used to describe examining different RF characteristics to determine signal strength, interference and noise, signal-to-noise ratio, frequency usage, and other factors that may affect the quality of RF transmission and reception. Spectrum analysis is also used sometimes interchangeably with the term "frequency analysis," although this isn't exactly correct. A spectrum analyzer is a device, or software used with a device, that can show the various RF characteristics for given frequencies and channels. From a technical perspective, the spectrum analyzer measures the magnitude of an input signal versus its frequency. The spectrum analyzer can give you a visual representation of how the physical radio frequency usage appears. Obviously, a spectrum analyzer works at the physical layer, versus the upper layers of the OSI model such as the network, transport, and application layers. To view traffic characteristics of the upper layers, you typically use wireless traffic or protocol analyzers (also called sniffers), such as Wireshark or the Aircrack-ng suite. Figure 5-14 shows a Linux host sniffing wireless traffic using the popular Airodump-ng wireless tool.

Spectrum analyzers can be dedicated devices, or they can be a software package that you can use on your laptop or tablet, for example, with your existing wireless network card. There are several popular spectrum analyzers available, with some of the dedicated devices and enterprise-level software packages coming from vendors such as Cisco, Fluke Networks, SolarWinds. There are also open source or freeware analyzers available. In Figure 5-15, you can see an example screen from a Fluke spectrum analyzer program.

```
CH  2 ][ Elapsed: 1 min ][ 2013-08-11 18:16

BSSID              PWR  Beacons    #Data, #/s  CH  MB    ENC   CIPHER AUTH ESSID

EC:1A:59:01:B5:74  -44      52        85     0   1  54e   WPA2  CCMP   PSK  thunderbird
00:1F:33:BA:F9:DA  -65      61         0     0  11  54e   WPA2  CCMP   PSK  texaspride
C8:D7:19:AD:B1:D3  -67      31         0     0   6  54e   WPA2  CCMP   PSK  dd-wrt
00:1D:D6:0F:0C:20  -70      14         0     0   6  54e   WPA2  CCMP   PSK  gnyxzl
CC:A4:62:D8:45:60  -70      16         0     0  11  54e   WPA2  CCMP   PSK  HOME-4562
26:AA:4B:66:11:48  -70       7         0     0   9  54e.  OPN                comer-brown-guest
00:19:9D:4B:89:19  -70      10         0     0   6  54e.  WPA   CCMP   PSK  VIZIO
20:AA:4B:66:11:48  -70       3         0     0   9  54e   WPA2  CCMP   PSK  comer-brown
C4:39:3A:13:52:4B  -71      12         0     0   1  54e   WPA2  CCMP   PSK  <length:  0>
20:4E:7F:98:AB:BC  -70       5         0     0   2  54e.  WPA2  CCMP   PSK  NETGEAR45
CE:A4:62:D8:45:60  -70      11         0     0  11  54e   WPA2  CCMP   PSK  <length:  0>
C4:39:3A:13:52:49  -70      11         0     0   1  54e   WPA2  CCMP   PSK  <length:  0>
C4:39:3A:13:52:48  -70       8         0     0   1  54e   WPA2  CCMP   PSK  HOME-5248
C4:39:3A:13:52:4A  -68      13         0     0   1  54e   WPA2  CCMP   PSK  <length:  0>

BSSID              STATION             PWR   Rate    Lost    Frames  Probe

(not associated)   00:90:4B:D6:F1:89   -72   0 - 1       0        1
EC:1A:59:01:B5:74  24:77:03:B2:A5:34    -9   0 - 6e      0       10  thunderbird
EC:1A:59:01:B5:74  70:D4:F2:3F:C9:9B   -40   0 - 2e      0        3
EC:1A:59:01:B5:74  88:32:9B:D2:8C:F0   -57   0 - 1       0        4
EC:1A:59:01:B5:74  00:19:D2:85:8B:A9   -58   0 - 1e      0        7
EC:1A:59:01:B5:74  00:1A:EF:11:D9:7A   -65   5e- 5e      0       82
EC:1A:59:01:B5:74  00:13:E8:8E:19:13   -24   0 -12e      0       14
00:1F:33:BA:F9:DA  00:21:6A:7D:FF:DA   -72   0 - 1e      0        3
20:4E:7F:98:AB:BC  00:21:00:E6:C7:FF   -70   0 - 1      29        3
```

Figure 5-14   Sniffing wireless traffic using Airodump-ng in Linux

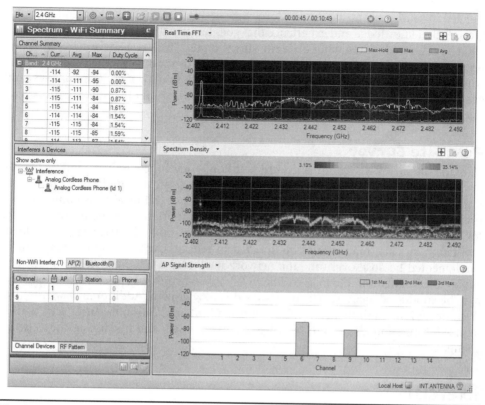

Figure 5-15 A software spectrum analyzer from Fluke Networks

Site Survey Documentation

After completing a site survey, maintaining the documentation generated is of utmost importance for a couple of reasons. First, the documentation may need to be referred to during later site surveys. Because it likely contains information that will be difficult to repeat without re-accomplishing the site survey, this will save a great deal of work during future surveys. Second, it also may be necessary to maintain some of this documentation for legal, safety, or regulatory compliance.

The main document produced by this effort should be a site survey report. This report, usually with multiple artifacts as attachments, includes an analysis of the wireless capacity, coverage, interference, signal strengths at various locations within the facility, and environmental factors (obstacles, machinery, and so on) that could affect the network. The report should also offer a solution for the network, in the form of network and facility diagrams laying out the proposed placement of access points, power settings, and even the recommended vendors and models for the equipment. Security recommendations, in concert with the organizations existing security policies, procedures, and infrastructure, should also be presented.

Keep in mind that this is a higher-level view of the site survey process, and there is a great deal of knowledge and experience that goes beyond the Mobility+ exam level that you should have before taking on a site survey alone. Usually a site survey, especially one conducted for a larger organization, is performed by several experienced individuals who have both the broad-based knowledge of wireless networks, RF principles, and general networking, as well as the specialized training on the possible solutions that may be required.

Post Site Survey

After the site survey is accomplished, the solution is developed, refined, and eventually implemented, although likely with some variations from the original recommendations. These variations may have resulted from budget or technical constraints, or even organizational business decisions. In any case, it's important that the documentation for the network be maintained in as current a state as possible. The site survey documentation and site map should be updated to reflect how the solution was actually implemented because over time the network may need to change and grow. The site survey documentation/site map will be invaluable when that happens because surveys may be conducted again to accommodate this new growth. Additionally, architecture documentation, equipment inventory information, application, security, and performance information all need to be collected, maintained, and updated as the network changes.

Exercise

Search the Internet for samples of site survey documents and templates that would give you insight as to what items are looked at during a site survey. Use these items to create your own checklist for any site surveys you may be performing in the future, and add any items that are relevant or specific to your organization.

Chapter Review

This chapter has provided an introduction to the common wireless technologies present in the commercial and consumer markets. You will see these technologies in your everyday work as a CompTIA Mobility+ certified professional, and you will definitely see them on the exam, so make sure you are very familiar with them.

A wireless client can be a laptop, desktop, tablet, cell phone, network server, or just about any other device that has a wireless network card. Wireless access points are network devices that clients connect to in order to access other devices, other networks, and the Internet. An access point can be a small home-use device, a simple SOHO (Small Office Home Office) device, or an enterprise-level device with several features that control access to multiple networks, including wired networks.

Wireless networks operate in two distinct modes of operation. The first type of network is the *ad-hoc network*; it consists of just a few devices that are connected together for the purposes of sharing files or gaming or Internet connection sharing. The other type of network is *infrastructure mode*; it is usually characterized by several client devices that connect to a central access point, usually a wireless access point, or WAP.

SSID stands for Service Set Identifier, and basically refers to the unique network name for the wireless network a client is trying to connect to. IBSS stands for Independent Basic Service Set, and describes the network in use in an ad-hoc mode network. ESSID stands for Extended Service Set ID, and is the network name of a larger infrastructure mode network that has multiple access points, and at some point, connects to a larger wired network. The BSSID, or Basic Service Set ID, usually a hardware address, describes the basic plain infrastructure-type of network, typically with one wireless access point.

Several organizations control existing Wi-Fi standards, including the Federal Communications Commission (FCC), Institute of Electrical and Electronics Engineers (IEEE), and the Wi-Fi Alliance. The IEEE 802.11 standards are a set of physical layer signaling and technology standards for implementing WLANs in the 2.4, 3.6, 5, and 60 GHz frequency bands, as allowed by the FCC, and promulgated by the Wi-Fi Alliance.

There are several standards, including the a, b, g, i, k, n, and many others. These are referred to as amendments to the original standard; they cover changes in technology, security, infrastructure, and more. They each have different data rates, physical signaling technologies, and other unique characteristics. Some are interoperable with legacy standards, while a few are not.

Bluetooth is an open standard for short-range radio frequency (RF) communication, and is used to support a wide array of data requirements, including exchanging contact information, streaming music, use of hands-free cell phone devices, and so on. The IEEE 802.15 standard covers devices that use Bluetooth technologies to connect to exchange data. WPANs can also use other technologies, such as the wireless Infrared Data Association (IrDA) standard, ZigBee, which is covered under the IEEE 802.15.4 standard, and Near Field Communications, or NFC.

Authentication is the overall process involved with identifying a user or device, and then confirming (or authenticating) their identity. Encryption ensures that the data transmitted and received between devices can't be easily intercepted and read. 802.11 wireless networks use several different types of authentication and encryption methods, depending upon which technologies are used. Wired Equivalent Privacy, or WEP, was the first attempt at securing wireless networks, and was introduced in 802.11b networks in order to provide the same security assurances that a wired network would provide. WEP is not very secure for a wide variety of reasons, including small key sizes, static keys, and a poor implementation of the RC4 streaming protocol.

WPA was developed primarily by the Wi-Fi Alliance and other interested industry parties, as a stop-gap method to secure wireless networks while the IEEE worked on a standardized improvement to security. WPA uses passphrases to allow users to create longer, stronger pre-shared keys. While WPA was used for some time, the IEEE finally ratified the 802.11i standard, which is a formalized version of WPA and is called WPA2. The primary differences between WPA and WPA2 are the encryption methods used.

802.1X is a port-based access control method, and has the ability to use a wide variety of authentication and encryption methods. It offers the capability to authenticate both users and devices to the enterprise network. 802.1X uses terms such as *supplicant*, which is a wireless client device; *authenticator*, which is the wireless access point or wireless LAN controller; and *authentication server*. EAP is the primary authentication protocol used with 802.1X networks. Because EAP is extensible, it allows a wide variety of authentication methods, including username and password combinations or certificate-based (PKI) authentication.

A site survey is an effort to understand the environment in which wireless networks operate for the purposes of determining capacity, coverage, and growth of the network, while also determining any factors that may interfere with the effective operation of the network, such as radio interference and other environmental factors. "Capacity" refers to the ability of the wireless network to handle increasing numbers of users and devices as well as the workload put on the wireless network itself.

"Coverage" refers to the distances involved between the access points and the clients; it can also refer to the signal strength between those clients and access points, as well as transmit and receive power, and bandwidth. Coverage typically addresses specific areas within the facility or organization. "Signal strength" refers to the strength of the transmission power from the transmitting device. "Receive signal strength" refers to the amount of power received from a wireless transmission; it is used by wireless LAN client devices. The receive signal strength is a factor of the power of the transmitter and radio frequency noise in the area. Signal-to-noise ratio (SNR) describes the difference between the received signal and the noise level. The Received Signal Strength Indicator (RSSI) on a device can show a value for the receive signal strength, but this is usually an arbitrary value assigned by the device itself and is not the true SNR value.

Spectrum analysis (sometimes called *frequency analysis*) is a term used to describe examining different RF characteristics to determine signal strength, interference and noise, signal-to-noise ratio, frequency usage, and other factors that may affect the quality of RF transmission and reception. A spectrum analyzer is a device, or software that can show the various RF characteristics for given frequencies and channels.

The site survey report includes an analysis of the wireless capacity, coverage, interference, signal strengths at various locations within the facility, and environmental factors (obstacles, machinery, and so on) that could affect the network. The report also offers a solution for the network, in the form of network and facility diagrams laying out the placement of access points, power settings, and even the recommended vendors and models for the equipment. Security recommendations are also included.

Questions

1. Which of the following two wireless standards operate in the 5 GHz range? (Choose two.)

   A. 802.11b

   B. 802.11g

   C. 802.11ac

   D. 802.11a

2. Which type of network usually consists of just a few devices that are connected together temporarily for the purposes of sharing files or gaming or Internet connection sharing?

   A. Infrastructure

   B. Ad hoc

   C. Peer-to-peer

   D. Centralized

3. Which IEEE standard covers Bluetooth networking?

   A. 802.11

   B. 802.15

   C. 802.15.4

   D. 802.11b

4. You are connecting your tablet directly to a friend's laptop wirelessly to share some files. Which term describes the network you are using in this ad-hoc mode network?

   A. IBSS

   B. ESSID

   C. SSID

   D. BSSID

5. Which of the following is a U.S. Government agency controlling the allocation and use of radio frequencies and technologies in the United States?

   A. Wi-Fi Alliance

   B. Federal Communications Commission (FCC)

   C. IEEE

   D. Federal Trade Commission (FTC)

6. All of the following are PHY signaling technologies used in 802.11 wireless networks EXCEPT:

   A. Frequency Hopping Spread Spectrum (FHSS)

   B. Orthogonal Frequency Division Multiplexing (OFDM)

   C. Time Hopping Spread Spectrum (THSS)

   D. Direct Sequence Spread Spectrum (DSSS)

7. Which of the following IEEE standards operates in the 2.4 GHz Industrial, Scientific, and Medical (ISM) range, and has a throughput of about 11 Mbps?

   A. 802.11b

   B. 802.11g

    C. 802.11ac

    D. 802.11n

8. Which technology uses Spatial Division Multiplexing (SDM), and is found in IEEE 802.11n and ac technologies?

    A. Direct Sequence Spread Spectrum (DSSS)

    B. Frequency Hopping Spread Spectrum (FHSS)

    C. Orthogonal Frequency Division Multiplexing (OFDM)

    D. Multiple-Input Multiple-Output (MIMO)

9. Which wireless security technology uses 104-bit keys with 24-bit initialization vectors?

    A. WPA

    B. WEP

    C. WPA2

    D. EAP

10. How long and what type of characters can a WPA/WPA2 passphrase be? (Choose two.)

    A. 8 or 63 ASCII characters

    B. 64 hexadecimal characters

    C. 6 or 10 ASCII characters

    D. 8 to 63 hexadecimal characters

11. How many bits is a TKIP initialization vector?

    A. 32-bit

    B. 24-bit

    C. 48-bit

    D. 128-bit

12. What is the 802.1X term for a wireless client device?

    A. User

    B. Authenticator

    C. Applicant

    D. Supplicant

13. Which of the following terms refers to the ability of the wireless network to handle increasing numbers of users and devices as well as the workload put on the wireless network itself?

    A. Capacity

    B. Coverage

    C. Throughput

    D. Bandwidth

14. All of the following are factors affecting wireless coverage EXCEPT:

    A. Distance

    B. Obstacles

    C. Encryption

    D. Placement

15. All of the following factors could affect transmission power levels on an access point EXCEPT:

    A. Manufacturer's restrictions

    B. Authentication protocol

    C. Local laws

    D. Type of antenna

16. Which of the following terms refers to the amount of power received from a wireless transmission by wireless LAN client devices?

    A. Receive signal strength

    B. Signal-to-noise ratio

    C. RF noise

    D. Receive power

17. All of the following are potential sources of RF noise and interference, EXCEPT:

    A. Microwave ovens

    B. Cordless telephones

    C. Pacemakers

    D. Industrial machinery

18. Which device can examine different RF characteristics to determine signal strength, RF noise, SNR, and other factors that affect the quality of wireless network transmission and reception?

    A. Wireless protocol analyzer

    B. Wireless sniffer

    C. ZigBee transmitter

    D. Spectrum analyzer

19. Which type of interference results from two 802.11 devices using the same frequency band?

    A. Adjacent channel interference

    B. Co-channel interference

    C. Non-wireless interference

    D. Locked-frequency interference

20. What type of documentation is completed after the site survey and usually contains an analysis and recommendations for the wireless network?

    A. Site survey report

    B. Site survey artifact

    C. Certification and Accreditation package

    D. Architecture diagram

Answers

1. **C, D.** IEEE 802.11a and 802.11ac standard devices operate in the 5 GHz Unlicensed National Information Infrastructure (U-NII) band.

2. **B.** Ad-hoc networks usually consist of just a few devices that are connected together for the purposes of sharing files or gaming or Internet connection sharing.

3. **B.** The IEEE 802.15 standard covers Bluetooth technologies.

4. **A.** IBSS stands for Independent Basic Service Set, and describes the network in use in an ad-hoc mode network.

5. **B.** The Federal Communications Commission (FCC) is a U.S. Government agency that regulates the use of radio frequencies and technologies in the United States.

6. **C.** All of these are PHY signaling technologies, but Time Hopping Spread Spectrum (THSS) is not used with 802.11 wireless technologies.

7. **A.** IEEE 802.11b operates in the 2.4 GHz Industrial, Scientific, and Medical (ISM) range, and has a throughput of about 11 Mbps.

8. **D.** Multiple-Input Multiple-Output (MIMO) uses spatial division multiplexing (SDM) and is found in IEEE 802.11n and ac technologies.

9. **B.** Wired Equivalent Privacy (WEP) uses either 40-bit or 104-bit keys, with 24-bit initialization vectors (IV).

10. **A, B.** A WPA passphrase can be from 8 to 63 ASCII characters (which are case-sensitive), or 64 hexadecimal characters.

11. **C.** A TKIP initialization vector (IV) is 48-bits.

12. **D.** A supplicant is an 802.1X term for a wireless client device.

13. **A.** Capacity refers to the ability of the wireless network to handle increasing numbers of users and devices as well as the workload put on the wireless network itself.

14. **C.** All of these are factors affecting wireless coverage, except for the use of encryption, which affects security, but not coverage.

15. **B.** All of these are factors that can affect transmission power levels on an access point, except for authentication protocol. This would affect security, but not transmit power.

16. **A.** Receive signal strength refers to the amount of power received from a wireless transmission by wireless LAN client devices.

17. **C.** Pacemakers are not normally considered sources of RF interference. All of the other answers are potential sources of RF noise.

18. **D.** Spectrum analyzers are devices that can examine different RF characteristics to determine signal strength, RF noise, SNR, and other factors that affect the quality of wireless network transmission and reception.

19. **B.** Co-channel interference results from two 802.11 devices using the same frequency band.

20. **A.** A site survey report usually contains an analysis and recommendations for the wireless network, and may include artifacts such as an architecture diagram and other documents. It may be used as an artifact itself in the organization's Certification and Accreditation package.

Planning for Mobile Devices

In this chapter, you will

- Examine mobile device infrastructure policies
- Determine a mobile device enterprise solution
- Explore disaster recovery in the mobile device infrastructure
- Learn about mobile device backup and recovery
- Learn how to plan for new mobile technologies

Up until this point, I have discussed the basic principles of networking, radio frequency theory, cellular principles, and wireless technologies, all from primarily a technical perspective so that you can understand how these technologies work. In this chapter, I begin discussing how to integrate all of the different aspects of mobile technologies into a unified approach to using and managing mobile devices in the business enterprise environment. This unified approach is called mobile device management (MDM), and essentially calls for a formalized infrastructure to be designed and built to manage the mobile aspect of information technology. This chapter discusses the planning aspect of centralized management, while Chapter 7 goes into greater depth on implementing MDM solutions.

The CompTIA Mobility+ exam objectives that we will cover during this chapter are 2.4, "Explain disaster recovery principles and how it affects mobile devices"; 3.1, "Explain policy required to certify device capabilities"; 3.2, "Compare and contrast mobility solutions to enterprise requirements"; 3.6, "Execute best practice for mobile device backup, data recovery and data segregation"; and 3.7, "Use best practices to maintain awareness of new technologies including changes that affect mobile devices." This is a great deal of information to cover in such a short space, so let's get started!

Basic Mobile Device Concepts

Mobile devices have become ubiquitous in our daily lives. With infiltration of these mobile devices into the workplace, organizations need to be able to manage them just as they would traditional desktops, servers, network devices, and other infrastructure pieces. In order to do this, a formalized structure has to be put in place to provide for policy, security, resource management, and long-term lifecycle planning for these devices. A concept called *enterprise mobility management* has emerged in recent years, which has slowly been bringing this formalized structure to the organization. Over the next few sections, you will examine the basics of mobile device management, including a little bit of history, terms and definitions, and general concepts that you'll need to know in order to implement a mobile device infrastructure in your organization.

History of Mobile Devices in the Enterprise

In the past dozen or so years, mobile devices have been infiltrating the enterprise. Although not the only player in the market, Research In Motion Limited's BlackBerry device was really the first major smartphone in the market to make a splash in the enterprise world. This is strictly from a managed perspective, of course, as early cellular phones, and then smartphones, were already starting to become more mainstream in the consumer market at that time. BlackBerry, however, infiltrated the corporate market at a significantly faster pace than any other device at the time. BlackBerry had the first devices that could be centrally managed by the IT department of an organization, and provided secure corporate email services to its users. At that time, mobile devices were really more of an extension of the normal IT paradigm that most employees were used to in their organizations. A mobile device was nothing more than an additional device, such as a desktop or laptop, which the user was issued and which was supported by the company. Its primary use was to contact the employee more efficiently and exchange secure corporate communications. There weren't really a lot of user-friendly apps or things the user could do on the device for entertainment, such as play videos or music.

With the massive flood of smartphones into the consumer space, however, a slew of devices eventually infiltrated the corporate arena because of their widespread use. Employees certainly didn't want to give up their personal devices simply because they entered the company doors, so many users eventually began to use their personal devices to check company email and so forth. Unfortunately, corporate IT departments realized only too late that mobile devices belonging to the user, but used to access organizational data, were here to stay, and most IT folks were likely ill-prepared to deal with it. However, over the past several years, a new paradigm in IT management has developed, not only in dealing with mobile devices (both personal and corporate-owned) in the enterprise, but actually embracing it to a large degree. Employers have found that they can actually use the introduction of mobile devices into the workplace to their advantage, simply because they are able to get more productivity out of users, and in some cases even cut IT costs by taking advantage of the fact that users want to use their own personal devices in the enterprise.

Several years ago, we might have thought of mobile devices simply as laptops, but of course, that's all changed with the smartphone revolution. Laptops are still a major part of the enterprise mobile device family, but the different devices that you see in corporate infrastructures today include not only laptops but tablets, such as iPads and Android devices, smartphones running Apple iOS, Android, and even Windows, and a host of other smart devices that are able to connect to an IP network and process data. No longer do these devices simply play music or download videos (referred to as *content consumption*); most smart devices have applications, called *apps* in the mobile world, that are able to actually produce work-related content (called *content creation*). Figure 6-1 gives you an idea of the differences between these two paradigms, and how advances in hardware, operating systems, and apps have really changed what we use mobile devices for.

One of the challenges, however, is that corporate data often is stored or processed on these mobile devices with little regard for security or control. Over the past several years, this has been addressed by the emergence of enterprise mobility management concepts, including mobile device management. Over the course of this chapter, I will introduce you to some of the common terminology relating to mobile device management, as well as some of the concepts you will need to understand, in order to implement an MDM policy and technical infrastructure in your organization.

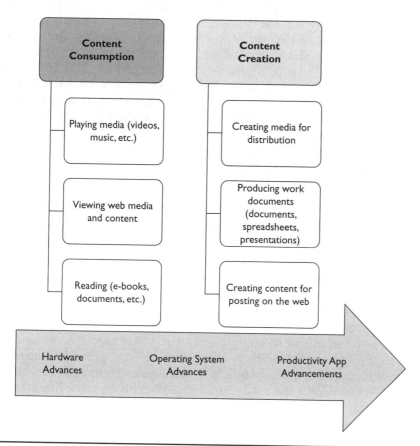

Figure 6-1   Content consumption compared to content creation

Comparing Apples to Androids

In this section I'm going to compare the different major mobile device platforms and vendors, so you have a basic understanding of the major characteristics of each. Keep in mind that there's no way I can cover every single device vendor and even every single major OS out there, but I am going to touch on the primary ones that I feel you should know about as a mobile device professional. That said, I'm going to cover Apple devices, Android devices, BlackBerry devices, and, to a smaller degree, Windows mobile devices in the next few sections. I'll talk about their development and implementation models, as well as some of their major features, including how their application stores work.

Apple

Apple devices began to enter the corporate market space only in the past four or five years. It has only been with the past few iOS versions that Apple devices could be centrally managed in a formalized structure within the enterprise. The Apple model of development is very monolithic in that Apple solely controls the development of the operating system as well as the hardware device and also maintains strict controls on app development for its platform. Apple has very strict development policies and controls for developers, and this is probably one factor that contributes to its high level of security. iPhone and iPad apps are almost exclusively installed and updated through the use of the iTunes software that Apple also produces. Because all developers have to come through Apple's app development model, there are no real third-party providers. An exception is providers of line-of-business apps specific to a particular organization. These internal development groups reside within an organization and can develop for Apple devices, but only for those that are under the organization's control. They still have to undergo a type of Apple partnering and enterprise licensing approval process, which I'll go into in more detail later in the chapter.

BlackBerry

BlackBerry can be considered the first big player in the enterprise market with smart devices. While other platforms easily penetrated the consumer market, BlackBerry focused on the corporate niche. This served them well for several years, simply because no other real contenders were available for enterprise-wide mobile device implementation. BlackBerry devices were at first really just an extension of the organization's email, as they previously did not come with a lot of different consumer-oriented apps. As Apple, Android, and Windows devices have also entered the business market, BlackBerry has had to rethink its strategy somewhat, and entered the consumer market somewhat late in the game. BlackBerry controls a significant portion of the enterprise market, particularly in the U.S. government, but that is slowly changing as other devices gain ground in that market space. Like Apple, BlackBerry has a monolithic, or vertical, development and control model, in that they control the operating system and the hardware device platform, as well as apps that are developed for the platform.

Android

In many different ways, Android devices can be considered almost the opposite of Apple and BlackBerry. This is because Android is an open platform, based on yet another open platform, Linux, and is owned by Google. The Android operating system is available for device manufacturers to alter or customize as they see fit, so there are some slight differences between the various vendors that implement it. The development model used by Android is more open, and is characterized by a wide variety of apps, as well as a wide variety of devices that run the Android operating system. While Google produces the basic Android code, vendors are able to customize it based upon their hardware device features. Apps are marketed and sold by a variety of sources, including the Google Play store, as well as other developers. It's not unusual to have a scenario with Google's Android OS, a device manufactured by a different vendor, such as Samsung, and apps that come from other stores, such as the Amazon Appstore for Android.

Windows

Microsoft struggled for years to break into both the consumer and the enterprise smart device market spaces. This is actually interesting, simply because for years Microsoft has already produced devices such as personal digital assistants, tablets, and so forth for both the consumer and enterprise markets. However, Microsoft always seems to trail the other platforms and vendors in both consumer adoption and market space with mobile devices. With the advent of Windows 8, and Microsoft's vision of one OS and one experience, regardless of device, this may signal a turnaround for Microsoft in the mobile device space. Like the other major platforms I mentioned, Microsoft mobile devices lend themselves to centralized management and can become an integral part of enterprise mobile infrastructure. Microsoft also maintains its own app store, but there are also third-party app providers. Microsoft primarily controls the OS portion of its platform, but naturally has developer requirements as well, although these are not as restrictive as Apple's. Figure 6-2 displays only a small sample of the many available devices from all four major platforms.

Figure 6-2
Apples and
Androids and
BlackBerrys and
Windows, oh my!

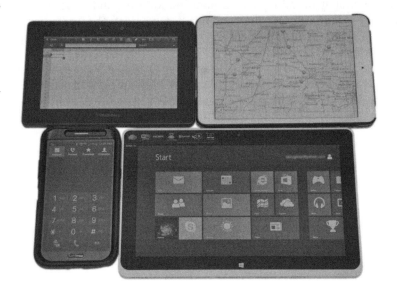

EXAM TIP While you won't be asked many details about specific hardware and OS vendors, you will be expected to know some particulars about their services, how their updates are performed, and so on.

MDM Concepts

As this chapter is concerned with showing you the basics of enterprise mobile device management, it's helpful to understand some basic concepts regarding MDM and its surrounding infrastructure. We've already discussed the more technical aspects of mobile technologies in the previous five chapters. Now it's time to discuss some of the concepts that relate to planning, integrating, and managing all of the mobile devices in your organizations.

Mobile Device Management

Mobile device management, as a unifying concept, basically means that the organization must develop a formalized structure that can account for all the different types of devices used to process, store, transmit, and receive organizational data. MDM can fit into the existing infrastructure right along with desktops, servers, and so on, but it also addresses unique challenges of centrally managing devices that don't always stay within the corporate walls. This may include unique software specifically designed to manage mobile devices that also integrates into existing infrastructure. It also may make use of software that manages applications on these same mobile devices (called Mobile Application Management, or MAM).

With MDM, what you're trying to manage is almost the same as what you would for traditional infrastructure: access control, secure identification and authentication, patch management, antivirus, and compliance with policy. Additionally, however, MDM manages things that traditional networks don't, such as provisioning a device and remotely installing and configuring software, and it can even be used to remotely wipe or lock a device in the event it is stolen or lost. The two major challenges with this, of course, are the mobile nature of the devices and the extent to which personal devices are used in the infrastructure.

The mobile aspect of the challenge is that devices can be removed from the organization's physical boundaries, and sometimes its logical ones as well, making it very difficult to manage in terms of security, patching, auditing, and so forth. When a mobile device is connected to the corporate infrastructure, it can be managed quite easily, but when connected to an external network (such as the employee's personal network at home, for example), it may be more difficult to manage without a direct, corporately controlled secure connection. The organization may, in some cases, have to simply manage the device as it connects to the corporate infrastructure.

The next major challenge is the extent to which personal devices are used in the infrastructure. I'll talk about that paradigm a bit more in the upcoming sections, but suffice it to say that employees are increasingly using their own personal devices to perform work functions, such as checking email, taking business calls, viewing sensitive data, and so on. Corporate data is being stored on personal devices, and this presents a challenge because personal devices can be more difficult to control by the organization. Absent a policy restricting the use of personal devices in the organization, the employee has the ability to do almost anything he likes in terms of organizational data being processed on his own device. Attempts to control these devices from the enterprise level can result in conflicts due to level of control and privacy issues. I'll discuss these issues, in particular, throughout this chapter and the rest of the book.

MDM encompasses the infrastructure, including servers, applications, security policies, and so forth, necessary to completely manage all the mobile devices that connect to the enterprise network. This could be devices that are completely owned by the organization, or personally owned devices brought in by the employee. In order to set up an MDM infrastructure, the organization has to use devices that lend themselves to centralized management (and not necessarily all do). The organization also has to have supporting infrastructure in place, which could include traditional network services, such as DHCP, DNS, email, security services, and so forth. Typically, having a separate set of common services such as those mentioned just for mobile devices, isn't practical or cost-effective. For the most part, these common core services will serve both traditional desktop and server devices, as well as the mobile infrastructure. However, the organization will have to implement some separate services specifically for mobile devices. One such service is the mobile device management software itself. This could be a commercial product purchased and licensed specifically for the organization, but it also could be in the form of cloud services that many organizations provide. One such example is that of MaaS360, a popular MDM cloud service that many organizations use, shown in Figure 6-3.

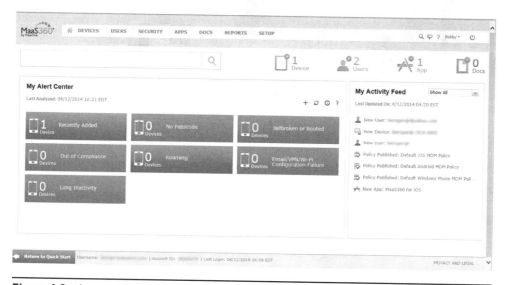

Figure 6-3 An example of a cloud-based MDM solution, MaaS360

Yet another component of the MDM infrastructure is policies (both in the management requirements sense, as well as the kind that are used to control configuration settings implemented on devices). The organization may have a centralized policy server as part of the MDM infrastructure that pushes security policy, as well as other types of policy, to the enterprise's mobile devices. The organization should use all of these different MDM components to initially provision devices, meaning that it will initialize, configure, and join the devices to the organization's infrastructure. Provisioning also includes setting up network services, such as email and Internet configuration, app store settings, and security settings. Beyond provisioning, mobile device management in the enterprise also means performing several actions on the device periodically, such as monitoring, patching the operating system, upgrading apps, and so forth. Over the course of this chapter, you will look at some of these particular features and characteristics of MDM and how they are implemented in the enterprise. While I will only briefly describe some of these in the following sections, many of these characteristics and concepts will be touched on throughout other parts of the book as well. Figure 6-4 shows a diagram of the overall MDM notional architecture.

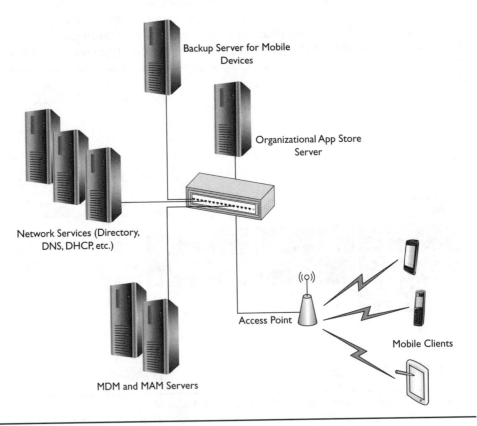

Backup Server for Mobile Devices

Organizational App Store Server

Network Services (Directory, DNS, DHCP, etc.)

Access Point

Mobile Clients

MDM and MAM Servers

Figure 6-4 A notional MDM architecture

EXAM TIP   Keep in mind that MDM is not only the management software, but also its supporting infrastructure and organizational policies.

Bring Your Own Device

The Bring Your Own Device (BYOD) war was one that was briefly fought and lost by organizations hoping to continue the long-held tradition that IT assets belonged to (and were strictly controlled by) the company, not the individual. As more and more mobile devices were brought into the infrastructure, however, IT folks realized that the genie was out of the bottle and they would never be able to completely control this new idea at all. In some cases, companies may be able to enforce a policy that prohibits the use of personal devices to access corporate data and resources, particularly those in high-security environments. At the other end of the spectrum, some companies allow (and even encourage) personal devices, as it saves corporate IT dollars and can contribute to a much happier employee. Most organizations, however, probably fall into the middle of the spectrum and have a mixed environment of both corporate-owned and personally owned mobile devices. In some cases the organization may institute a cost-sharing program, subsidizing an employee's personally owned device by offering a monthly phone stipend or through discount agreements with mobile device and telecommunications vendors. Regardless of the degree of BYOD in the organization, there are challenges that must be dealt with.

One challenge is, of course, device control and how much control the corporation has versus the individual. If corporate data is processed or stored on the device, then rightfully so, the organization should have some degree of control over the device. On the other hand, if the device also belongs to the employee, then obviously the employee should have some control over it. This conflict is probably best solved by policy, which I will discuss later in the chapter. Another challenge is who pays for the device and its use. If the organization allows the user to use her own device for company work, does the organization help pay for the monthly bill or compensate the user for its use? Again, this issue is probably better solved by defined formal policy and procedures. Yet another and equally important challenge in a BYOD environment is employee privacy. If policy allows the organization some degree of control over the device, what degree of privacy does the user maintain on her own device? Can the organization see private data or have the ability to remotely access or control a user's personal device and its use? Again, different aspects of these issues will be discussed not only later on in this chapter but throughout the remainder of the book.

 EXAM TIP   The two critical issues with BYOD are personal data privacy versus protection of corporate data, and level of organizational control versus individual control.

Policy and the Mobile Infrastructure

As I emphasize throughout the book, policy controls everything you do in the enterprise infrastructure, to include security, equipment acquisition, provisioning devices, setting up users, and so forth. Policies are promulgated down from the organization's senior management to be followed and implemented by everyone in the organization. Where policy states what is required in terms of compliance, procedures are used to supplement and support policies by explaining how something is done. Procedures must be developed to support any mobile device policies in place, such as requirements for encryption or authentication, for example. Before mobile devices are purchased and provisioned, the organization must develop a solid set of policies for them. That's what I'm going to discuss during the next several sections.

Organizational IT and Security Policies

Information technology policy not only applies to mobile devices, it applies to every IT asset in the organization. However, because mobile devices present unique challenges and issues, special attention should be paid to developing good mobile device policy that balances the needs of the organization with available resources, device functionality and use, and, of course, security. Information technology policy, combined with other organization policies, such as acquisition or resource management policies, will drive what types of mobile devices are accepted and acquired for use in the organization. The organization may have a policy, for example, that states that only iPhones or BlackBerry devices are to be used for official company business. The company may allow users to bring their personal iPhones, or may purchase and issue them to its employees. The same set of policies may have requirements regarding who in the organization gets to use company devices, such as senior management or the sales force, for example. In any case, the acquisition and use policies for mobile devices should spell out how these devices will be acquired, who will be allowed to use them, what they will be allowed to be used for, and other conditions for use.

Policies should also dictate terms concerning the extent of use of personal devices in the enterprise, either supporting or restricting a BYOD program. In the event the organization allows the use of personal devices to store or process company data, then the policies must also state what the responsibilities are of both the organization and the user in protecting any company data on these devices. Policies should dictate the level of control the organization has over the device, especially if it is a personally owned one. Elements stipulated in this policy may include the organization's ability to restrict network connections or applications used on the device, or even control over the data itself, including the user's personal data that may reside on the device. These policies should not be taken lightly and created without considerable thought, because of the ramifications that can occur when dictating corporate control over personally owned devices. Users must be required to agree to these policies in order to use their devices in the corporate infrastructure, so naturally they should be well briefed on the policy and the consequences of abiding by it. To be fair, users should also be given the option of not using a personally owned device to access corporate data, if they desire. Figure 6-5 gives you an idea of how policies relate requirements to implementation.

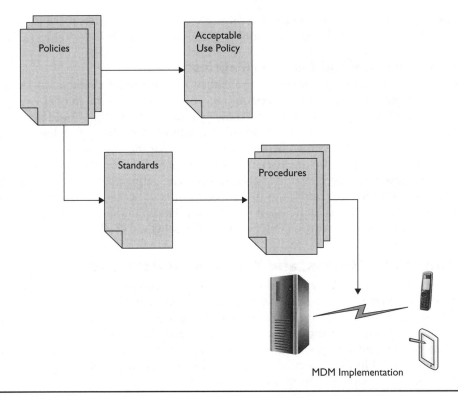

MDM Implementation

Figure 6-5 Relationship of policies (requirements) to standards, procedures, and implementation

Balancing Security with Usability

One important issue with centralized management of mobile devices is one that is similar to managing traditional desktop devices—the issue of security versus functionality. In this case, I'm looking at how to balance security with usability of the device in order to get the right degree of data protection and still afford users the right amount of functionality so they can do their jobs. With traditional desktop devices, administrators balance security and usability by allowing only certain applications to be used, allowing the user to access only certain Internet sites, and so on. This has been effective simply because traditional desktops typically remain inside the company when the user leaves for the day, and typically only connect to the corporate infrastructure. Mobile devices, on the other hand, are more ubiquitous and personal for most users, and have not been traditionally subject to centralized control.

The question of how much security is enough versus how much usability the user should be allowed to have really depends upon the organization's policies and its tolerance for risk. Obviously, the more secure a device is, the less usable it may be, and the reverse is also true. With more usability comes greater risk, but with more security may come less productivity. The organization must take a true risk-based approach to security and functionality, not only with traditional devices, but with mobile devices as well.

Backup, Restore, and Recovery Policies

As you'll see from the earlier discussion and in the upcoming chapters on security, policy plays an important role in determining what's required from an organizational perspective. Managing data backups and recovery is no exception and should be covered by organizational policy that is developed and approved at the higher levels of the organization, and implemented by those responsible for the day-to-day operations of the mobile device infrastructure. Backup policies, which also should cover data restoration and recovery, dictate how often backups should be performed, to what degree they should be performed in terms of the amount of data that is backed up, and how quickly backups must be restored in the event of data loss. Backup policies should also dictate what types of backup are run, as well as the backup schedule itself.

Vendors, Platforms, and Telecommunications

In planning a mobile device infrastructure, there are concerns with standardization across the enterprise. The organization may choose to have a uniform standard for devices and operating systems, to include only specific vendors' platforms, or only specific versions of a given operating system. For organizations that have a more homogenous mobile device hardware population, it still may desire standardization for each OS and version used in the enterprise. There are also considerations with app stores and OEM vendors, as well as telecommunications providers. Obviously, the more differences in operating systems, versions, providers, and so on, the more difficult it would be for administrators to centrally manage and keep up with the configuration on those devices, as well as manage challenges with billing, updates, interoperability, and supportability. Standardization, as much as the organization can attain, is a significant goal in mobile device management in the enterprise. This, again, can be affected by organizational policy, which can set standards for different devices, platforms, and providers.

One other interesting aspect of policy that relates to telecommunications vendors and carriers is the issue of expense management. The organization should have a defined policy that addresses who pays for monthly carrier expenses, especially if the users are in an officially sanctioned BYOD environment. The company may help offset the users' expenses incurred while conducting organizational business on personal devices, for example, by subsidizing monthly bills or even paying the charges outright. In an environment where the organization owns the devices, this is a bit more clear-cut; likely the organization will bear all (or at least the majority) of the expenses. In any event, policy should be created to handle these issues.

OS Vendors

As I discussed earlier in the chapter, several major operating system and platform vendors are available for an organization to choose from. The choice of platform vendors depends upon several factors, most of which I've already discussed, and include interoperability with existing infrastructure, the ability of the organization to support the device platform, resources available, and, of course, security. Each different device platform has its own advantages and disadvantages that the organization must weigh in deciding what the standard will be. In some cases, the organization may not settle on a particular vendor or platform as a standard, especially in an environment where there already are many different types of devices and operating systems in use. Even when settling on a particular hardware platform or OS vendor, the operating system vendor may have different OS versions available for a given device or platform. These all may be supported by the OS vendor, but for the sake of standardization, the enterprise should attempt to use the same version for all its devices to the extent possible. This may be problematic in some cases because there are different versions of the same device. For example, a given version of iOS may be the enterprise standard, but because of different versions of iPhones purchased in different fiscal years, the organization may be required to support different iOS versions until it periodically refreshes those legacy devices. Careful acquisition planning, as well as IT lifecycle planning, should be considered so that devices are upgraded or refreshed on a periodic basis, keeping in mind operating system version and device supportability as well.

OEM

OEM (Original Equipment Manufacturers) vendors provide additional hardware, accessories, and even apps for mobile devices. Some of these may be optional, and some may be purchased by the organization out of necessity. For example, a given enterprise time-keeping app that comes from an OEM, or specific piece of hardware designed to plug into smartphones and scan credit cards for POS purchases, may be acquired for business purposes. Decisions to acquire these types of software and hardware from OEMs are also influenced by factors I discussed previously: interoperability, supportability, security, and cost. In planning a mobile device infrastructure, the different requirements the business may have with OEM hardware and software must be carefully considered.

Telecommunication Vendors

The selection of a particular telecommunication vendor for the enterprise's mobile devices is also very important. In an infrastructure where the enterprise maintains tight control over its devices, and standardizes device hardware and operating system platform, the carrier used by the organization will likely also be standardized. The factors that go into selecting a carrier are based upon different elements such as pricing structure, mobile device volume supportability, and so on. In a less structured environment, such as a BYOD environment, or somewhere in between those two extremes, there may be several different communication vendors used by the organization to provide data services to its mobile devices. In addition to cost and supportability, other factors that affect standardizing with a particular telecommunications vendor include data rates, throughput, coverage areas, and so on. All of these elements can vary with different carriers, and should factor into the organization's decision.

Additionally, each different telecommunications vendor can add its own unique changes to different hardware devices and operating systems. There may be slight differences in operating systems, apps, and even supported features found on the device, which can be attributed to the telecommunications vendor and the changes they may make to the device and its software. As an example, both Verizon and AT&T bundle different applications or additions to the operating system in some of their devices. Some of these modifications may or may not significantly affect the functionality of the device, but are typically used to personalize the device for the carriers' networks or infrastructures. One significant example of a carrier-specific change that did, in fact, affect functionality of a device was when AT&T previously imposed a limitation on iPhone devices that restricted the ability of the device to tether (share its Internet connection) to another device. As this example illustrates, the organization should completely investigate all of its carrier choices, as well as how the carrier alters devices, apps, and platforms, to ensure that they choose the right telecommunications vendor.

 EXAM TIP   Policy is a critical aspect of planning and implementing an MDM, and includes a wide variety of policies that cover security, standardization, interoperability, acceptable use, and many more.

Enterprise Mobile Device Infrastructure Requirements

In order to implement a mobile device infrastructure on an existing corporate network, there are several requirements that should be met. Obviously, careful planning is required, and involvement from all levels of management and technical personnel. The organization first has to decide on a policy, of course, to include privacy issues, the use of personal devices, standardization of devices, apps, and so on. The organization also has to have the infrastructure that can support a managed mobile device implementation. This includes the existing services that are provided to non-mobile clients, such as desktops, but also includes new infrastructure required to successfully integrate and manage mobile devices. Over the next several sections, I discuss more characteristics of mobile device management and mobile application management, to give you a general idea of how these two important concepts work. I also discuss some other considerations that apply not only to the entire enterprise, but also to mobile devices specifically.

Security Requirements

Security should be one of the major requirements that an organization considers when implementing a mobile device infrastructure. For the most part, mobile devices are probably more secure these days than their traditional desktop counterparts were when they first flooded the market a few decades ago because mobile device manufactures and operating system vendors, as well as app developers, have all learned from the mistakes of the past. Unlike some of the big security issues the desktop operating systems suffered through their evolution, mobile devices, their operating systems, and their apps, are much more secure. Even having said that, security is still a big issue in the mobile device world. Of course, there are still some of the same issues with OS vulnerabilities, poorly secured apps, and so on, but these have become very minimal because of some of the stringent requirements that vendors impose on operating system and app developers. Some of the bigger security problems you see in the mobile device world relate to data loss from mobile devices because of the way they're used sometimes, or because they are not securely configured. When developing security requirements for implementing a mobile device infrastructure in the enterprise, several items should be considered. I'm going to cover a few of those key items in the next few sections, but please see the chapters devoted to security for additional information.

Device Groupings

Device grouping allows administrators to place devices in certain groups, based upon logical user groupings or other requirements. This would allow administrators to apply different policy options to the different groups of devices, such as security policies, software or patching policies, or any other updates based upon the requirements of that particular group of devices.

Devices can be grouped by organizational unit, geographical location, and class of device, such as laptop, tablet, smartphone, and so forth. They can also be grouped by user function, such as administrators, sales personnel, and so on. MDM software can also be configured to group devices according to organizational specific requirements. For example, suppose you wanted to group devices by vendor platform or an operating system version. You may want to do this in order to push certain patches down to devices that have a certain version of operating system on them, but not to others. For example, you could also group them by VLAN or IP subnet if you wanted, for segmentation purposes. Device grouping would allow you to do that, and this should be an important management tool used in any mobile device management infrastructure. Figure 6-6 shows one example of how device groupings could help an organization from a security policy perspective.

 TIP Device groupings may be accomplished by using structures (such as groups) in particular MDM software, but it could also be implemented by using structures in directory services, such as the organizational units found in Active Directory.

Administrative Permissions

Administrative permission to devices is almost a necessity in a centralized management setup. In order to effectively manage mobile devices, the organization has to have the elevated permissions necessary to perform privileged actions, such as installing software, patches, and upgrades; audit device use and access logs; and configure security settings on the device. In an environment where all devices are centrally controlled and managed by the organization, this is not difficult to implement. However, in an environment that has adopted a BYOB paradigm, this may not be easily accomplished, because once again, that relates to the organization's control, versus the users' control of a device.

Figure 6-6
Device group-
ings and different
policies

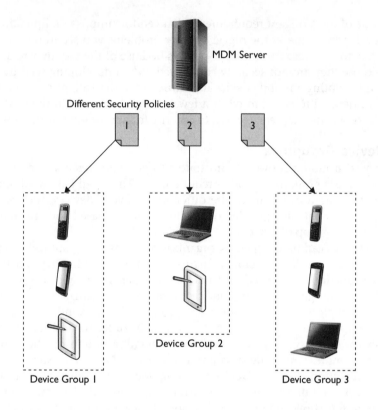

EXAM TIP As with traditional devices, you should ensure that administrative permissions are only granted to those personnel that have a valid need
for them. Policy should also state this as well as part of the principle of least
privilege.

Password Strength

Password strength for mobile devices works similarly to the way it does for traditional
desktops and devices within the enterprise. As with all things security-related, password
strength comes from security policy. The enterprise policy for mobile devices should
closely mirror that of traditional desktop devices and other infrastructure devices such
as servers, routers, and so forth. Only when legacy mobile devices cannot support the
complex passwords required in the organization should policy be different for these
types of devices. In general, however, password strength should be a combination of
length and character key space. The longer the password (more characters), the stronger
the password is. Additionally, the more characters that can be used (including numbers,
lowercase letters, uppercase letters, and special characters), the stronger the password

is. You should set the password strength requirements in the security policy according to the requirements of your organization. By and large, the same password policy used for traditional devices on the network should be used for mobile devices as well. Keep in mind, however, that many mobile devices may use a personal ID number (PIN) in addition to passwords for different apps. I go into more depth on password complexity requirements in Chapter 9, when I discuss mobile security technologies.

 CAUTION Some individual apps on a device may have different password strength capabilities, in addition to the capabilities offered by the enterprise-level apps. Be cognizant not only of using different passwords for these apps, but also the fact that some may be weaker passwords because of limited capabilities of non-managed apps.

Remote Wipe

Remote wipe refers to the ability of the organization or the individual to access the device via wireless or cellular signal, and institute a complete erasure of the device's storage. Normally, a remote wipe would be used in the event the device is lost or stolen. A remote wipe ensures that sensitive data on the device is not compromised when the user no longer has positive control over the device. Most modern mobile devices, particularly smartphones, allow for some type of remote wipe capability, either through the MDM infrastructure, or using a remote wipe app provided on the device and accessed via a Web or desktop app. One major issue of remote wipe is that the data will be lost permanently unless it has been backed up; however, this may be preferred to losing data due to unauthorized access.

Remote Lock/Unlock

Yet another feature related to security is the ability to remotely lock or unlock a phone for a user. This, of course, has to be accomplished through some type of network connection, whether it's cellular or Wi-Fi, and can be a function of an app on the device or, preferably, through the MDM infrastructure. An administrator may want to remotely lock a phone if it has been temporarily lost or is not currently under the control of the user. The organization may want to do this if it could reasonably believe the user will get the phone back under their control, but needs to make sure that no one can access data on the device in the interim. Remote unlock is a feature that the company may want to have in the event the user locks the phone and can't remember the passcode, for example. In this case, if the user inputs the wrong passcode too many times, it may actually wipe the device, if it is configured as such in policy. This offers the user a way to get the device unlocked without risking a device wipe. Again, this has to be configured on the device and in the MDM infrastructure.

Captive Portal

A captive portal is usually an administrative or management function, possibly implemented as a web site, or through a network access control device, whereby the user is prevented from entering the network infrastructure until they are properly authenticated or verified as having sufficient need to connect to the network. A captive portal can also ensure that devices meet certain requirements in order to connect to the corporate infrastructure. Such requirements may include antivirus updates or security solutions installed on the device, proper authentication and encryption mechanisms implemented, connection to a specific subnet or VLAN, or any other requirements the organization wishes to impose on a device before it connects to the network. Typically, a captive portal will prevent the device from connecting until it meets those requirements and the user is properly authenticated. This may be something that happens when the user connects to a self-service portal or when the device is first provisioned and connects to the network for the first time. One example of a captive portal is that of an open Wi-Fi network, for example, that requires a web site authentication from the user in order to proceed to connect to resources on the Internet or other internal assets. A captive portal can also be used for guest networks, for example, if the user doesn't have full authorization to access most resources on the internal network.

 EXAM TIP   A captive portal is a security measure that is often implemented by a network access control (NAC) device.

Monitoring and Reporting Capabilities and Features

I discuss mobile device monitoring and reporting in great detail in Chapter 8, but for now you should know that these two functions are very important to centralized management of mobile devices. Monitoring and reporting allows you to ensure that devices are being used according to policy, that the security posture of the device is being maintained in terms of configuration and patching, and that the actions of the user are appropriate. Monitoring and reporting can tell you if the patch level of the device is up-to-date, what applications are being used, which policies are being followed, and if there are any security issues on the device. Most MDM software allows you to implement monitoring and reporting on devices, as long as the device operating system supports these functions.

Interoperability and Infrastructure Support

Another consideration in planning an MDM implementation is the infrastructure support the organization already has or is willing to add. Obviously, implementing an MDM program from scratch would require a significant investment in time, money, training, and other resources. Leveraging existing infrastructure wherever possible is likely a good idea if it is feasible. For example, using existing services, such as DNS, DHCP, email, and security services, is not only feasible, but definitely recommended for integration purposes and economy of use. Adding additional infrastructure, such as MDM servers, cellular signal boosters, wireless LAN controllers, and so forth, should be done with seamless integration in mind. If these new pieces of infrastructure are added to an existing network that can't support them, due to legacy equipment issues or bandwidth issues, for example, then the existing network may have to be upgraded first before the MDM pieces are introduced.

Interoperability

I've mentioned interoperability already as one area where you must ensure that your mobile device infrastructure works with the other technologies used by the enterprise. When planning a mobile device infrastructure, you should examine your existing infrastructure and compare products, protocols, services, network requirements, and application requirements, as well as user requirements, to make sure that the mobile devices you implement in the infrastructure interoperate with existing technology. That's not to say that things will always work perfectly; you will often have to make adjustments to the existing infrastructure, including equipment configuration, topology, addressing schemes, and so on, when installing new pieces to support mobile device management. However, services, such as DHCP, DNS, and Active Directory authentication and resource location should be checked for compatibility with your planned mobile device infrastructure.

Self-Service Portal

A self-service portal is the function of a mobile device management infrastructure that allows mobile device users to connect to the network, possibly to a shared folder or self-service web site, and perform many different functions for themselves without the need to interact with mobile device administrators. Self-service functions may include installing certain common software apps, such as antivirus updates, trusted certificates, proxy settings, or even application and operating system updates. Many different functions can be accessed through a self-service portal by the user, but some will require intervention from the administrative staff, usually those involving administrative permissions, such as significant configuration changes and major operating system upgrades.

Device Platform Support

Device platform support is another consideration in the broader mobile device management infrastructure, as well as mobile application management. The organization has to determine what types of devices it will support, including from which vendors, which operating systems, and from which application stores it will allow users to download apps. Factors that affect the decisions regarding the device platform support include supportability, cost, flexibility, level of integration into the existing network, and even how deeply embedded existing devices already are in the infrastructure. For example, forcing all of the employees that have previously used Apple devices to switch to Android devices may affect the ease in which the organization can support those particular platforms. For the most part, an organization may settle on one major platform and try to steer all of its users to that platform, especially in a situation where the organization owns all the mobile devices. However, in a BYOD implementation, this is likely not possible because the users will purchase and use whatever device they like the most. In this case, the organization should be prepared to support multiple platforms.

NOTE Support for a particular device platform depends largely on whether the organization is standardized with a few specific devices it uses, or if it employs a BYOD program where users can bring almost any device. The organization may also need to specify which devices, platforms, and operating systems it will support, as a matter of policy, in a BYOD program.

On-Premise vs. SaaS

I mentioned previously one example of a cloud-based mobile device management service, and this actually brings up the discussion regarding on-site provisioning and support versus cloud-based support. Remember that cloud services include Software as a Service (SaaS), Infrastructure as a Service (IaaS), Platform as a Service, (Paas), and, as previously mentioned, even MDM as a Service (MaaS). Other types of services can also be outsourced, including, oddly enough, security. Regardless of the services that are outsourced by your organization or used as a cloud service, there are several considerations involved with both on-premise and cloud providers.

There are several advantages and disadvantages to both of these paradigms, not the least of which are cost, ownership and control, security, and responsiveness. All these factors affect the decision to host services on-site or to contract them out to cloud providers. The organization must look at these factors from the perspective of cost savings, naturally, which would include the necessity to buy, provision, and maintain equipment, train personnel to operate the services on-site, as well as do the same for supporting infrastructure. From a cost perspective, a cloud provider may be the right way to go. However, from a security and control perspective, the organization may have limited control over a cloud provider's handling of the organization's data and services. The organization should do a cost-benefit analysis to determine whether on-premise support is more cost-effective and secure than contracting out services to a third-party provider. One way to ensure that there is a good balance of cost, security, control, and so on, is to have a well-written, well-vetted service-level agreement between the organization and the cloud provider.

EXAM TIP The decision of whether to employ on premise support versus cloud-based "as-a-service" support is one that includes control over data, cost, and service agreements as the primary factors.

Multi-Instance

Multi-instance refers to the use of multiple instances (meaning multiple installations) of software configured for specific groups of users or even having different configuration settings per installation. The MDM solution you choose to implement may provide the ability to have multiple instances of software or settings based upon different configuration requirements in your organization. Some individual software packages, such as Microsoft's SQL Server, also provide this capability built-in. There are a number of different reasons that an organization might want to use different versions of software. User groups may have different mobile hardware devices or even different versions of an operating system on the same hardware platform. In addition, different groups of users within an organization might have unique security requirements and require different instances of software to support their requirements.

Multi-instance software can be managed by the MDM software itself, if the software supports multiple instances. In the multi-instance model, the software itself may execute on a remote server and allow the mobile device user to interface with it through the user's browser, or through a front-end app that connects to the instance. Cloud-based services, such as those that provide SaaS, typically provide for multi-instance use, as that is part of their business model. Licensing would be one issue you would have to consider when deploying multiple instance software, as well as network configuration. Some software licenses may permit the use of multi-instance use, but others may not. Each separate instance may also require different ports or IP addresses. It also may be licensed on the basis of a certain number of users that can simultaneously access the software instance at a time. In the case of database access, software may require separate database stores. These are all considerations when deploying multi-instance software.

 EXAM TIP   The primary concern with multi-instance software is licensing. Make sure the licensing model for the software you wish to use in a multi-instance environment supports using multiple instances of it.

Location-Based Services

One major feature of mobile devices is the ability to track the device's location through GPS, cellular, and, when needed, Wi-Fi connections. After all, because they are mobile devices, location information is very helpful for a wide variety of reasons. Obviously, users rely on location services to conveniently find things near them, such as stores, restaurants, banks, and so forth. Various apps use location services in different ways, but location-based services are also important to an MDM implementation. I discuss two of those ways in the next two sections.

Geo-Location

Because of the mobile nature of personal devices, and the built-in location-based technologies that mobile devices usually have, geo-location services are a must for MDM programs. In addition to helping users find out where they're at and where points of interest are near them, geo-location can be used by the organization to help keep track of devices in the event they are lost or stolen, to aid in their recovery. Aside from that, geo-location can also help make sure that users are compliant with policies that may prohibit the use of their mobile devices in certain locations. There are many features of both end-user apps and MDM that require the use of geo-location services, so it's a good idea to ensure that the service remain enabled on the device. This can be accomplished with configuration policies pushed to the device during its initial provisioning, or even afterwards. For personally owned devices, this may be a little bit more problematic, and the organization should set policy accordingly.

Geo-Fencing

Geo-fencing is an interesting way to keep track of both mobile devices and users. Geo-fencing uses the geo-location capabilities of the mobile device to locate it and keep track of its location while it is within a pre-defined electronic perimeter. This electronic perimeter is set up using electronic sensors around a physical location, such as an office building, warehouse, or even a business campus area. For a few examples of how this might be useful to an MDM infrastructure, picture being able to determine which mobile devices are currently in the building, and, by extension, their users. Or think about being able to detect when an organizationally controlled smart device enters the building. The system could detect it and send notifications to you or even the device. If the device is a tablet used for inventory or sales, for example, the system could pick it up and automatically download its data from the day's work. Also consider how valuable this system would be to prevent mobile device theft, or even unauthorized use, if the device is not allowed to leave the property.

As I mentioned previously, you could conceivably use this capability to track user locations as well, and this might not sit too well with users. Employees might consider this a form of workplace surveillance, and in some cases may rebel somewhat at using geo-fencing for this particular use. In the most benign cases, they may simply leave the device somewhere on a desk and leave the building anyway, or, in the most serious instances, they may seek legal advice and consider bringing litigation to the organization for invasion of privacy. Depending upon how geo-fencing is used to track employee movement within the bounds of the employer's property, there may be legal ramifications to using it to track workers. It's a good idea to research the legal issues of using geo-fencing to track employees, as well as intelligently discuss the merits and pitfalls from an employee satisfaction perspective, and try to get a realistic view of what benefit you may or may not actually get from this practice.

Mobile Application Management

Mobile application management (MAM) is a concept related to MDM, but on a different scale. With MDM, the idea is to reach out and control a device in its entirety through the corporate infrastructure and policy. MAM is limited to simply controlling the applications on the device itself, whether owned by the organization or the employee. There are several different ways this is possible, including controlling individual apps, controlling the source of the apps, controlling the security features of the apps, and controlling the app's data. MAM usually isn't a solution by itself; it's typically used in conjunction with MDM to varying degrees.

The next few sections go into a bit more depth on MAM and describe some of its features and characteristics, so you can get a better idea of how it works. Organizations have the ability to implement MAM as part of their overall MDM infrastructure, but it isn't necessarily a requirement to implement MDM in order to have MAM. Organizations often choose to simply manage applications on mobile devices, particularly in a BYOD environment, rather than manage entire devices, their updates, their security, and so forth. Two of the reasons an organization may choose to simply manage applications on the device include the size of the infrastructure and cost; if an organization has too few users to warrant implementing a significant MDM infrastructure, but yet has a desire to remotely manage a few key apps the employee uses in relation to corporate data, then this might be a good reason to implement MAM without MDM. Another reason may be the organization's tolerance for risk; the corporate decision may be to accept any risk incurred by users managing their own devices while connected to the corporate infrastructure or accessing and storing corporate data on their devices.

There are a few different ways to implement MAM. First, absent centralized management software specifically used for managing mobile apps, the organization could build management capabilities into the app itself. This could be accomplished by pre-configuring the app to communicate with certain corporate servers, using particular authentication and encryption methods, and restricting permissions on the app. This would effectively control what the app could do on the device, and how it could interact with the device and the user, as well as other apps and their data. The problem with this approach is that the organization would have to manage this for every single app it wanted to control. MAM is more of a centralized solution, and requires the use of specialized enterprise software either combined with an MDM solution or implemented separately. Another problem may be that the app, especially if it comes from the device vendor or another third-party developer, may not lend itself to be manageable to the degree the organization would like.

Another way to manage apps on mobile devices is simply to use integrated MDM and MAM software to control the devices' operating systems, which would in turn be used to control apps on the device. Regardless of whether an organization chooses to implement MAM with or without MDM, however, there are still some characteristics of MAM that require centralized policy, planning, and management. I'll discuss these over the next few sections. Figure 6-7 illustrates the differences between using MAM as part of an MDM solution, and implementing MAM separately.

Application Store

Although the focus of this discussion is really on application management by the enterprise, it's worth it to discuss some major differences between both the vendor

Figure 6-7
MAM employed both independently and as part of an MDM solution

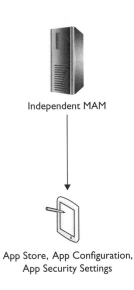

Independent MAM

App Store, App Configuration, App Security Settings

MDM and MAM

Device Configuration and Security, App Store, App Configuration, App Security Settings

app stores and how mobile devices receive apps from organizational app stores. While Apple tightly controls its app store and how apps are introduced into the Apple marketplace, for instance, Android users can install apps from other sources in addition to the Google Play store (a process popularly known as *sideloading*). These apps may come from independent app developers or enterprise-specific app stores created to develop applications specifically for the mobile users of a particular organization.

Apple actually has several different ways to distribute iOS apps. Obviously, the most common way is through iTunes, which is Apple's own app store. Developers use the iOS Developer Program, where they submit an app to Apple for approval to be included in the app store. Apple has very stringent requirements, for both quality and security, so this can be a difficult and rigorous process for developer, particularly if the developer wishes to distribute the apps to anyone else. Figure 6-8 shows an example of the ubiquitous Apple iTunes storefront.

Android app stores, such as Google Play, also have strict requirements on apps developed and marketed through them, but these requirements are not necessarily enforced in the same ways. There are guidelines for developing secure code, ensuring that there is no malware in the app, and so forth, but it still a lot simpler to get an app into an Android store than it is into Apple's store. This way of doing business in no way means that Android apps are less secure or of lesser quality, however.

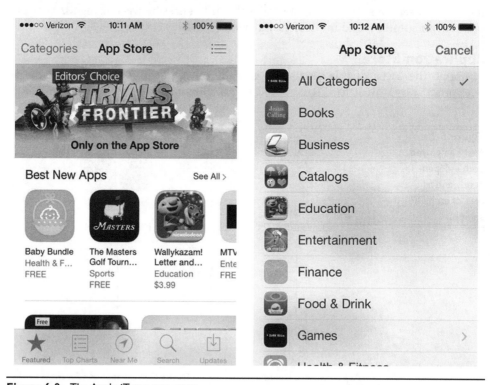

Figure 6-8 The Apple iTunes app store

While most users of smart devices are familiar with a vendor application store, such as iTunes, or Google Play, for example, there are third-party app stores, depending upon the platform used, such as Android or Microsoft devices, from which users can acquire apps. The focus of our concern, however, is app stores that are provisioned and managed by the organization themselves. An organization might develop its own app store to deploy apps specifically for its enterprise users. Given the distribution and approval model used by stores such as iTunes and Google Play, an organizational app store is often a better alternative for several reasons.

First, the organization can control the developmental process and ensure that quality apps are produced for its users and their devices. Second, the organization can make sure that apps include the level of security controls necessary for use in the enterprise. Control over which apps are installed and used on the device is yet another reason for employing an organizational app store. Another reason is cost. For larger organizations that have in-house developers, it may be more cost-effective to simply develop an app that is specific to the enterprise, versus the cost of buying a volume license for apps that may not fully meet the organization's requirements. In any event, establishing an organization app store and managing it through the MDM and MAM infrastructures may be an integral part of the centralized management of mobile devices in the enterprise. Keep in mind that at its most basic level, an app store is really nothing more than a repository for software, so mobile users can simply click on a link or go to a network share to find and install apps.

In order for an organization to distribute Apple iOS apps to its employee users, the organization has to participate in the iOS Developer Enterprise Program, enabling them to develop and distribute apps specific to their own needs. These apps are digitally signed by the company, ensuring that they are legitimate and have been designed around Apple's rigorous standards. They can, however, only be distributed to users associated with the organization. One interesting fact about apps developed by an organization using this model is that they have to use what is called a provisioning profile. The provisioning profile contains metadata about the app itself, as well as information about its developer. The provisioning profile is pushed to an Apple device via the device's cellular or wireless connection. Provisioning profiles can also be included in the app itself to provide for ease of installation by the user. Because Android employs a more open method of getting apps into the hands of the users, enterprise app stores don't have to go through any of the hoops that they would for iOS apps.

 EXAM TIP Understand the ways that the different app store vendors develop and distribute apps for their respective devices.

Default Applications and the Enterprise

For ordinary users with their own devices, the default vendor applications that come with the devices may sometimes be enough. Most people, however, usually shop through a vendor's app store specific to their device in order to find apps that are more useful to them than the default apps sometimes are. In a corporate environment, the default vendor applications may not be what the organization wants to use for several reasons. First, the organization could have a standardized app it wants to use for a given function, such as email, for example. Second, as security is a much more important issue in the enterprise space, default applications may not be configured securely or may not offer security features the organization needs. There also may be some default applications that are not approved for use on devices owned or managed by the organization, such as those used for social media, for example. Policy and standardization will typically determine what apps are used on centrally managed devices, and they may need to be located in the enterprise app store for users to download and install.

Pushing Content

Most mobile app management is done through a *push model*. In a push model, applications centrally managed by the enterprise app store are typically automatically installed on the device, as are updates to both the device's operating system and the different apps that reside on the device. Occasionally, the MDM infrastructure may poll the device to determine what version of apps and patch levels the device is currently using, and may install an upgraded version as needed.

Devices can be notified of updated or new apps via *push notifications*. Push notifications are messages sent to the device from the app store, informing the device about a new version of an app or, in some cases, about new configuration settings. One important item about pushing content from the application store is that, regardless of whether or not the organization has its own enterprise app store, in the case of iOS or Android apps, these push notifications don't actually come from the enterprise app store at all; they actually come from the OS vendor. For example, the Apple Push Notification service (APNs) requires that the enterprise register its app with Apple and will send notifications through Apple servers for forwarding to the device. Similarly, Android devices work the same way, using the Google Cloud Messaging (GCM) service. The main difference between Apple and Android notifications is that GCM notifications can hold more data, but both similarly interact with individual apps on their respective devices.

 EXAM TIP Understand how APNs and GCM work, including their associated port numbers you learned in Chapter 1.

Disaster Recovery Principles and the Mobile Infrastructure

Disaster recovery is an issue that affects not only mobile device infrastructures but also every part of the organization, to include its entire infrastructure, and beyond that, all of its business processes. There are really two main focuses of disaster recovery. First of all, there is *business continuity*. Business continuity is concerned with keeping the business going after a major event or incident. For our purposes here, an event or incident can be a wide range of either man-made or natural disasters, to include hacking attempts, thefts, terrorist attacks, tornadoes, hurricanes, floods, fires, and so on, even to include events that indirectly affect the organization. For example, an incident that cuts the major communications lines for an upstream Internet service provider (ISP), could significantly affect the business itself, even though no damage has occurred to the organization's computing assets. For the most part, a disaster could be considered anything that affects the organization in a negative way, with a serious impact. Business continuity requires careful planning and actions, such that the business can successfully recover from an event and continue to operate, fulfilling its mission and goals.

Disaster recovery, on the other hand, is concerned with all the activities immediately following an incident or disaster that are designed to help the organization and its personnel recover to the state that business can continue. Disaster recovery could be considered a subset of activities in the overall business continuity effort. However, disaster recovery is focused on a different set of planning activities and actions. For our purposes here, we're going to look at the different general disaster recovery and business continuity principles that your organization should adhere to in order to keep its business going and recover from any type of incident.

The next few sections cover major disaster recovery principles and how the organization reacts to disasters and maintains business continuity. In later sections, I discuss how these principles apply to the mobile device infrastructure specifically.

 EXAM TIP Understand the differences between business continuity and disaster recovery. Remember that business continuity is concerned with the planning and execution activities involved in getting the business back up and operating, while disaster recovery focuses on recovery after a major incident or disaster has occurred.

Business Continuity and Disaster Recovery

An entire book could be written on business continuity and disaster recovery, but I'm going to cover just the important basic elements for you to become familiar with for both the exam and for the real world. Throughout your career as a mobile infrastructure professional, you will likely learn a great deal more about business continuity and disaster recovery. In this chapter, I discuss business continuity principles and disaster recovery principles separately, but understand that there is a significant overlap in the overall business continuity planning world between these two disciplines.

Business Continuity Principles

As I mentioned in the preceding sections, business continuity concerns itself with the organization's ability to maintain its mission after a significant negative event has taken place. Whether this event is man-made or natural, the business must be restored to some level of operation or it risks shutting down completely and not surviving the event. Business continuity planning is all of the careful planning and activities that ensure a business can survive an event and continue in its mission. While disaster recovery typically comes before the business can resume operations, understand that the business continuity planning function is a long-term, on-going activity in the organization, and should be happening long before a disaster actually occurs. Business continuity planning is very much connected to risk management. While I discuss risk management several times throughout this book, risk management is an entirely different discipline that requires a great deal of experience and training to become proficient at. For our purposes, understand that business continuity planning is part of the overall risk management process.

One of the first steps in business continuity planning is to identify the organization's critical processes and business operations. These processes and operations, if lost, would significantly affect the organization's ability to conduct its business. The organization should determine criticality for each of these processes, and how much protection each of them requires in order to maintain them after a negative incident or event. The second step the organization needs to take is to inventory all of its critical assets, such as equipment, data, and yes, even people, to determine how these assets relate to these critical processes and operations. The organization must determine how much protection each of these assets should be afforded to protect them in the event of a negative event or disaster. The organization should also prioritize these different processes and assets for protection and recovery after an event. These first two steps in the process are commonly called a *business impact analysis,* or BIA. In addition to identifying and prioritizing the different processes and assets, this part of the planning should also show what the impact would be if that particular article process was lost or that asset destroyed, for example. Table 6-1 shows an example of how an organization might list and prioritize these different processes and assets for business continuity planning. Keep in mind that this is a very simplistic example, of course, but could be used as a starting point in your business process analysis. For a more in-depth view on conducting a BIA, you should take a look at the National Institute of Standards and Technology (NIST) Special Publication 800-34, Contingency Planning Guide for Federal Information Systems, which offers some really great information on planning and conducting a BIA, to include useable templates.

Once the business impact analysis is complete, the organization should commit resources and planning to protect those processes, and its critical assets. Business continuity planning should also include a step-by-step process for bringing these critical processes back online after a disaster or other negative event. The plan should include an analysis of how much data the organization can afford to lose or how much time the organization can afford to be out of commission before bringing operations back up to an acceptable level. In the next few sections, I discuss parts of business continuity

Process or Asset	Value	Replacement Cost	Organizational Priority (1–10)	Impact If Lost (High, Med, Low)
Order intake process	$10,000 in sales per day	$20,000	10	High
File server	$5,000	$3,500	6	Med
Administrative Computer	$2,500	$2,500	3	Low

Table 6-1 A Sample View of a Business Impact Analysis

planning that help support bringing these critical processes back into operation as quickly as possible with minimal data loss.

 EXAM TIP Understand the purpose of a business impact analysis (BIA).

Disaster Recovery Principles

As mentioned previously, disaster recovery planning and related activities are designed to become effective when an event or incident has actually occurred. Careful planning has to go into disaster response and recovery, to include establishing a response team and chain of command, assessing the situation, and ensuring personnel and critical equipment are protected during a disaster. Disaster recovery also includes restoring equipment and data, and supporting infrastructure so that business continuity activities can proceed and the organization can resume operations. The most important part of disaster recovery planning is the recovery plan itself. The disaster recovery plan, or DRP, is designed to provide the organization a set of concrete, step-by-step actions and activities the organization will take in the event of a disaster, in order to preserve lives, prevent injuries, and save critical equipment and data. The DRP should address the different range of possible events that could happen to the organization, on a reasonable basis, of course, and plan the organization's response to each of those possible events. For example, part of the DRP should cover how the organization will respond to a serious weather condition, such as a tornado, for example. This may cover alerting personnel, shutting down equipment, securing the facility, and evacuation if necessary. Obviously, the most important aspect of planning is protection of human lives, and emphasis should be given to this aspect during the planning process. Beyond protection of personnel, planning should cover protecting facilities, equipment, and data. During a disaster that immediately threatens human lives, such as serious weather conditions like tornadoes, flooding and hurricanes, or fires, priority should be given to saving lives and there may not be time to secure equipment or data. However, if the threat is not immediate, or if there is some time leading up to the event (such as, for example, a few days' notice before a severe storm or hurricane), the DRP should cover the orderly backup, shut down, and security of systems and data, as well as the facilities if at all possible. Activating the disaster response and business continuity plans in advance of an event, wherever possible, is also a good idea to give the organization additional preparation.

 EXAM TIP Remember that the most important consideration in disaster recovery is saving human lives and preventing injury.

Other normal business operations, such as backup processes, for example, serve to supplement the disaster recovery and business continuity planning, and should be utilized with that in mind. For example, routine backups of the servers should be performed with a quick restoration and minimal data loss as the goal, and possibly should be located off-site in the event a disaster occurs. Other processes, such as alternate work locations and so forth should also be created with disaster recovery planning in mind.

Disaster Recovery Locations

Often, a disaster will completely render the primary business location unusable. This is definitely a possibility in the case of fire, tornado, hurricanes, or flooding. In this event, the business may not be able to restore operations back in its primary site, and should plan in advance on an alternative location for restoring operations. There are different ways that this could be accomplished. Owning or leasing a separate facility located some geographic distance away from the primary site is one way of ensuring an alternate processing location. Cloud services provided by third parties could be another solution, especially if most of the organization's business processes occur online or via the Internet. It's still likely, however, that there will need to be an alternate physical location for employees to work at and restore business operations to.

There are three types of alternate locations that provide differing levels of readiness or support for restoring business operations after a disaster. Each of the sites has different advantages and disadvantages directly related to how fast the business needs to restore operations, as well as how much of an investment the business can put into the alternate location. The first type of location is the *cold site*. The cold site is usually nothing more than a bare facility with empty space for equipment and offices. There's usually some limited level of utilities turned on for the site, such as heat, water, and electricity, but nothing more than that, to include Internet or communications access. In the event of a disaster, the organization would have to take a great effort in physically relocating people, equipment, supplies, and so forth to the cold site. In a large scale natural disaster, this might prove to be difficult due to road conditions, lack of vehicles, disrupted public utilities, and so forth. Disaster recovery planning, when using a cold site, should take all of this into account, as well as the time and effort required to set up the cold site for business operations. One key advantage to using a cold site is that there is probably very little investment required. Basically, the organization only has to lease or own an empty building, for example. The key disadvantage to using a cold site is that the time required to restore operations could be excessive, and it's possible that an organization would not have the manpower or tools to successfully relocate personnel or equipment over to the cold site very effectively.

The next type of alternate business site is called the *warm site*. A warm site provides the next level of readiness for business operations restoration above a cold site; in addition to workspace and basic utilities, it also could provide communications links, such as Internet and phone service. A warm site also usually has some basic level of equipment already installed at the site. This could include simple office furniture, such as desks and tables, but also likely includes equipment such as servers, workstations, and other types of equipment used to bring the business back into operation. This equipment could be turned off and simply waiting for someone to flip the switch and turn the power on, and it could also require a quick data restoration from organizational backups to ensure that it is using the most current data available from the organization's business transactions. In some warm sites, organizations often restore data from backup on frequent basis, so that the business can come back to operations much more quickly. The key advantage to a warm site is that the time to recover business operations is much less than it would be when using a cold site. The key disadvantage is that a warm site requires much more of an investment of time, money, and resources. The organization may find, however, that the extra expense is worth it if the amount of money and business lost in the event of a disaster would be much more than what it invested in the warm site over time. So, business criticality and the need to restore operations much faster would be a deciding factor in maintaining a warm site.

The next level of an alternate processing site is the *hot site*. As you can imagine, the hot site is capable of providing business operations much faster than a cold or warm site. In a hot site, not only do you have workspace, utilities, communications services, and equipment, but this equipment is usually maintained in a high state of readiness, powered on and ready to process data. Backups from the main processing site may be transferred to the hot site and restored very often, even on a daily basis. This would ensure the hot site has the ability to pick up processing quickly with minimal loss of data in the event of a disaster. Obviously, a hot site requires much more expense of resources on the part of the organization in order to maintain this high state of readiness. A hot site would be used in business operations that are extremely critical, and when anything beyond a very small amount of downtime in business processing is intolerable to the organization. In maintaining a hot site, not only would an organization have to maintain a completely separate facility in a high state of readiness, but also all of the redundant equipment and supplies used in the facility would be almost a duplicate of anything in the primary facility.

 EXAM TIP Understand the differences between hot, warm, and cold alternate site locations and how they affect disaster recovery.

Network Device and Server Backups

I've mentioned backups a few times throughout this discussion over the past few sections, so it's probably very appropriate that I discuss it in depth at this point. This discussion assumes that the organization already has a routine backup strategy in place, and follows best practices by having different levels of backups, to include full, incremental, or differential backups, or even organizationally customized backups for specific data sets, such as transactional or specific data backups, to restore different levels of data at different required restoration points. This discussion also assumes that backups follow the best practice of being stored at an off-site facility, in order to protect them from disasters that could happen at the primary site. These are important best practices that most organizations cognizant of business continuity will follow, but they are worth mentioning here as well.

Network device and server backups are important because they are the first line of protecting the organization's entire infrastructure data. This can include all the business process data, financial transactions, employee data, and any other relevant organizational data, on a massive scale. Server backups include not only data related to the business, but also operating system data as well. Without backing up this type of data, it would be almost impossible to restore the organization to a functioning capability in the event of a disaster or serious incident. Network device backups, on the other hand, usually consist of backing up the operating system (if applicable), and any configuration files or settings, so the device can be restored to an operational state very quickly. While some data processed by network devices may not necessarily be backed up (i.e., real-time traffic passing over a device) because it will never be restored, per se, other peripheral data created by the device; logs and other audit trails, for example, must be backed up for a variety of other reasons, including security auditing, regulatory compliance, and so on.

Directory Services

There are several critical business and information technology processes for an organization, but few are as critical as directory services. The reason for this is that modern IT infrastructures rely so heavily on directory services for security, resource location, and many other critical services. For example, imagine what the workday would be like if first thing in the morning, when most of the users came into work and started logging into their workstations, they could not authenticate to the network because the directory services were down. In addition to network authentication, many network services, such as shared folders and DNS, for example, rely on directory services to do their job. Because of this criticality, directory services must be maintained in a high availability state for users in the network. In the case of Microsoft Windows Active Directory (AD) services, the distributed data stores that make up AD are often spread across several servers for redundancy purposes, in the event that one server fails or becomes unavailable, for example. In addition to this load-balancing type of set up, the Active Directory database is typically backed up several times a day in order to ensure that a current copy of the directory is maintained. Like directory services, several other infrastructure type services are also critical and should be protected to the use of backups and redundancy. Some examples of these services may include the domain name service, the DHCP database, different business critical Web services and related databases, financial and accounting transaction databases, and even key data from users' workstations.

 EXAM TIP Understand why critical services, such as directory services, must be backed up and recovered. These services are what make authentication, encryption, name resolution, resource location, and other network functions happen. Without them, you will not be able to restore your network to its previous operational state.

Frequency of Backups

How often and to what extent an organization backs up its data depends upon several factors. The first factor is how critical the data is. When doing business continuity planning, the organization should take a good hard look at its data assets and determine how critical they are. In some cases the organization has to prioritize different types of data to determine how critical they are and how much protection they should be afforded. Obviously, more critical data should be afforded better protection, which could cost the company more in terms of equipment and money. For example, maintaining backups on critical company servers to process customer financial data is probably a higher priority than maintaining equipment and the necessary backups to provide for data protection for an administrative assistant's workstation.

Another factor, aside from data criticality, that affects both frequency and degree of backups, is how quickly data would have to be restored, given a catastrophic event that destroys the primary data source on a server, for example. Certain types of backups take longer to execute than others, and certain types of backups restore more quickly than others. Essentially, the amount of data backed up is the determining factor in the trade-off between a quick backup and a quick restore, with some consideration, of course to the speed and efficiency of the backup equipment used. With that said, it's probably a good opportunity to talk about the three primary types of backups that most organizations perform.

Most information security best practices describe three basic types of backups: full, incremental, and differential. Obviously, each of these different types of backups is chosen based on the amount of data the organization wants to back up, as well as the resources required to execute each type, such as media capacity and cost, and efficiency of the backup equipment. However, each type also varies in both its backup time and the time required to restore the data from the backup. Let's discuss each of these in a bit more detail to clarify.

A full backup basically does exactly what you would think; it completely backs up the entire hard drive or an entire data store or whatever else the administrator specifies as a complete backup of an asset. In addition to data, this could also mean the operating system and applications on the server's hard drives. That way, if the server's drives fail, the complete contents of the hard drives, including the operating system, applications, and data, could be fully restored back to the point that existed before the failure. Because a full backup can get every piece of data an organization needs to completely restore a server, it would seem that this would be the preferred type of backup every time a backup is executed. The problem with this, however, is that a full backup can take quite a while to back up as well as restore. How fast exactly depends upon the hardware involved as well as the amount of data. If the server crashes in the middle of the day and has to be restored from a full backup, it may take more time than the organization can afford to spend in downtime. Most organizations perform a full backup on an asset on a regular basis; once per week is a fairly normal schedule for executing a full backup, but it really depends on the organization's policies and its tolerance for data loss and/or downtime.

The second type of backup is an incremental backup. An incremental backup is a type of backup that will back up any data that has been added or changed since the last full backup or the last incremental backup. The incremental backup is able to determine what data has changed through the use of the *archive bits* set on data files. Most file systems have the capability to set different bits on files that determine whether or not they have been backed up recently, or have changed. If the file has changed, the archive bit is turned on, and this lets backup software know that the file needs to be backed up. Once the incremental backup has occurred, the archive bit is turned off. So if a full backup occurs, and then a file changes, an incremental backup will back up that file and then turn the archive bit off. Assuming the file doesn't change any more, the next incremental backup will *not* back up that file. It will only back up files that have the archive bit set to on, meaning they have changed since the last full or incremental backup. An incremental backup does not take very long to back up the data because it backs up only a subset of the files on the media. However, an incremental backup can take a long time to restore, simply because the full backup must be restored first, and then every single incremental backup performed since the full backup was executed must be restored, in order, until all the data has been successfully restored. For example, let's say that a full backup of a server was performed on a Sunday night. On Monday, Tuesday, Wednesday, and Thursday, an incremental backup was performed on each of those nights. If the server's hard drive crashed on Friday, then the full backup from Sunday night would have to be restored first, and then, in sequence, the incremental backups from each of the succeeding nights would have to be restored in order to make sure that all of the data changes were accounted for. Depending upon the amount of data involved, this may take a while, and for a critical server, this amount of downtime required to restore data from a backup may be unacceptable.

The third type of backup is the differential backup. A differential backup is similar to an incremental one in that it does not back up every piece of data from the media. Like incremental backups, a differential backup also relies on the archive bit on a file to determine what data has changed since the last full backup. If a full backup is performed on a Sunday night, for example, and then data changes on Monday during the day, a differential backup will back up that data because the archive bit has been turned on for the file. However, unlike an incremental backup, a differential backup does *not* reset the bit and turn it off. The archive bit remains turned on. So, the next time a differential backup is run, it backs up all the data with the archive bits turned on, including all data that has changed since the last full backup. So its backups of data are inclusive of all changes since the last full backup. Because of this, the first differential backup in a series doesn't take very long to perform. However, as time passes and more data changes, each subsequent differential backup takes a bit longer. The advantage of this is that because backups are inclusive of all data changes since the last full backup, restoring data requires less time. For example, let's say that a full backup was performed on the server on a Sunday night. Files changed on Monday during the business day, and a differential backup was performed on the server Monday night. It would back up all files that had changed since the last full backup because the archive bit on those files would be turned on for those files. Because the differential backup does not turn off the archive bit, the next night the differential backup is run (Tuesday night), it backs up data that changed on Tuesday as well as the same data that changed on Monday, because the archive bits are still turned on. It does this every single day, increasing the amount of data that it backs up, as well as the time it takes to perform that backup. On Friday, the server hard drive crashes, and the full backup is restored from Sunday night, but the only differential backup that is required to be restored, in addition, is the last differential backup performed on Thursday night. This is because it contains all of the changes to all of the data since the last full backup. So while differential backups may take progressively longer to perform, they are usually faster to restore.

Obviously, the examples I've given you for the scenarios involving full, incremental, and differential backups are intentionally simplistic, so you can understand how they work, but in the real world, it can get much more complex than that. Most organizations differ in how they perform their backups, but they usually use a combination of full, incremental, and differential backup techniques in order to make sure they maximize the amount of data backed up, as well as reduce backup and restore times to the best extent possible. Obviously, speed and efficiency of backup hardware contributes to backup and restore times.

 EXAM TIP Be able to identify and explain the three basic types of backups for the exam, and to know the advantages and disadvantages of each.

High Availability

High availability is an interesting concept in the world of information technology. To average users, high availability means that their data is always there whenever they need it, and this is a fair presumption to make. To IT professionals, however, it means a little bit more than that. It means there are several technologies that are used to make sure the user's data is there and available when they need it, of course, but there are more detailed considerations as well. There are several different ways of assuring high availability of data and services. Redundancy is one such way, both of technology and in data. Redundancy contributes to high availability by providing additional technologies and services that are available in the event of primary source of services becomes unavailable. One such redundancy technology is *clustering*. Clustering involves using multiple devices, usually servers, connected together in a cluster, such that all of the servers act as a single logical node, all communicating on the network with each other and with clients. In the event that one server fails, another server, usually with identical data stores, takes over in answering client requests. This is usually completely transparent to the client. Load balancing is another way to ensure availability. Load balancing involves similarly using servers in a cluster; however, in this case different servers in the cluster could answer a particular request. There's no primary or alternate server that takes its place if a primary fails; they all equally can answer a client's request for data or services.

Similarly, redundant data can affect high availability. This means that on a near real-time transactional basis, data is backed up and duplicated to multiple sources, so that if a connection to a data source is lost or if the data source itself fails, the client requests can be redirected to a duplicate instance of the data. Such redundancy requires very high-speed controlled communication between data sources as well as clients to manage requests almost instantly.

There are several different technologies that can contribute to high availability, including load balancing and clustering, as I mentioned earlier, but also redundant communication lines, high-speed fiber connections, and so on. Virtualization can also contribute to high availability, in that virtualizing servers and data stores can be restored easily by provisioning a new, identical virtual machine in case a primary server or other virtual machine fails in some way.

High availability is often measured in terms of "the nines." For example, an availability of 99.9 percent might be considered very high for the average user. When you look at it in terms of 365 days per year, 99.9 percent availability actually gives you only 8.76 hours of downtime per year—but this totals out to an entire work day. Even if most employees don't work a continual 365 days per year, businesses today that use e-commerce models are subject to being accessed online and may sell products and services every day of the year, so the loss of business even one day (especially during a peak season, for example) due to system downtime may be completely unacceptable. Most IT professionals would rather have what's referred to as "five nines" of availability, which is 99.999 percent. It doesn't seem, at first glance, that this would be too much different from our 99.9 percent availability, but when you do the math, it actually equates to only slightly over 5 minutes of downtime per year—a significant difference for a business that relies on its online assets being available to paying customers 24/7. Figure 6-9 shows the different measurements of availability in terms of "the nines."

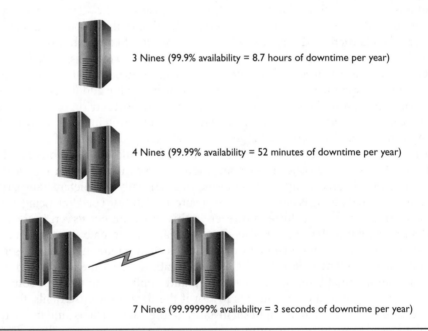

3 Nines (99.9% availability = 8.7 hours of downtime per year)

4 Nines (99.99% availability = 52 minutes of downtime per year)

7 Nines (99.99999% availability = 3 seconds of downtime per year)

Figure 6-9   High availability, with regards to "the nines"

Best Practices for Mobile Device Data Protection

Previously I discussed business continuity and disaster recovery concepts and how they apply to the overall organization. In this section, I extend the discussion on BCP and DRP, examining how they would apply to the mobile device infrastructure, as well as mobile devices themselves and the data that is contained on them. I also continue the discussion on backup and restore of all data from a perspective of mobile devices, corporate data, and personal data.

Mobile Device Backups

Mobile device backups happen on a different level than the infrastructure and server backups. A device backup likely includes the user's individual device, and also probably includes not only the operating system itself but also corporate data, and sometimes personal data from the device. There are a couple of different ways to back this data up, but all of them should come from the perspective of centralized control and management. In other words, you could conceivably rely on a user to back up their own individual device, but this might not be the best solution because it couldn't be managed and controlled such that the backups would occur on a complete and routine basis. So backing up individual devices from a corporate management perspective is probably the better idea. Typically, backups would occur for each individual device on a recurring basis, according to policy. These backups could be sent to a centralized backup server, such as a network attached storage or NAS, or a storage area network (SAN), if there is enough data to warrant it.

Corporate Data Backup

Corporate data is organizational data that must be protected and separated from personal data. On almost every mobile device that the organization allows to connect to its infrastructure, especially personal devices in the case of BYOD implementations, there exists some level of personal data collocated with corporate data. The organizational concern should be corporate data obviously, and there needs to be a formal policy as well as management infrastructure designed to back up corporate data from mobile devices. As I have mentioned, the backup policy driving corporate data backups should be such that corporate data is backed up to organizational assets, such as backup servers or officially sanctioned third-party providers, specifically used for that purpose. Corporate data, and its backup, should not be entrusted to individuals for mobile devices simply because there's no way that the organization can ensure that it gets done on a periodic basis. There are a few different ways to implement backing up corporate data to centralized infrastructure devices. First, regularly scheduled backups could be implemented through policy pushed to the device, so that it backs up critical data through the MDM infrastructure in a defined time period. You could configure the device to back up its data immediately upon connection to the corporate infrastructure, as part of its connection process enforced by network access control (NAC) devices. You can also enforce backups when not connected to the corporate infrastructure, making sure that they are securely backed up using encryption and only when there is a strong signal, such as a wireless connection available. One key aspect of backing up corporate data is security, of course, and encrypting backups as they occur is a highly recommended best practice. Additionally, another issue is how much data to back up from a mobile device. It may not be necessary to back up the entire device, including the operating system and applications, on a regular basis. Critical corporate data may be all that's required on a regular basis, with full backups spaced out with a little bit of time between them. I discuss data containerization in later chapters, which is one method of keeping corporate data and personal data separate. With data containerization implemented on a mobile device, it may be possible to confine the backup set to a particular set of data, and backing up all relevant corporate data in the container.

Personal Data Backup

Backing up personal data from a mobile device is not much different than backing up corporate data from a technical perspective, except that personal data may or may not be transferred to the corporate infrastructure backup devices. If data containerization (discussed in Chapter 7) is in effect on the mobile device, then it could be quite simple to separate personal data from corporate data, and ensure that personal data is left alone by the organization's backup system. There may be reasons for backing up personal data from the mobile device, and that's more a matter of organizational policy than technology. I discuss the legal and ethical ramifications of corporate access to personal data on a mobile device in Chapters 7 and 9, but suffice it to say for now that there are good reasons for keeping personal data and corporate data separate, even in backup policies.

For the most part, users should be responsible for backing up their own personal data, especially on devices that are employee-owned and used in the organization as part of a BYOD infrastructure. Employees can back up their own personal data to their own computers, of course, but one valid option that is becoming more and more prevalent is the use of third-party backup solutions. In some cases this may be the use of cloud services offered by the different device and operating system vendors. For example, Apple offers its iCloud services for iPhone and iPad users, which allow users to back up their personal data to an Internet-based storage device. While I won't debate the security or privacy issues of this practice here in this text, suffice it to say that it is a decision that each user should make based upon their feelings of control over personal data and their trust in the third-party provider.

Local Device Backup

One common method of backing up data, whether it is corporate or personal data, is backing up to the local device itself, in addition to any backups that are made to the network infrastructure or a cloud provider. Most devices, such as tablets and smartphones, have a finite amount of internal storage provisioned with the device. It's not unusual to find devices with 8, 16, 32, and even 64 GB of internal storage on the device itself. While backing up data to the device's local storage is one option, it should be considered a short-term option in the event the user can't access any other backup devices, such as computers or corporate servers. The primary reason for this, of course, is that if the device is lost, then the backup is also lost with it. Another option available to both corporate and personal users is the use of removable storage media, such as the ubiquitous SD card for backups. There may be advantages to using an SD card in a mobile device for backups, especially for corporate users. Corporate data can be backed up to the removable SD card, and then removed from the mobile device when the user leaves the facility. In this way, corporate data can be protected when the device leaves the facility. Personal data could also be backed up to a separate SD card and maintained by the individual.

Yet another method of backing up data to the device itself is the use of the SIM itself that is located inside the device. The SIM can hold certain data for the user, including contacts messages, and so forth. This data is persistent when the SIM is removed from the device. While not the ideal form of backup, use of the SIM is one method of storing data on the device itself. Whenever possible, backups should be kept in an encrypted form, regardless of media, while on the device.

EXAM TIP   Understand the best options for backing up a mobile device. Corporate data should be backed up to the infrastructure or contracted third-party provider. Personal data should be backed up to a computer or user's third-party provider.

Data Recovery

The next few sections cover concepts essential to data recovery. Once data has been backed up and stored, obviously there's a concern in restoring it, both from a time perspective, as well as an integrity perspective. For example, even if a backup for particular server goes successfully, how does the administrator know the data can be successfully restored without being corrupt or that it can be restored within a certain amount of time? These questions can be answered through carefully planning the backup and restore process, as well as testing backups occasionally to make sure that they work properly. Additionally, there are considerations with restoring both corporate and personal data to mobile devices.

Testing Backups

As mentioned above, the restoration of data from a backup incurs questions of integrity. Tapes deteriorate over time, CDs and DVDs become scratched while in transport or storage, and even issues with mass storage devices connected to the network can occur, whether it's failure of the device itself or issues with the network connection. In any case, data integrity during the restoration process is very important to an organization. It does no good to get a good data backup if it can't be restored fully intact, and within an acceptable amount of time. For this reason, backups should be tested for integrity using tools designed specifically for that purpose. Hashing, discussed later in the book, is one method of ensuring integrity of data, in that the hash or cryptographic sum is computed for the data both before and after the backup. If the cryptographic sums match, then integrity is ensured and the data is assumed to be intact. If the cryptographic sums are different, however, then it can be assumed that the data is corrupt or has been changed in some manner. Likewise, when the data is restored, the cryptographic sum can be computed again, and a match with the same number computed after the backup would indicate that the data has not changed and is still intact. The property of data being intact, without being modified in any manner, is called *integrity*.

Yet another way that the backups can be tested is to actually perform a restoration from them to ensure that the data can be quickly restored and that integrity has been maintained. It's a good idea to test backups by restoring them from time to time to ensure that the backup process is working the way it should. It's probably not a good idea, however, to wait until an emergency or incident happens to test your backup process and see if the restore process works the way it should. A much better idea would probably be to restore backups from time to time, possibly even to a server designated for that purpose, instead of to the original server, so as to prevent interruption of services during the restoration. Restoring a backup from time to time can give the organization confidence that the backup and restore processes are working properly. Backups that can't be restored or persistent problems with restoring data would indicate that there is a failure somewhere in the backup process, either with equipment, network connectivity, or even the backup application software.

 TIP   In the real world, you should randomly restore data backups periodically to ensure that you have a good backup set; this can validate the effectiveness of your backup and restore processes.

Restoring Corporate Data

Restoring corporate data, particularly to mobile devices, will likely require that the device is somehow connected to the corporate infrastructure. If there's a suspected reason to restore corporate data to the device, then obviously the user would need to bring the device to the organization's mobile device technicians so they can restore the data appropriately from the centralized backup source. If the operating system and application data need to be a restored as well, then restoring the device completely from a full backup is in order, followed by restoral of the latest backup of the corporate data.

Restoring Personal Data

Restoring personal data again is not much different than restoring corporate data, except that the backup source may be different. While you could restore personal data from a corporate device from a corporate device, it's more likely that you'll be able to restore from a user's personal computer, or even a third-party provider, such as a cloud backup type of service. The types of data restored from backup may include individual files, or compressed or containerized data formats unique to the application used to perform the backup. If the backup is being restored from a third-party service provider, the user will obviously need appropriate credentials and connectivity to a third-party provider so the data can be restored. Although the user is likely going to be responsible for storing their own personal data, it may be wise to get the mobile device technicians involved as well, to prevent the restoration of personal data over corporate data or violating the security of the device with the restoration process.

New Technologies and the Enterprise Mobile Infrastructure

Exam objective 3.7 requires that you have an understanding on how to use best practices to maintain awareness of newer technologies, including changes that affect mobile devices, and by extension, the enterprise. In order to maintain awareness, as mobile infrastructure professionals, we must keep up with technology changes, new capabilities, and even new business practices that quickly adapt to rapidly evolving technologies. We can maintain our proficiency and technical awareness through a variety of sources, including trade journals, socializing with other professionals, and keeping up with new technologies through vendors, developers, and integrators. Some of these sources are discussed in the text that follows, along with the technical security awareness that comes from watching for new threats and vulnerabilities in software and hardware.

OS Vendors

Operating system vendors are a good source of keeping up with technological advances because, from a business perspective, one of their goals is to continue to improve their market share by introducing new and innovative products to businesses and consumers. Vendor publications, web sites, and even trade shows are excellent sources of keeping up with new technologies in the mobile market. Some of the key things to keep track of from the operating system vendors are new versions of existing operating systems, or even completely new operating systems. Additionally, new features, particularly those related to productivity or security, should be carefully watched as well. When examining some of these new technologies from OS vendors, you should carefully consider how they will integrate and interoperate with your existing technologies and infrastructures. In some cases, simply trying to bolt the new technology into the existing infrastructure won't work, and other pieces of infrastructure may have to be changed or upgraded as well in order to accommodate the new platform changes. Conducting a limited test or pilot program when considering new operating system platform advances is usually one first step in successfully integrating a new operating system into the enterprise infrastructure.

Hardware OEMs

Hardware or device vendors are another forceful driver in bringing new technologies to market. New technologies and features often cause advances in software or platform advances and may bring additional features to the enterprise as well. An organization must keep up with hardware and device changes in order to make sure they maintain interoperability with existing networks and related infrastructures. Hardware, software, operating systems, and network infrastructures are the key drivers in technology changes in the mobile device world. Better and faster hardware obviously requires updated operating systems. More advanced network infrastructures and signaling methods in turn require more advanced hardware, and again, more efficient and feature-driven operating systems.

Telecommunications Technologies

As mentioned, network infrastructures are frequent factors in driving hardware and software advances. For example, the changes over the years in network speeds and throughput from 2G all the way to 4G technologies brought about significant hardware changes with increased processor speeds and expanded storage capacity, which in turn drove vendors to produce more advanced operating systems that were ever more feature-rich and secure. The different telecommunications carriers have contributed to the new technologies that have been introduced into the devices and software itself. Some of the organization's infrastructure may have to be adapted to these new technologies, especially if they are legacy network switching, routing, and management technologies. Consider these factors as well when evaluating new carrier technologies for the enterprise.

Third-Party Application Vendors

Application vendors, particularly those representing third-party developers not necessarily connected to the operating system vendor, also drive technological changes that must be considered. As developers produce more complex apps that are full of new and exciting features, both hardware and operating system developers must ensure their respective products keep up in order to continue to offer support for the apps with these features that consumers and businesses want.

As you can see, this is a continually cycling process. Consumers require more features, speed, and data consumption with their applications. This in turn drives hardware and operating system vendors to create better, faster, more efficient devices to support this increased demand for data and features. Although the consumer market has caused the mobile device demand to increase at an astounding rate during the past decade, the business sector has quickly caught up, causing organizations and their employees to demand the same features and technological advances for the workplace. Technology managers in organizations have to keep up with all of these advances and determine how best to adapt and integrate them to the enterprise environment, while ensuring the security of enterprise data and interoperability with the existing infrastructure. Figure 6-10 demonstrates the relationship between apps, hardware, the OS, and the consumer.

New Risks and Threats

In addition to technology advances, unfortunately, there is also new risk and threats with every new piece of technology that gets implemented on a widespread basis. In Chapter 8, I discuss security risks and threats in depth, but at this point in the text, you should at least be aware that those are the pitfalls that come with new technologies and how they are integrated into the existing infrastructures. New technologies often change existing security policies and configurations settings, creating additional risk. When considering integrating these new technologies into the enterprise, a risk assessment should be performed in order to ascertain threats and how they affect existing corporate assets. Determining the impact on the corporate infrastructure from new technologies will help you to determine how much resources need to be spent reducing that risk, as well as how to mitigate any risks that can't be further reduced.

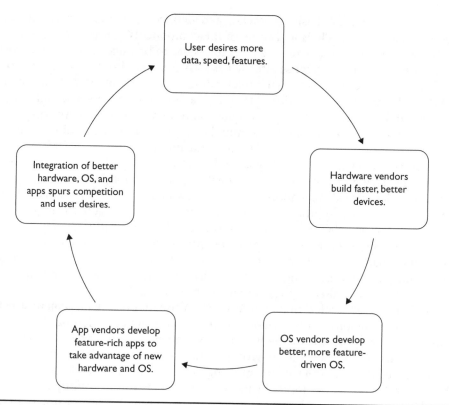

Figure 6-10 Relationships between hardware, OS, and mobile apps

 EXAM TIP Remember that there are several important factors to consider when evaluating new technologies for implementation in the enterprise; the most important are interoperability, cost-effectiveness, and of course security and risk to the existing infrastructure.

Chapter Review

In this chapter, I covered a great deal of material over several examination objectives. First, I covered a little bit of history of mobile devices in the enterprise, and talked about how the massive infiltration of these devices into the consumer market spurred their adoption in the corporate world. I then also compared a few of the many brands of devices available to the enterprise user. I described the most popular, including Apple devices, Android devices, Windows platforms, and BlackBerry devices.

I then went into the basics of MDM, and defined what the purpose of mobile device management really is and some of the components that you may see with an MDM infrastructure. I discussed the need for MDM to be integrated with the existing infrastructure that the enterprise already has in place, and also discussed some of the challenges organizations face when implementing MDM. An important related concept to MDM, the bring your own device (BYOD) paradigm, where users actually bring in user-owned devices to access corporate data and resources, was also covered.

From there, I went a little bit deeper into the mobile infrastructure, discussing policy as one of the primary factors in planning and implementing an MDM program. I discussed several elements of policy, including IT and security policies that drive the extent of balancing security and usability with regards to mobile devices. Backup and restore policies were also mentioned as a critical set of security policies that affect the organization's ability to protect data. Other elements of policy affect external factors, such as vendors and their default applications, the different platforms available to the organization, and even telecommunications providers. The need for standardization across the board in terms of mobile device platforms, apps, and carriers is also important. I also discussed the need for interoperability with the existing equipment and network services already present in the infrastructure.

Requirements are critical in planning an enterprise mobility management structure. Some of these requirements include security requirements, including the need for administrator permissions on mobile devices, and password strength requirements. I also discussed the benefits of grouping devices by particular users or security requirements, in order to facilitate applying granular policy down to specific classes of users or devices. Other topics included in my discussion of security requirements included the ability to remotely wipe a device in the event it is lost or stolen, in order to prevent unauthorized access of data, as well as the ability to remotely lock or unlock a device. I also mentioned the benefits of using a captive portal to control mobile device access to the network infrastructure. I then briefly touched upon something very important to security requirements in the mobile enterprise: the ability to monitor mobile devices and report their status back to the central infrastructure.

Later on, you learned about the importance of interoperability and infrastructure support. As I mentioned previously, interoperability is extremely important in terms of implementing an MDM in a network infrastructure that is already configured and working properly, to make sure that seamlessly integrates with network services, applications, existing equipment, and even user requirements. I also looked at the usefulness of a self-service portal as a means to allow users to perform common functions for themselves without the need of an administrator. In addition to infrastructure support, device platform support in the organization is a very important aspect of the MDM infrastructure. Finally, to conclude the discussions on infrastructure support, I looked at the benefits and detriments of using cloud-based services with MDM, such as Software-as-a-Service, and even MDM-as-a-Service, as well as the benefits of using multi-instance software in the enterprise.

Location-based services are a product of built-in technologies such as GPS, cellular, and Wi-Fi, and are used to pinpoint a mobile device's location. Apps use location-based services to provide many different useful features for a user, including the ability to locate points of interest near the user. Two important functions that use location-based services are geo-location and geo-fencing. Geo-location allows the organization to track the position of the mobile device at all times, which can prove to be useful in the event the device is lost or stolen. Geo-fencing allows an electronic perimeter to be set up around the organization's property, which is useful for tracking devices within its boundaries, communicating with devices, and preventing theft or misuse.

Mobile application management is a subset of the functions that can be performed in an MDM setting, except that instead of managing the entire device, particular applications are managed. This may include imposing configuration or security settings on the application, and restricting its use and interaction with other applications and data. MAM can be implemented separately from MDM, or as part of it. Application stores are important to the management of mobile devices from several different perspectives. Vendor-specific stores can be used to access commonly used apps, such as email and productivity apps. Enterprise-level app stores can also be created to deliver apps specific to business functions of the organization to mobile devices. Both apps and notifications are typically pushed to the device by the organization or the vendor.

The next major topic I covered was disaster recovery principles in the mobile infrastructure. Disaster recovery and its counterpart, business continuity planning, apply to the entire organization and I discussed several principles from an organizational perspective. Business continuity planning and disaster recovery principles entail preparing the organization for disaster, as well as determining how an organization will resume business functions after it has recovered from a disaster or incident. In discussing the basics of business continuity, I covered how important a business impact analysis, or BIA, is in managing the risk involved with disasters. I also discussed the importance of a disaster recovery plan and mentioned some key concepts surrounding developing and implementing the plan. Disaster recovery locations are alternate sites designed to recover the organization and its assets to those locations to begin the process of getting back in the business. Hot sites are more expensive to obtain and maintain, but provide almost immediate recovery and restoration capabilities. Warm sites are little bit less expensive, but don't always provide immediate way of recovering the business back to an operational state. Cold sites are the least expensive sites to obtain and maintain, but they provide almost no capability, other than physical space and utilities, for the organization to resume critical business functions.

Closely related to business continuity is the concept of backing up data, and being able to restore it quickly and efficiently. I discussed the importance of backing up servers, devices, and critical network services, such as directory services databases. The basics of how to conduct backups, and how often they should be run, were also covered. I discussed the pros and cons of three basic backup techniques, including full backups, incremental backups, and differential backups, with regards to how effective each are for backup and restoration purposes. High availability is another related concept in protecting data, and we covered the importance of uptime in terms of a percentage of the total time per year that the service or device is available to the user.

In terms of best practices for protecting mobile device data, we looked at protection from three perspectives: mobile device backups, corporate data backups, and personal data backups. Corporate data is best backed up to an enterprise infrastructure device, such as a backup server or data store, or even to a contracted third-party provider. Personal data is best left to the individual to back up, particularly from a personally owned mobile device. In either case, data should be backed up to a location other than the device itself because the possibility of device loss or theft is always present. For temporary purposes, backing up to the device itself—using the built-in storage or a removable media, such as an SD card, or even to the device SIM—is a good method, provided that it's only temporary and that the backup is encrypted. Data recovery is an essential part of both day-to-day business and disaster preparation, and must be considered in terms of speed, efficiency, and data integrity. Testing backups is one way to assure data integrity and to gauge speed and efficiency of backup methods and technologies. Restoring both corporate and personal data to a mobile device requires considerations such as restoring particular files, certain apps, or the entire operating system to the device from a full backup.

I discussed the necessity of enterprise administrators and information technology personnel to maintain awareness of new and emerging technologies, and how they affect not only the mobile device infrastructure in an organization, but also the entire enterprise. Advances in technology, new versions of operating systems, and new device features are areas that MDM personnel need to keep track of in order to effectively plan on integrating these new technologies into the existing infrastructure. Along the same lines, IT security personnel have to also keep up with new and emerging threats that seem to always come with those new technologies.

Questions

1. Which of the following was the first centrally managed device platform used in the enterprise?

   A. Apple

   B. Android

   C. Windows

   D. BlackBerry

2. The ability to produce work-related content on a mobile device is called:

   A. Content creation

   B. Content consumption

   C. Content production

   D. Content encryption

3. Which of the following is a goal of MDM? (Choose two.)

   A. Decentralized management

   B. Content creation

   C. Centralized management

   D. Acceptable use

4. Your manager is contemplating initiating a BYOD program in the enterprise and asks you what the disadvantages are to this type of implementation. Which of the following are the primary issues the enterprise must deal with in instituting a BYOD program? (Choose two.)

   A. Employee privacy

   B. Mobile application management

   C. Mobile device control

   D. App licensing

5. Which of the following is the most important consideration in the development of mobile device policy for the enterprise?

   A. Cost

   B. Employee satisfaction

   C. Security

   D. Using the latest technologies

6. Which of the following can assist administrators in pushing security policies and other configuration settings to only certain devices on the network?

   A. Containerization

   B. Device groupings

   C. BYOD

   D. MAM

7. Which of the following should be considered in developing strong passwords? (Choose two.)

   A. Length

   B. Ease of remembering

   C. Key space

   D. Dictionary words

8. A user has called you in a panic because he has left his company smartphone at a relative's house and won't be able to retrieve it for a week. Which of the

following steps should you implement to prevent unauthorized access to data on the device?

A. Remote backup

B. Remote wipe

C. Remote unlock

D. Remote lock

9. Your organization employs several groups of developers, some of which work off-site, and use different types of mobile devices to aid in their development. One group requires an enterprise database installed and configured specific to their needs, while another group requires the same enterprise database, but configured much differently. Which of the following solutions will meet the requirements of both groups?

A. Separate installations of the same database software

B. Separate licenses of the same database software

C. Multi-instance installations of the same database software

D. Using one installation of the database software, configured to the specifications of the more important group

10. Which of the following services should not have to be modified to integrate with your mobile device infrastructure? (Choose two.)

A. Network access control

B. DNS

C. Legacy authentication services

D. DHCP

11. Which of the following technologies uses geo-location to help prevent device theft and misuse?

A. Geo-fencing

B. MAM

C. Containerization

D. Geo-tracking

12. MAM is used to manage all of the following EXCEPT:

A. App store configuration

B. Device configuration

C. App security settings

D. Push notifications

13. Which the following types of alternate sites allows an organization to recover business operations, but may have only utilities, communications, and some equipment set up?

    A. Hot site

    B. Warm site

    C. Co-located site

    D. Cold site

14. Your manager has asked you to write a business continuity plan for the organization. What is the first step in developing this plan?

    A. Data backup policy

    B. Alternate site selection

    C. Disaster recovery plan

    D. Business impact analysis

15. You perform a full backup of a server's hard drive on Monday, followed by a specific type of backup on each of the following nights until the next full backup. Each of these backups turns the file archive bit to off. Which type of backup are you running?

    A. Differential

    B. Sequential

    C. Incremental

    D. Partial

16. All of the following are used to ensure high availability of data and services for users EXCEPT:

    A. Load balanced servers

    B. Clustered servers

    C. Standalone server

    D. Redundant data

17. Which of the following is the preferred solution for backing up corporate data from a mobile device?

    A. Back up to device internal storage

    B. Back up to corporate servers

    C. Back up to user's computer

    D. Back up to user's cloud storage account

18. Which of the following can help assure integrity of data backups?

    A. Using cryptographic sums on backups

    B. Making copies of backups

    C. Password protecting backups

    D. Setting strict permissions on backups

19. To whom should backup of personal data from a mobile device be entrusted?

    A. The operating system vendor

    B. A third-party provider

    C. An enterprise administrator

    D. The mobile device user

20. What is the primary issue when evaluating new technologies to be integrated into the existing enterprise network infrastructure?

    A. Data throughput

    B. Authentication mechanisms

    C. Interoperability

    D. Encryption strength

Answers

1. **D.** BlackBerry was the first major device to be centrally managed in the enterprise.

2. **A.** The ability to produce work-related content on a mobile device is called content creation.

3. **C, D.** MDM has goals that include centralized management of mobile devices, as well as enforcing security policies, to include acceptable use, on mobile devices.

4. **A, C.** Two key issues in implementing a BYOD program in the enterprise are employee privacy and level of organizational control of the device.

5. **C.** Security is an important consideration in developing mobile device policy.

6. **B.** Device groupings allow an administrator to categorize groups of users or devices and easily push security policies to those particular groups.

7. **A, C.** Two key factors in developing strong passwords are password length and key space.

8. **D.** Remote lock would be the best solution in this case because the user expects to regain control of the device within a certain amount of time.

9. **C.** Multi-instance installations of the same database software will meet the requirements of both groups because each instance can be configured to the needs of each group without violating licensing agreements.

10. **B, D.** Network services, such as DNS and DHCP, normally do not have to be modified to integrate with a mobile device infrastructure, as both services support mobile devices.

11. **A.** Geo-fencing can help prevent device theft and misuse by establishing an electronic perimeter around the organization's property, which can help track mobile devices within the perimeter.

12. **B.** MAM is normally not used to manage the device hardware settings but rather the individual applications that reside on the device, as well as their security and configuration settings.

13. **B.** A warm site can allow an organization to resume operations and usually has basic utilities set up, as well as communications and some limited amount of equipment installed.

14. **D.** The first step in developing a business continuity plan is performing a business impact analysis, which prioritizes all critical processes and equipment, and determines the impact on the organization if they are lost.

15. **C.** An incremental backup sets the file archive bit to off.

16. **C.** A standalone server does not contribute to high availability of data and services.

17. **B.** Backing up corporate data to corporate backup devices is the preferred solution.

18. **A.** Using cryptographic sums on backups, both before they are performed, as well as after they are restored, is one way to ensure data integrity.

19. **D.** The mobile device user should be personally responsible for backing up their own personal data, regardless of backup method or location.

20. **C.** Interoperability is the primary concern when evaluating new technologies to be integrated into the existing corporate network.

Implementing Mobile Device Infrastructure

In this chapter, you will

- Learn about installing and deploying mobile device management solutions
- Understand mobile device on- and off-boarding processes
- Be able to explain how to manage mobile device operations
- Learn about configuring and deploying mobile device technologies

Now that you've learned how to plan for a Mobile Device Management (MDM) infrastructure in Chapter 6, it's time to get involved in how to actually implement MDM in your enterprise. I can't stress enough the value of planning and developing requirements for your MDM solution, to the maximum extent possible, before you actually install the first piece of software or provision the first device. Planning is crucial in determining what you want out of the solution and how to get it. Establishing your requirements is an important part of this process as well, simply because you can't implement the solution unless you know what technical, organizational, management, and security solutions you require. Chapter 6 helped you to understand this planning and requirements process, and this chapter will help you understand how to implement your solution based upon those requirements. The Mobility+ exam objectives covered in this chapter include 3.3, "Install and configure mobile solutions based on given requirements"; 3.4, "Implement mobile device on-boarding and off-boarding procedures"; 3.5, "Implement mobile device operations and management procedures"; and 3.8, "Configure and deploy mobile applications and associated technologies."

Install and Configure Requirements-Based Mobile Solutions

This part of the chapter discusses installation and configuration of the MDM solution the organization selects based upon its requirements. Installing and configuring the MDM infrastructure is a great deal more than simply inserting a disc into a server and installing software. There's that part of it as well, to be sure, but what I'm going to discuss during the next few sections deals essentially with tasks you must complete prior to installation, as well as during the pilot and deployment stages. Additionally, these tasks span the entire enterprise—not just those that require you to sit down at the server. Some of this may seem to be the non-technical pieces of the installation as well, but they are equally as important (if not more so in some ways) because they usually have to be done correctly the first time. Failing to implement some of these tasks may result in a bumpy installation at best, or even an installation that has to be rolled back and restarted at worse.

Pre-Installation

Several considerations go into installing the MDM infrastructure in an organization. To some degree, it depends on your user base, the infrastructure size and complexity, and the interoperability requirements you may have in the organization. Generally, the larger the scale of the deployment, the more the complex the deployment will be, as well as the planning and tasks that have to be accomplished. Having said that, there are some pre-installation activities that you will likely need to address, regardless of scale. Many of these are discussed in the upcoming sections, and include considerations such as coordinating the installation with your infrastructure team, creating user profiles, preparing your directory services for the new mobile infrastructure, establishing certificate policies, and even reviewing end-user licensing agreements. You'll also have to determine both the policy and technical aspects of setting up security features such as containerization and sandboxing.

Group profiles are another aspect of the mobile infrastructure you'll need to plan for and set up in advance, as they include groupings of particular types of users such as executives, consultants, and so forth. You should also consider conducting a pilot test program and evaluating it for possible changes to your overall implementation plan. You'll also need to gain final approval for the implementation from upper management, as well as document the policies, plans, and procedures for the implementation. This preparation phase also includes training both users and administrators on your new implementation to ensure that it goes smoothly. Finally, you'll also want to consider the program's life cycle, in terms of equipment and service acquisition. All of these items will be discussed in the upcoming sections and should be carefully thought out and prepared for prior to implementing the new infrastructure. For the purposes of the exam, I'll discuss most of these items from a large enterprise deployment perspective, rather than from the view of a small organization with a limited user population.

Infrastructure Support and Coordination

As a best practice, you want to make sure you have everything in place that supports your MDM infrastructure. This may mean buying additional servers or provisioning additional bandwidth to accommodate the added capacity you'll require. It may mean upgrading server operating systems or adding security devices. In any case, it's also a good idea to test all of these additions, as well as your existing infrastructure, to make sure they can handle the increase in network traffic and server load that will come with adding MDM to the organization. You'll want to schedule downtime to install and test additional components, such as the MDM server, and possibly other devices, such as

NAC devices, wireless LAN controllers, and so on, to ensure that they integrate with your network.

Downtime should be scheduled at a time that is convenient for your users, of course, but should also be scheduled such that you can take the time you need for the integration activities. If any components require switching from a legacy device or service to a new service, you'll want to schedule cutover testing to make sure that there is a seamless transition between legacy and new services. You may also want to re-baseline your network after installing the new infrastructure components. This may include measuring latency, bandwidth, throughput, and so forth. All the integration and testing activities will require coordination with management, other technical areas, and sometimes even users to ensure that critical business functions aren't interrupted.

 EXAM TIP Infrastructure interoperability and support are critical pieces of an MDM deployment. Make sure you understand this for the exam.

Directory Services Setup

As most MDM solutions, as well as mobile devices, integrate to some degree with the organization's existing directory services structure, there may be changes you want to make to directory services in order to successfully integrate MDM. Assuming your organization uses some sort of Lightweight Directory Access Protocol (LDAP) structure, such as Microsoft's Active Directory, you may want to take a look at your directory containers and policies. For example, because you will likely want to administer specific groups of devices and users as distinct units, you should examine your existing domain groups to see if any changes need to be made to their memberships or to the permissions that are given to them. You also may have to create additional groups to account for different categories of devices or mobile users. If you intend to delegate administration down to certain business units, you may want to create or modify organizational units in Active Directory to administer them as different groupings.

Because you will likely push different organizational policies down to mobile device users, even creating new Group Policy objects (GPOs) for the security requirements of mobile users may be something you should consider. Some of these GPOs may contain different device or security settings for mobile users. Regardless of the changes you make to the directory services structure, keep in mind that the entire enterprise depends upon directory services, so you should carefully plan your adjustments to it to prevent negative impact to the rest of the infrastructure.

Initial Certificate Issuance

I cover digital certificates and the Public Key Infrastructure (PKI) a great deal starting with Chapter 9, so I won't address those topics in depth here. However, you should begin the process of determining how you will issue digital certificates for mobile devices and their users prior to rolling out MDM. If you don't already have a PKI set up, this may be something you want to do before implementing MDM. If you already have a stable PKI, then you likely only have to develop further policy regarding mobile device and user certificates. You can use existing policies and processes for user verification and certificate enrollment. You should also consider setting up an automatic enrollment process using a self-service or captive portal, or even third-party enrollment if you use outsourced certificate services.

Initially getting certificates to mobile devices is part of the provisioning process and will be discussed later in the chapter. Mobile devices will need, at a minimum, the organization's certificate installed. The user's personal certificate will also need to be installed on the mobile device, as well as the device certificate itself, if it is used to authenticate itself to other devices.

End User Licensing Agreements

This may seem mundane and of little value, but reviewing your end user licensing agreements (EULA) is very important in making sure that you are in compliance with laws and contracts you have with software vendors. From the infrastructure perspective, you'll need to review licensing agreements that cover enterprise and volume licenses, as these determine the number of users or devices that can use software purchased for the enterprise. You'll also want to take a look at multi-instance software licenses to make sure you can run the authorized number of instances for the numbers of users and devices you need.

On the client side, both you and your users should understand licensing requirements so that you are both aware of what is required for compliance, and understand how end-user apps and software can be legally used. You don't want users to try to use apps on additional corporate devices, or on personally owned devices not permitted by the license. You should make mention of license use in your organizational policies, and implement technical measures, such as license servers, to keep track of organizational and user compliance with those policies.

 EXAM TIP   Make sure you are familiar with the requirements to review and adhere to licensing requirements for the exam.

Sandboxing

Sandboxing is a technique that is used to keep apps separate by allowing them to run in their own restricted memory space, with restricted resources and limited access to other apps and hardware. Before implementing MDM, in addition to defining your policy on sandboxing, you want to make sure that the technical infrastructure you need to implement it is in place. Chapter 8 covers sandboxing in more detail.

Containerization

Containerization is a method of separating data, and is commonly used to keep corporate and personal data separate on both organizationally owned as well as personally owned devices in a Bring Your Own Device (BYOD) environment. Containerization is discussed further in Chapter 8, but you will need to prepare for it prior to rolling out MDM. You need to make sure that the MDM solution you are going to use supports containerization, as well as the mobile devices that you will be managing. End users also need to be trained on how containerization works and how they can use it to protect their own personal data and keep it private, as well as protect corporate data that resides on the device. Of course, developing organizational policy on how containerization will be implemented, what data will be segregated, and so on is also important at this stage. Figure 7-1 shows the differences between the concepts of sandboxing and containerization.

Device and Group Profiles

A profile is a collection of configuration and security settings that an administrator has created in order to apply them to particular categories of users or devices. A profile can be created in several different ways, including through the MDM software, or in a program such as the Apple configurator, for example. Profiles are typically text-based files, usually in an eXtensible Markup Language (XML) format, and are pushed out to the different devices that require them. Profiles should be developed based upon the needs of the organization. You can develop a profile that is device-specific, and applies to only certain platforms or operating systems, so that a particular type of device will get certain settings.

You can also develop profiles that are specific to different user categories or management groupings. For example, if you have a group of mobile sales users, you might create a profile that would contain certain settings for security, apps, network

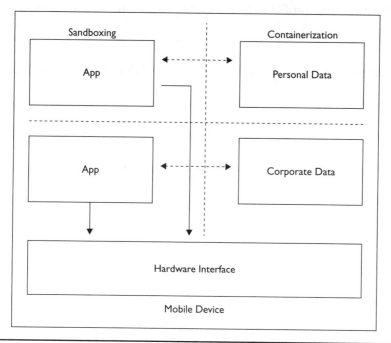

Figure 7-1 Illustration of sandboxing and containerization concepts

connections, and so forth. Your senior organizational executives may have a specific profile that is applied to them based upon their unique requirements. Frequently, senior managers or executives may be allowed to do more with a corporate device, in terms of what's acceptable to the organization. There also may be unique apps or connection requirements for these executives. Likewise, middle managers may require unique profiles that have configuration settings applicable to what functions they need to perform on the network using a mobile device. These managers also may have specific apps such as human resources or payroll-related apps installed on devices.

There are also group-specific profiles that may apply to external users, such as consultants or business partners, for example. These users may require limited access to organizational resources using their own mobile device, their organization's mobile devices, or even mobile devices issued from your organization. A group-specific profile applied to these external users may give them particular network configuration and security settings so that they can access a business extranet, for example, or use specific VPN settings. They may also require access to particular enterprise or business-to-business (B2B) apps hosted on your organization's servers. In any case, both device- and user-specific profiles can be very helpful in managing larger groups of users, delivering uniform security and configuration settings to their devices based upon different mission or business requirements. Figure 7-2 displays configuration and security settings that can be included in a profile.

Depending upon your organizational needs, you could conceivably apply several different profiles to a device at once, based upon platform, user group, and so forth. It's possible that some profile settings will conflict with those in a different profile. For

Figure 7-2 Configuration and security settings for a profile

example, some restrictive settings for a device profile may not be consistent with some less-restrictive ones in a group or user profile. When both are applied to the device, the different configuration settings may conflict and overwrite each other so you may want to pay special attention to profile precedence when applying them to the device. You may decide to configure settings precedence in the MDM server to resolve conflicts based upon a number of criteria, including user group membership, or security requirements, for example.

As far as devices go, in addition to vendor or OS platform–specific profiles, you should also develop profiles that may apply to corporate-owned versus personally owned devices. A profile applied to a device in a BYOD environment may be considerably different than one applied to an organizationally owned device. This would, of course, be based upon policy settings affecting privacy, acceptable use of the device, and so on. Figure 7-3 shows how you can conceptually apply different profiles to different device and user groups.

 EXAM TIP Understand the reason for, and use of, different device and user profiles for the exam.

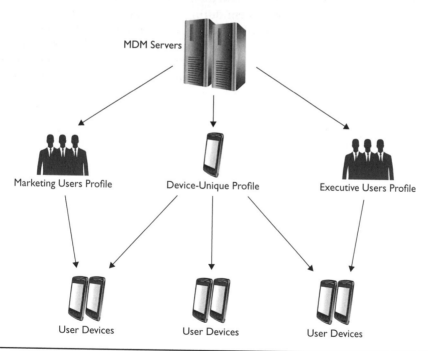

Figure 7-3 Applying profiles to different device and user groups

Conducting a Pilot Test and Evaluation

One of the smart things you can do before rolling out your full MDM implementation is to conduct a pilot program. In a pilot program, you would implement your MDM solution on a small-scale test basis to a select group of users and devices. The benefit of a pilot program is that you can try out the rollout before you actually implement it on the broader enterprise scale. You can see what problems there may be lurking in implementation and solve them before rolling it out to the general population. You can also use the pilot as a kind of dress rehearsal for the major rollout, testing your processes, procedures, and infrastructure to make sure they work as planned.

In conducting a pilot, you likely want to be very selective in the types of users and devices that participate in the test. You'll also likely want to get a sample of the general enterprise population that represents most use cases in business use, technical ability, and even device variety, as the case may be. You want to spread out your participation to a variety of users so that you can cover as many deployment scenarios as possible with the small test group. Some of the participants may be business executives, administrative personnel, sales force members, as well as technical users. You also should include some non-technical users as well, to see how effective the rollout will be in terms of addressing user needs and questions.

The participants should be carefully selected and then briefed on what the expectations are of the pilot test. You'll want them to use their devices for all the different things they are expected to do with them during the course of a normal business day, such as accessing the routine types of data and internal resources they need, so that you can see how effective the program is during normal use. The test group users should be instructed to report any issues or problems, either with connectivity, security, configuration, and so on, to the lead administrator for the pilot test, so these issues can be resolved quickly and efficiently. The pilot test should normally last as long as it takes to get a really good idea of how the actual rollout will go; depending upon the scale of the pilot test, this may take a month or two. For example, if it takes you a full week to roll out to a small pilot test group—to include device provisioning and issuance, certificate enrollment, profile configuration, and so on—then you could reasonably estimate the time required to roll out the solution to the general population, based upon the proportional number of users and devices involved. Of course, there will be bumps in the road when you implement the program with the pilot group that you'll be able to solve before rolling it out to the larger population in the enterprise. This will enable you to actually learn from the experience and be able to implement the MDM solution more efficiently during the actual rollout.

You should also brief the users in the pilot group that they should be prepared to offer feedback, both formally and informally, to administrators and management on the success of or issues with the program. This feedback will be used to gauge the effectiveness of the rollout, report any issues or problems test users had during the pilot, and offer suggestions for improving the process. Once you have received feedback from all the participants, you should summarize the results and present the report to management. There may be some action items that result from the evaluation and feedback process, including some changes to the process and maybe even some configuration changes to the MDM software or infrastructure.

EXAM TIP   Understand the reasons for conducting a pilot test, and how it can be used to more efficiently implement the MDM solution in the enterprise.

Documentation

The importance of documentation before your major MDM rollout can't be understated. Much of this documentation may be produced before your pilot, as you plan for your solution, develop requirements, document the architecture, and so on. Some of the documentation will be produced as a result of the pilot test, such as revised procedures, rollout processes, problem resolution, and so forth. It's important to document your solution from beginning to end, to assist in troubleshooting any issues that arise, as well as maintain good visibility on the entire MDM infrastructure. Documentation should include architecture documents, procedures, configuration and security settings, individual device information and inventory, and so on. Some of this documentation could also be used to enhance the overall architecture documentation for the enterprise, including the fixed desktop, server, and network device documentation. Keep in mind that documentation should be updated as needed, based upon changes in the infrastructure.

Launching the Program

Once you have completed your pilot, tested your solution, and made appropriate changes to the plan and infrastructure, the time will come when you actually have to launch the new MDM program. If you have planned carefully, and made sure that the organization is ready for it in terms of management and user buy-in, infrastructure upgrades, policies and procedures, and so forth, then everything should go smoothly, right? It more than likely will, assuming you've done all of these things I just mentioned, but there will always be little bumps in the road in any major scale infrastructure implementation. Part of your planning process is to make sure that you prepare for those as well, as much as you can. Getting approval from management is one of those things, not only for implementation itself but for contingencies and adjustments that will need to be made to the plan and infrastructure when necessary. This may mean approval for extra money in the budget if something needs to be purchased on-the-fly; it also may mean approval to make configuration changes in the existing network when implementing the solution. You should also consider getting management approval in advance for any additional hours that may involve implementation personnel working overtime.

In addition to management approval, training is also a key aspect of launching the MDM solution. Everyone in the organization should be trained on the policies that will have to be followed because of centralized management, of course, because it will affect how they use their mobile devices from that point forward. There may also be technical training required that is specific to device use, especially if users are migrating to a different platform or OS. There should also be training for the administrators who will manage the program. Training can come in the form of user manuals, handouts, short presentations, intranet sites, or even full scale vendor training classes for administrators whose daily job it will be to manage the MDM infrastructure. Regardless of training level, this piece of the deployment should also be resourced and planned to help ensure a smooth implementation.

Deployment Best Practices

Even with thorough planning and consideration to all the aspects I previously discussed, there are still factors that can make your implementation easier or more difficult, depending upon how you approach them in your organization. For example, sheer numbers of users or devices can definitely affect how quickly and how effectively you implement a mobile device infrastructure, and there are some different approaches that may help you manage this aspect of it better. I discuss those best practices in dealing with large-scale deployments in the next few sections.

Scaling Numbers of Devices and Users

Two important considerations in deploying your mobile device management solution are the numbers of users as well as the numbers and types of devices that you must provision and manage. In a new implementation, there's a temptation to rush everything, especially if both management and users' expectations are high, and there is pressure to get the ball rolling on the program. If you're dealing with a small number of users and devices, then deploying to everyone at once may not necessarily be a big deal in your organization. If, however, you are in a large organization, it may not be a good idea to deploy to everyone at once. Even with the best planning and a good pilot program, there are likely going to be problems that crop up during a large-scale MDM implementation. Deploying to everyone at once could cause even smaller problems to become enterprise-wide issues and require a lot of attention from the limited number of technical personnel you may have available for the rollout. This would slow the implementation down because everyone would have to stop what they are doing to solve problems with the entire workforce. You may even have to backtrack in your implementation and rethink some policies or technical details.

In large populations, you'd likely want to deploy to small groups first, as part of a phased deployment approach. There may be several different ways to deploy the program initially to smaller groups, but that's likely going to depend upon several things. From a managerial perspective, there may be certain groups of users who have priority, based upon their status or level in the company. For example, management may want certain employees, such as managers or key personnel, to be the first to receive new devices and be enrolled in the MDM infrastructure. From a technical perspective, however, there may be other groups that it may make more sense to roll the solution out to first. These groups may include administrators or more tech-savvy people who can help make sure that things work the way they should before the program gets pushed out to the general population of users. Still, yet another perspective is the business angle. There may be sound business reasons for pushing the solution out to the sales force first, for example, because that may provide faster return on investment in terms of increasing productivity for that group. Whatever the decision is, it may be a good idea to scale your rollout incrementally to different smaller groups within your entire user population. This way, you can determine what major problems exist more quickly with smaller groups and try to solve them before you implement the solution on a larger scale. You can also get good feedback from different groups of users with different levels of technical ability, which may tell you whether any of their issues are with the technical solution, user functionality, or even with the deployment process itself.

Similar to the scale of users, there's also the scale of devices. While this may simply mean sheer numbers, typically the number of devices will roughly equal the number of users, so you would take care of those numbers the same way, and at the same time, you would with users. But scale of devices could also mean a larger *variety* of different types of devices on the network that need to be initially provisioned and managed. If you're using a single-vendor solution, this may not be a problem because you are going to be managing the same types of devices, whether it's ten devices or a thousand devices. This makes the deployment much easier and more consistent, and limits your set of potential problems to a specific platform. This would also mean that your solutions may be easier to implement because they could apply to most devices. If, however, you are deploying many different types of devices from different vendors and using different operating systems, then your scale has just increased beyond simple numbers. Like deploying to smaller groups of users first, you may want to consider initially deploying to smaller groups of device types first. You can group these in terms of vendor, such as Android devices, for example, and deploy only to them first.

So you can see, there are several ways you could slice and dice deployment groupings in terms of to whom you would deploy to first, and in what order. Typically, these decisions are not only going to be technical; they're also going to be managerial and business case related, and in some cases, even political. It's going to take a coordinated effort between the technical personnel and managerial personnel, and even some coordination from the user base, to help make the right decisions as to what the most effective and economic approaches are to the problem of deployment scale. Figure 7-4 shows a conceptual way to deploy an MDM infrastructure based on scale.

Figure 7-4 A conceptual way to deploy MDM based on scale

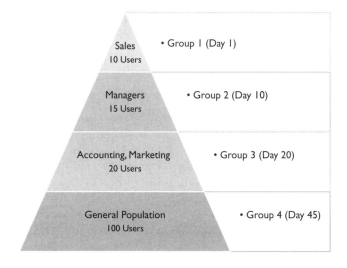

EXAM TIP Know how the concept of scale, in terms of simple numbers of users and devices, as well as types of devices and level of control, affects an MDM deployment.

Level of Control

Assuming you've already made the policy decisions regarding level of control the organization has over the mobile devices versus the level of user control (especially when it affects BYOD implementations), deploying your solution may be driven by this aspect of MDM as well. Users who have already been using mobile devices, either corporately owned or personally owned, prior to the formal deployment, may get a little bit of "sticker shock" when the solution is first deployed and they figure out that they don't have as much control over their device or how it's used as they previously did. Although most people may simply accept it, you may have some issues with people that may require you to scale back the level of control over devices temporarily and increment it slowly. This may not only be due to the users' resistance to change, but may also may come from a practical perspective of not having line-of-business applications or business processes ready for a stricter level of control and security requirements. Depending upon what the organization's tolerance for risk that affects either security or functionality is, a temporary reprieve from tighter control may or may not be a good idea, simply because once you give someone a certain degree of control or freedom over their device, it's typically really hard to rope that back in and impose tighter control over it later. There also may be other technical reasons, however, for initially loosening control, as you may not have the proper infrastructure in place to manage that level of control over devices. For example, you may choose to not initially impose content filtering on devices if you don't have the right content filtering infrastructure in place when the rollout occurs. The best solution is to have that infrastructure already in place, but this isn't always possible due to resource constraints or even organizational issues. In any case, scaling back the level of control when you initially deploy a solution may be something your organization considers for various reasons, but the path forward for eventually exerting that level of control should be decided upon and made clear to everyone.

Mobile Device On-Boarding and Off-Boarding

On-boarding and off-boarding are two of the most common processes that organizations have to go through on a fairly regular basis. Employees and other members of the organization come and go, and when they do, there are a lot of checklist items that must be accomplished. Among these are getting their mobile devices set up or, in the case of those that leave the organization, turned in and deprovisioned. It's important to establish a defined, documented process for these two important activities so the organization can ensure that employees get their devices set up correctly and securely when they arrive in the organization, and that employees are able to turn in and secure their devices in an orderly fashion when leaving. The next few sections cover the important aspects of on- and off-boarding employees and their mobile devices. These are all important considerations to add to your organizational process checklists, as well as to know for the exam.

Device Activation

As part of the provisioning process, at some point devices will need to be activated, if they have not already been. If a previously used device is being issued to a new user, the device may simply have to be reprogrammed with a new telephone number and reloaded with the baseline operating system and apps. If the device is brand-new, it likely requires some sort of over-the-air (OTA) programming on the part of the carrier. For the most part, this is a simple process and can be performed by the user or by the mobile device administrator. Devices can also be activated on a larger scale, with coordination from the carrier as the devices may be preprogrammed and activated prior to shipment to the organization.

Device Activation on Cellular Networks

As mentioned previously, for the most part, activating a mobile device on a cellular network is an easy process. In some cases, the user needs only to dial a number to get the phone activated with the carrier. This wasn't always the case, however. Just a few years ago, for example, iPhones in particular had to be activated using iTunes software with a cable connected to a computer, but this has changed over the past couple of years. Nowadays, just like their Android and other platform counterparts, iPhones can be activated simply by dialing a number, which initiates an over-the-air programming process. There are other things that iTunes can still be used for during the initial setup for devices not centrally managed, such as synchronization, restoring from a backup, and app and content management, but in an enterprise situation, this should all be done using MDM software and techniques. This applies to most other vendor platforms as well.

Obviously, in order to activate a mobile device on the cellular network, the device must be able to get a clear signal in order to receive programming. If the device can't be activated, there may be issues with the device itself, the Subscriber Identity Module (SIM), the account, or even the cellular network. The troubleshooting steps covered in Chapter 12 can help you to solve these problems if this is the case.

Other Mobile Hardware Facilitating OTA Access

Aside from activation on a cellular network, devices can be activated and provisioned as well using different types of mobile communications hardware that can be used for OTA activation, provisioning, and programming. These types of hardware include built-in or removable (SD or USB) wireless cards and cellular cards, for those (rare) devices that don't have those capabilities included. Most of the scenarios involving OTA access using connectivity other than cellular technologies usually involve the organization versus the telecommunications carrier.

TIP Remember that device activation can be a separate activity from provisioning the device in the enterprise, especially if it's a device in a BYOD environment. Sometimes activation and provisioning go hand-in-hand and are done together, but this is not always the case.

Provisioning

I've mentioned provisioning several times throughout this chapter already, and I'll go a bit more in depth on the provisioning process over the next few sections. Provisioning a device means to initially configure it according to the corporate mobile device policy and make sure that it has the proper network, app, device, user, and security settings. It also entails installing the baseline set of apps the user requires for their job. There are several different ways to provision a device; it can be done by the administrators before the device is even issued to the user, or it can even be done by the end user when he or she first uses the device.

Provisioning a device doesn't necessarily stop when the device is first issued; there are several points during the life of the device that additional provisioning actions take place, in the form of patches, operating system upgrades, installing new apps, and so on. These actions will usually take place over-the-air, when the device is connected to the corporate network, but also may happen by allowing the user to connect to the vendor's app store, if that is how the organization chooses to permit upgrades, patches, and so forth. Over the next few sections, I'll discuss provisioning a device using several different methods, and how on-boarding and off-boarding is accomplished with mobile devices and users.

On-Boarding and Provision Process

The on-boarding process entails assigning a device to a user, activating the device, provisioning the device based on the corporate policies and requirements, and making sure that the device functions as it needs to for the user. There are a few different ways to make this happen, including having the mobile device administrators perform this function, or even the users themselves. Some of the particular on-boarding and provisioning items that must be accomplished include setting up app stores, both vendor and enterprise controlled, as well as corporate email services. You can even set up network connections, such as VPN connection profiles. Digital certificates used for authentication to enterprise services will also have to be provisioned to the device. Backup services, including apps and policies, should also be deployed during the provisioning process.

The device will also have to be entered into the MDM inventory tracking process, as well as any other organizational equipment inventory systems. For corporate-owned devices, the user should also have to physically sign for the device and take responsibility for it so that the user can be held accountable for loss or damage to the device. The user should also be briefed on his or her responsibilities, both in the care and the acceptable use of the device. For personally owned devices, corporate policies will dictate what level of provisioning and on-boarding activities take place.

IMEI, ICCID, and IMSI Numbers

There are three particular identifiers you will need to understand both for the exam and for real-life management of mobile devices. The International Mobile Equipment Identity (IMEI) number is a 15-digit number used to uniquely identify a mobile device, typically a smartphone or other device that connects to a cellular network. IMEI numbers are unique to the GSM family of devices, including current devices that descended from GSM technologies (including current day 4G LTE and LTE-Advanced). You can typically find this number printed inside the battery compartment of the mobile device, but you may not necessarily need to take the device apart, as some operating systems will allow you to find it inside the device configuration settings. The IMEI number can be used to identify a specific device and even to block that device from accessing the carrier's network. So, if the device is lost or stolen, the user can notify their carrier, and the carrier can make sure that the device can't be used on the network.

The ICCID number, which stands for Integrated Circuit Card Identifier, uniquely identifies a SIM. Remember from the discussion in Chapter 4 that the SIM contains information unique to the subscriber (the owner of the phone), and is used to authenticate the subscriber to the network. SIMs can be moved from phone to phone, usually with no problems.

The third number, while not listed in the exam objectives, is equally good to know for your professional endeavors. This is the International Mobile Subscriber Identity (IMSI) number. It is also included on the SIM, but represents the actual user associated with the SIM. There may be some cases where you might want to record these numbers for each managed device in the enterprise, for inventory purposes or so that you have them handy when you work with your MDM software. Typically, during the device provisioning process, those identifiers, as well as other information particular to the device, such as telephone number and MAC address, are collected by the server and stored in inventory for you. Figure 7-5 shows how IMEI and ICCID numbers are listed for a newer Android device in the device settings.

 EXAM TIP Remember the differences between the IMEI and the ICCID numbers for the exam. The IMEI represents the device, and the ICCID is tied to the SIM.

Profile Installations

As I discussed previously, a profile is a set of configuration settings applicable to a particular device or group of users. A profile can contain security information, such as authentication and encryption settings, network settings relating to connecting to a particular MDM server or to a VPN, for example. It can also contain settings relating to apps, app stores, email configuration, and so forth. Provisioning a device involves sending these profiles to the device, and can be done using a variety of methods. Administrators can provision a device with the appropriate profiles, or the user can actually do it himself under certain circumstances. I discuss these methods in the next few sections that follow. Before we do that, however, let's talk about profiles and how they are constructed.

Figure 7-5
IMEI and ICCID
numbers

Profiles are typically small configuration files formatted as an XML file. These files can be written manually, of course, but it's probably easier to create them using the MDM software or other types of configuration generators. Apple has a very good utility, called the iPhone configuration utility, that can be downloaded and used to create configuration profiles. Other vendors, of course, have their own configuration utilities as well; however, for the sake of consistency across the enterprise and the ability to generate profiles for multiple device and user groups, it's probably more effective to use MDM software to create device and user profiles. Figure 7-6 shows an example of an XML configuration file for a device, and Figure 7-7 shows an example screen from the iPhone configuration utility that anyone can use to create their own iOS configuration profiles.

Device profiles can be sent to a device using several different methods; again, I'll discuss these methods in the sections that follow, but they include sending them via the organizational intranet and a website (assuming the mobile device has a good cellular or Wi-Fi connection), or even installing them to the device via a USB cable or SD card.

Manual On-Boarding and Provisioning

Manual on-boarding and provisioning is typically done by the administrator, who may have the device connected to a USB cable and a computer. The administrator could also provision a device by manually installing an XML file via an SD card, which contains all the configuration settings needed to get the device initially set up for the MDM infrastructure. This can be a bit of a time-consuming process if there are a great many users to provision at the same time, but it may be necessary for the occasional user who is having difficulty getting the device set up.

```
<plist version="1.0">
<dict>
    <key>ConsentText</key>
    <dict>
        <key>default</key>
        <string>This profile is mandatory for all MH mobile
devices.</string>
    </dict>
    <key>PayloadContent</key>
    <array>
        <dict>
            <key>PayloadDescription</key>
            <string></string>
            <key>PayloadDisplayName</key>
            <string>Restrictions</string>
            <key>PayloadIdentifier</key>
            <string>
MH_Baseline_Profile.restrictions1</string>
            <key>PayloadOrganization</key>
            <string>McGraw Hill</string>
            <key>PayloadType</key>
            <string>com.apple.applicationaccess</string>
            <key>PayloadUUID</key>
```

Figure 7-6 An example of an XML configuration file

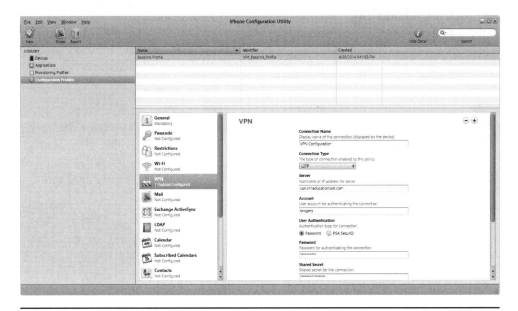

Figure 7-7 The iPhone Configuration Utility

Self-Service Provisioning

Self-service provisioning is a method by which the user assists in the provisioning process by going to a website and clicking on a link provided by the mobile device administrator. This link could be provided via email or even SMS text message. The link would then initiate the provisioning process for the user, who may have to answer a few prompts or perform a few actions, such as powering down the device and powering it back up again. The user could also self-provision using a pre-installed app on the device, which would take the user through the various provisioning steps that the user must provide input for, such as their own username and password or passcode. Figure 7-8 shows the URL and temporary passcode as sent via email to the user for a new self-service device enrollment.

Batch Provisioning

Batch provisioning is what an administrator might do if they wish to provision groups of mobile devices at once. This could be done by sending mass emails to users, which contain URLs and other settings they need in order to connect to a centralized provisioning website. It can also be accomplished by running scripts on the MDM server that contain the device IMEI or ICCID numbers for the applicable devices. Another way might be to send bulk SMS messages out to the users. Typically, batch provisioning might be done for a particular set of devices or users requiring a specific device or group profiles.

Remote Provisioning

Remote provisioning is likely how you'll eventually update all of your mobile devices, whenever possible. This is, of course, over-the-air provisioning and involves sending provisioning profiles to the remote device using cellular or Wi-Fi technologies. XML

Figure 7-8 URL and temporary passcode sent to a user for device enrollment

files provide the data necessary to provision a device, and can be sent in small bursts as individual files or as one large file. Even a typically medium-size XML file shouldn't require a great deal of time or bandwidth to push to a smart device using current 4G or Wi-Fi technologies. Remote provisioning may sometimes require that the administrator temporarily assume control over the device, via an app or configuration feature set up specifically for that purpose. Remote provisioning may be more useful after the initial provisioning has been accomplished, to provision updates or changes to the configuration settings or profiles.

Secure Device Enrollment (SCEP)

Secure device enrollment is, of course, an issue that you will want to pay attention to when administering an MDM infrastructure. One way to enroll devices is through a secure portal or other method that uses the Simple Certificate Enrollment Protocol (SCEP). This protocol was designed to deliver secure, encrypted settings to devices (both mobile and traditional) so they can securely communicate with the infrastructure. Both Apple and Microsoft make use of SCEP, albeit somewhat differently. SCEP is used to push a secure profile to a device that contains authentication credentials for the infrastructure, usually the management server in the case of mobile devices, so the device can securely connect to the network and complete their enrollment. SCEP contains certificate information from the server so that the device and the server can negotiate secure communications.

One important item of note is that SCEP is still only a draft standard, published by the Internet Engineering Task Force, or IETF, so it is not fully developed or implemented in many infrastructures or with all vendors. Another interesting note that you should investigate before using SCEP is that there are security issues with it that have been identified by various reputable security researchers. Using SCEP is not the only way to securely register a device and provision it over the air; other protocols can be used as well, including good old SSL. Figure 7-9 shows an example of the SCEP configuration screen in the Apple iPhone configuration utility.

 CAUTION You will need to research how your organization uses various implementations of SCEP to securely provision your devices, as some implementations may not be fully secure.

Off-Boarding and De-Provisioning

Just as on-boarding and provisioning are important parts of the process of bringing a new employee into the organization and equipping them with a mobile device, off-boarding and de-provisioning a mobile device with an employee who is leaving the organization is equally as important. Even if an employee is not leaving, sometimes a device is de-provisioned so a replacement can be issued. Regardless, there are some processes that need to occur to orderly get the device from an employee, transfer its content, sanitize the device of remnant data, and either dispose of it or return it into the active inventory. I'll discuss these activities in the next few sections.

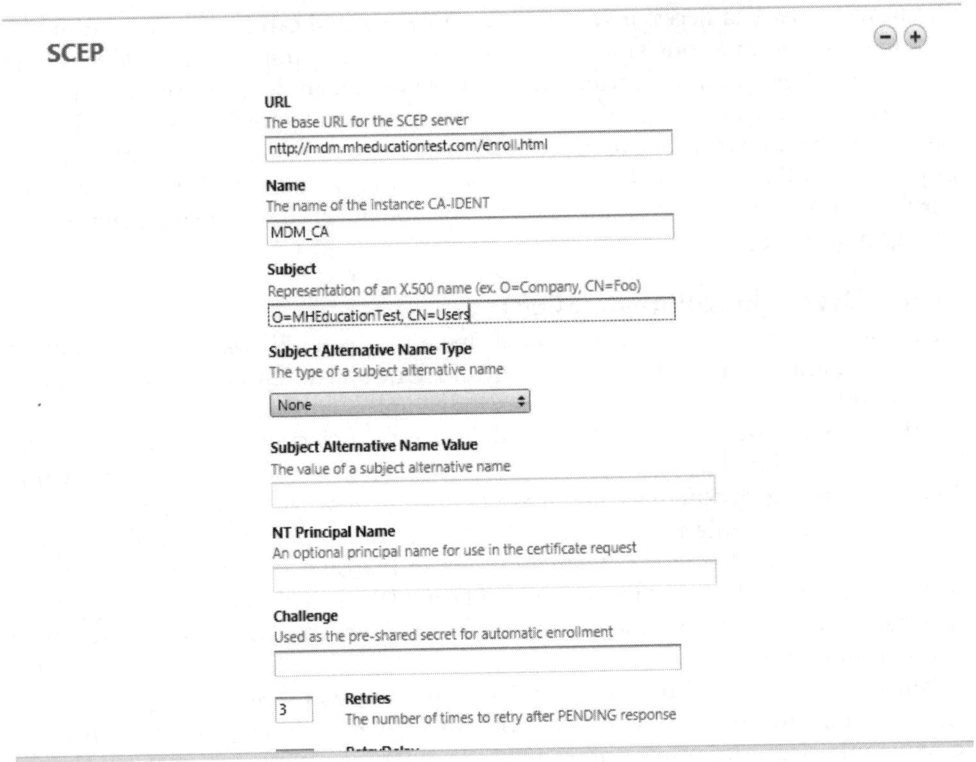

SCEP ⊖ ⊕

URL
The base URL for the SCEP server

http://mdm.mheducationtest.com/enroll.html

Name
The name of the instance: CA-IDENT

MDM_CA

Subject
Representation of an X.500 name (ex. O=Company, CN=Foo)

O=MHEducationTest, CN=Users

Subject Alternative Name Type
The type of a subject alternative name

None ⬍

Subject Alternative Name Value
The value of a subject alternative name

NT Principal Name
An optional principal name for use in the certificate request

Challenge
Used as the pre-shared secret for automatic enrollment

3   **Retries**
The number of times to retry after PENDING response

Figure 7-9   SCEP configuration screen in the Apple iPhone configuration utility

Employee Terminations

As a fact of life, employees come and go in the organization. Sometimes this is due to retirement, transition to another job, or, unfortunately, due to cause. When an employee leaves the organization, there is usually an orderly process for terminating her access to the network, retrieving any organizationally owned equipment from her, and managing any data she may have that the organization needs in order to continue business. The organization usually has some sort of checklist, published by the human resources department, which details all of the steps that both the employee and managers need to take in order to affect an orderly exit from the organization. One of the steps in that checklist should be for the mobile device administrator to take possession of the device, checking to make sure it has not been damaged and is still operational, transferring any organizational data from it, and wiping the device so it can be disposed of or reloaded. If there's any personal data on the device, arrangements should be made to get the data from the device and return it to user.

Once the organization has the device, the device should be reloaded back to its approved baseline and reissued to another member of the workforce. If the device is a smartphone, it may be a good idea to have the device phone number changed to a new one, so that the previous employee will not continue to receive calls on the device. This is also a good time to upgrade the operating system, as well as any apps that the new user needs. The device may need to be re-provisioned to suit the needs of the new user, to include any necessary device or user profiles.

Migrations

Aside from turning a device into the administrator because he is leaving the organization, an employee may need to turn it in to upgrade or replace the device with a new one. If this is the case, then the data should be migrated to the replacement device. This typically means backing up the data separately from the operating system and apps on the old device so that it can be restored to the new device. If there are any incompatibility issues that may affect the data, such as apps that require the data in a new format, you should consult with the app vendor to see if there are any vendor tools or instructions available to migrate the data to the new app. Additionally, settings such as contacts, files, media, URLs, and other information should be migrated to the new device, via a data backup, or using vendor tools created just for that purpose. MDM software may also have the functionality needed to migrate user data from an old to a new device.

Applications

As part of the off-boarding and de-provisioning process, there are some considerations with applications on devices that are decommissioned. Uninstalling an app on a device that is to be de-provisioned may free up available licenses in an environment where there are limited volume licenses available for users. For some applications, particularly multi-instance software, the administrator may need to go in to that software and remove the device from the list of allowed devices, if the software maintains such a list. Another consideration is that when an organization replaces a device, if licenses are tightly controlled, the app may need to be uninstalled first from the old device before it can be reinstalled on a new device. Additionally, this may be a good opportunity to upgrade apps if there are any new versions available that the organization would like to include in its mobile device baseline.

Content

Any data or content on the device should be examined to determine if it is corporate property or personal data. If the data is organizational in nature, it should be backed up to a centralized location so it can be restored to another device if necessary, retained, or destroyed. If the data or content is of a personal nature, then the user should be given the opportunity to retrieve it before the device is wiped and replaced. In some cases, the organization may find content on the device of a prohibited nature because the content may violate policy or laws. In this event, the organization would need to determine whether it needs to retain the data as part of an investigation or simply delete the data from the device.

Deactivation

Finally, if the device is not going to be reused by another member of the organization, it should be deactivated and otherwise stored or disposed of. The deactivation process may depend upon the device or telecommunications vendor with whom the organization has their contract. In some cases, the device may be returned to the vendor for credit. Deactivation may require that the organization remove the SIM, if applicable, and return or destroy it, as well as contact the carrier to inform them that device is being deactivated and should not be allowed to connect to the carrier's network. As telephone numbers can be ported from device to device, this is another consideration when deactivating a mobile device because the number should be ported first before the device is deactivated.

 EXAM TIP   Keep in mind all of the factors that should be considered when de-provisioning a device, including deactivation, data backups, and so on.

Implementing Mobile Device Operations and Management

Now that I've covered the critical aspects of deploying the MDM infrastructure to the enterprise, let's discuss the day-to-day tactical level of managing the MDM solution in the organization. Common, routine activities that an administrator may perform include those relating to managing applications and content, and getting them out to the mobile users, as well as remotely managing the organization's mobile devices. Beyond the day-to-day management activities, implementation also includes considering both the near- and long-term aspects of managing the infrastructure's life cycle, such as managing certificates, upgrading equipment, implementing changes in the infrastructure, and even managing the retirement and disposal of devices and equipment. Our discussions in the next few sections won't be very technical in nature because every organization conducts its day-to-day operations differently, using a wide variety of platforms, technologies, and equipment. But you should be able to easily apply these concepts to your organizational infrastructure to help you manage your routine operations.

Centralized Content and Application Management and Distribution

One of the major issues that organizations are encountering in the mobile device world is that not only are employees bringing their own devices and accessing corporate resources (networks, services, data, and so on) on their mobile devices, they are also actually exfiltrating (removing data) from the corporate infrastructure and storing it on devices or in the cloud via personal services, such as Dropbox or similar services, so it can be easily accessed later. For the most part, removing this data from its corporate boundaries isn't necessarily done with malicious intent (although that is out there also); it's done simply to facilitate a legitimate business need. For example, employees who must work on large files while away from the office at home or on travel often upload company sensitive documents to their personal cloud storage or download them to their own devices. This is often more efficient than simply emailing oneself a document because of email attachment size and file restrictions. Usually, when users resort to these unapproved means of content transfer and distribution to get their jobs done, it means that the organization hasn't done its job in developing approved methods of content distribution and use. Unfortunately, these unofficial methods used by employees put the organization at great risk because of the possibilities of data loss, unauthorized access, malicious or unapproved content, and so on. Additionally, it creates practical issues with versioning control because there may be multiple instances of documents and content that is being created or edited by different users.

Likewise, employees also obtain apps and other content from sources other than those officially approved by the organization. I'll cover enterprise app stores a bit more later in the chapter, but it brings up interesting discussions on the organization's ability to properly control both inbound (apps and approved content) and outbound (company sensitive data) content from mobile devices. The solution, of course, is an enterprise solution for content management and distribution. Of course, apps can be distributed through a centralized MDM and MAM solution, through the initial provisioning of the device, by using device profiles, and by providing either on-demand or forced installation, usually through an enterprise app store. Other content, however, may need to be considered, and the organization may need to implement a centralized content management and distribution solution to meet those requirements.

As with all enterprise solutions, content management and distribution systems should be integrated with the existing infrastructure, leveraging methods and technologies already in place. Mechanisms—such as Active Directory's centralized authentication and resource access control, as well as content filtering, device and user profiles, and other centralized services—can be used together to help make content management work. There are several out-of-the-box solutions that are designed to effectively integrate with those mechanisms and provide centralized content control and distribution, but I'm going to focus more on how they work than what they are over the next few sections.

Distribution Methods

There are several common methods of distributing content in an MDM infrastructure. On-premise models usually involve servers that contain content in shared folder locations, and can be, at the simplest level, accessed through simple share mapping techniques. This content can be files such as documents, spreadsheets, databases, slide presentations, media, and even executable files. There are also the more sophisticated models that use enterprise software suites designed for just such a purpose, such as Microsoft's SharePoint, which serves as both a content repository and access interface, providing location, indexing, and searching methods and structures. On-premise server-based solutions can also use custom-built web applications to manage and distribute content to its users, who are connecting with both mobile and traditional devices.

Cloud-based services are also viable distribution models for enterprise content. If the organization contracts with an approved cloud vendor, then employees can access business content via mobile apps and web sites. Data security and availability, as well as other issues, are concerns with organizational data stored in a third party's infrastructure, such as a cloud service provider, so the organization must be careful to ensure that those concerns are addressed in contracts and service-level agreements (SLAs).

Content Updates and Changes

In addition to managing aspects of content that include storage, security, and delivery to the user, content management also includes controlling the content itself. This aspect covers content change and updates, as well as versioning control between those changes. This ensures that not only are files kept up-to-date and records of changes maintained (either as metadata or as a separate record), but that file inconsistency and concurrent use issues are prevented and resolved when they occur. This could be implemented using permissions, file locks, or other mechanisms.

Application Management

Again, I'll discuss the particulars of enterprise apps and their control through in-house development and app stores shortly, but for this discussion, we could also consider vendor apps or those purchased and licensed through a third-party provider. These may also be controlled by the enterprise app store in conjunction with or as part of the content management system. In the case of vendor or third-party apps, content management and distribution solutions could centrally manage which users and devices get which apps, of course, but could also be useful in maintaining multiple versions of apps if these are required for supporting business processes. The solution could also be used to maintain changes and updates for vendor apps and ensure that devices get the most current version when they need it, either through push or on-demand distribution methods. The system could also assist in managing licensing requirements because there may be limitations on the number of concurrent or total licenses in use that the organization may have through their agreements with the vendors that provide those apps. Figure 7-10 shows an example of listing and managing the applications on an

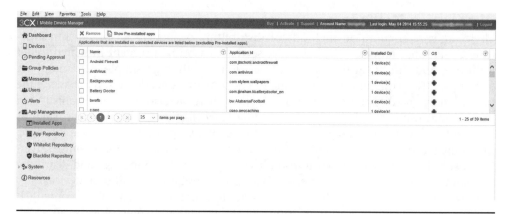

Figure 7-10   Listing and managing apps on an enrolled device

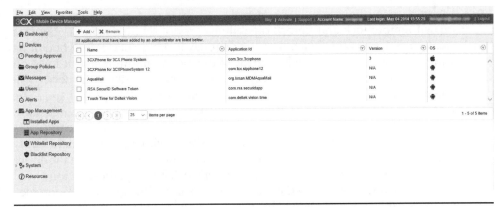

Figure 7-11   An example of an app store repository for an organization with required apps

enrolled Android device. Note the options in the left-hand pane that allow different management tasks. Figure 7-11 shows how an enterprise app store might be configured in an MDM structure, with required apps in its repository.

Security and Access Control

Another aspect of content management and distribution systems that's important to understand is content security and access control. While many content management systems provide built-in security controls to apply to content and storage structures, most integrate with existing security mechanisms, such as LDAP authentication, HTTPS encryption, PKI, and file permissions to secure content. Like most other networked resources, content should be secured by those supplemental mechanisms based upon need-to-know and job requirements. Additionally, other restrictions designed to control access to content may be applied, such as time-of-day restrictions, role-based access control, and granular permissions (editing versus reading, and so on).

Implementing Remote Capabilities

One of the most useful features about mobile devices is that they are, of course, mobile. Unfortunately, one of the most troublesome security issues with mobile devices is that they are, of course, mobile. As mobility can be both a blessing and a curse, mobile device administrators must be able to manage the device using over-the-air methods for a wide variety of tasks, including security, configuration, updating, and other day-to-day administrative reasons. Over-the-air management is one of the primary means that administrators use to configure and secure devices, through cellular and Wi-Fi connections, because it's fairly impractical to have the user bring his device in on a recurring basis to connect it to a USB cable and a computer. In these next few sections, I discuss remote administration capabilities, to include methods and tasks that administrators can perform on mobile devices.

Lock/Unlock

I've discussed remote lock and unlock capabilities previously in Chapter 6, but it is worth mentioning here again. Remember that remotely locking a device is a good thing to do if the user has temporarily lost control of the device, meaning that it's not on her person and she can't get to it at the present time. Remotely locking the device will prevent unauthorized personnel from accessing it while it's out of the user's hands. The scenario assumes, of course, that the user will be getting the device back at some point in the immediate future. Remotely locking a device could also be used in those instances where the user is using the device in a manner inconsistent with organizational security policy. Through monitoring and logging, the organization could determine that the device is being subjected to unauthorized use, and remotely lock it to halt those actions.

Remotely unlocking a device, on the other hand, is something the organization may want to employ if the user has forgotten her passcode or personal identification number, and the device is locked. Because many devices have security policies implemented on them that will wipe a device after so many times if an incorrect passcode is entered, the ability to remotely unlock a device would be useful to prevent that scenario from occurring. While remotely locking a device is used to protect data confidentiality and prevent unauthorized access, remotely unlocking a device can be used to protect data availability.

Remote Wipe

While it's a drastic step to take, remotely wiping a device is done to prevent unauthorized access to data, should the device be irretrievably lost or stolen. Typically, the data on the device is so sensitive in this event that unauthorized access or disclosure would be detrimental to the organization. This could also be the case if the data is of a personal nature and could result in harm to an individual or identity theft, for example. Of course, unless the device has been backed up recently, any data on the device is going to be lost forever once it is wiped. Remote wipe can be easily accomplished if the device has been enrolled in the MDM infrastructure and the policy is set up to allow for remote device wipe.

Remote Control

There are several reasons an organization may want to have the ability to remotely control a device. First, the most obvious reason is to help a user who is having issues with the device or apps on the device. In this case, administrators can remotely access and control the device to help the user solve a problem with connecting to corporate resources or using an app. Another reason for using remote control features may be to configure the device or change its settings for some reason. There are several different ways that an administrator could remotely control the device, some of which involve putting an app on the device and enabling settings that allow remote control. Typically, remote control would be done over a TCP/IP connection, provided that the device has cellular data access or a Wi-Fi connection. Another way may involve pushing remote control policies down to the device from the MDM server, and using built-in capabilities in the MDM software.

Location Services

I discussed location services in Chapter 6 as well, and how these location services make use of the device's built-in GPS features, as well as cellular and Wi-Fi connections in some instances. In addition to the user being able to use location services to pinpoint her own location, as well as use location-based services and apps, mobile device administrators can use the same location services to locate a lost or missing device, or track its previous locations to ensure that it has been used in accordance with policy. Geo-fencing, also discussed in Chapter 6, is another use of location services a mobile administrator may employ in order to track and manage mobile devices that are on the organization's premises. Figure 7-12 shows how MDM can locate and remotely manage a device. Notice the menu items that allow you to remotely lock, unlock, and wipe a device.

Reporting

Reporting on a mobile device's status is one way that mobile administrators make use of remote capabilities so that they can keep track of a device's location, its use, and its security and configuration settings. I discuss monitoring and reporting more in Chapter 11, but you should know that this is one remote capability that an administrator must have in order to ensure compliance and security for mobile devices. The MDM software used may have remote monitoring and reporting features, but this is also device-dependent to some degree. In some cases, the type of device used may not provide for remote monitoring and reporting, or it may require the installation of a small piece of software, called an *agent*, on the device, or even an app that can provide remote monitoring and reporting capabilities.

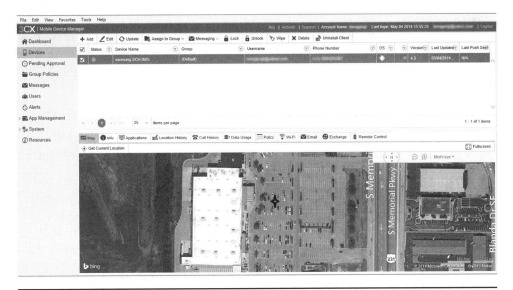

Figure 7-12   Locating and remotely managing a device

TIP   Make sure your MDM solution and your mobile devices are set up correctly to enable you to remotely manage your devices, as this is a major consideration in mobile device administration.

Life-Cycle Operations

Like the rest of your infrastructure, you must manage the MDM life cycle. The life cycle refers to the beginning-to-end process of planning, acquiring, developing, implementing, maintaining, and disposal of both MDM systems and related software in the enterprise. Both software and systems development life cycles (SDLC) provide a framework and structure for managing assets from the time they are newly acquired or developed, to when they are introduced into the infrastructure, all the way to the time when they are replaced, retired, or otherwise disposed of.

Software/System Development Life Cycle

All of the different phases of the life cycle should be considered at various times throughout the usable life span of mobile hardware and software assets. This process requires that management maintain visibility on all assets in the enterprise, including purchase date, maintenance history, testing, and other relevant data. For software, this could include versioning control, upgrades, updates and patches, and eventual obsolescence or end-of-life support. The organization should develop a process for periodically examining the different aspects of the MDM infrastructure, such as devices and other equipment, as well as software and apps, in order to ensure that those assets are still fulfilling their required functions and are not obsolete. For example, the organization may need to develop a policy that states that it will review devices for upgrade or replacement every three years. Obviously, you would not want to replace every single device at the same time every three years, so you might stagger device replacement such that you are replacing one-third of your inventory every year. This will enable the organization to use its financial resources more wisely, as well as more efficiently manage its inventory. Figure 7-13 shows a conceptual diagram of mobile device life-cycle management.

TIP   The organization should develop a life-cycle management plan for managing all of its mobile assets (both hardware and software) throughout the organization.

Certificate Expiration/Renewal

Installing certificates was discussed earlier in the chapter; however, at some point, certificates will need to be renewed, simply because they will expire after a certain point in time. While I won't go into the particulars of the certificate life cycle at this point because that is discussed in depth in Chapter 9, you should still consider it as part of your overall life-cycle management. Certificates must be renewed periodically, and

Figure 7-13
Conceptual
diagram of mobile
device life-cycle
management

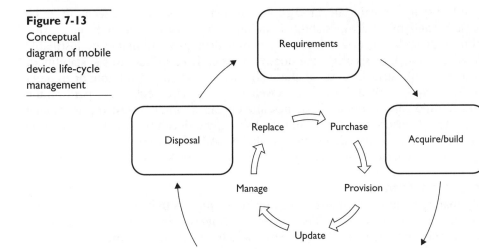

there is a process that you must develop and follow to do that. It may be via an internal self-service enrollment web site, or it may be a process whereby you simply provision the device remotely when the time comes. Certificates can be sent via profiles to the device as well, as part of its normal update process. This may be via a push action from the MDM servers, or even using a pull method from the device itself, when the certificate is within a certain amount of time of expiration. Troubleshooting certificate issues is also covered in depth in Chapter 11.

Updates, Patches, and Upgrades

Part of the life cycle for mobile device software and hardware includes updating software with new versions, as well as applying patches between updates. Updates tend to offer new features or fix existing issues with functionality and security, whereas patches may also include fixes for small issues that don't warrant a full update. Upgrades tend to be major version revisions, often leading to a major new version of the operating system, software, or even in the device firmware itself. Upgrades not only usually offer fixes where needed, but also add new functionality to software and hardware.

All of these must be managed from the life-cycle perspective, simply because the changes to the operating system, as well as the apps and the device itself, should be made based upon functional or security necessity. While the latest upgrade or patch may be a necessity to restore broken functionality or to fix a security issue, the organization must keep track of what updates, patches, and upgrades actually do for them, and not necessarily rush to install the latest and greatest of them simply because it seems the thing to do. Device and infrastructure stability is an important consideration in the enterprise, and constantly updating, upgrading, and patching devices when there is no justifiable business, functional, or security case doesn't necessarily contribute to the stability of the enterprise infrastructure. All of these should be considered for interoperability with existing equipment and software, as well as the resources required to perform the updates. Additionally, all updates, patches, and upgrades should be tested before being implemented to ensure that they don't inadvertently reduce functionality or security for the devices they are intended to be installed on. All of these considerations directly relate to the concept of change management, which is discussed in the next section.

Change Management

Change management is a formalized organizational process that is used to ensure that any changes in the device or infrastructure baselines are appropriately considered, approved, and documented. The degree to which change management is formally implemented in the organization depends upon several factors. The most important factor is the level of change and the level of impact it may have on the organization. Typically, a change control board (CCB) is formally created and responsible for over-seeing the change management process, and usually consists of both technical person-nel and managers, who evaluate proposed changes to the infrastructure for impacts on interoperability, availability, and even security. The change is normally tested first, in a nonproduction environment, to ascertain what effects the change would have on the existing infrastructure. When the change is evaluated, the board is responsible for approving the change and directing its implementation. If approved, the change is implemented and documented as part of the infrastructure architecture.

Whether or not all levels of changes make it to a change control board depends upon the organization's tolerance for risk and level of concern for changes that affect the mobile device infrastructure. For example, changing a setting on one user's device probably isn't an adequate enough reason to submit a formal change to the process. However, instituting a major enterprise-wide update or upgrade to an operating system or to a group of devices may warrant such a formalized request. This is because the change could negatively impact the operations of the organization, and the purpose of the CCB is to carefully consider these impacts before approving the change. Like all other aspects of managing the enterprise infrastructure, change management should be implemented according to policy developed and approved by senior management.

 EXAM TIP   Remember that although it may be structured and named differ-ently in various organizations, the change control board (CCB) is the formal authority for evaluating and approving major changes to the infrastructure.

End-of-Life

Managing the end-of-life cycle for mobile platforms in an MDM infrastructure is also important because it involves changes to the infrastructure and the user experience. As devices and operating systems transition to a legacy state, they often lose support from their manufacturers and vendors, as it's not very cost-effective to keep maintaining patches and updates for legacy platforms. This can often force the organization to transition to newer devices and upgrade operating systems and applications. Additionally, new features and new technologies frequently require moving from legacy systems to new ones. How organizations manage the end-of-life cycle for platforms is critical to making a smooth transition to new technologies.

Operating systems will have to be upgraded from older versions to new versions, and this often requires moving to newer hardware in order to support new OS versions. Likewise, moving to new applications may require also moving to new operating systems and device hardware. As you transition to new technologies and platforms, you'll need to plan on how to deal with the older hardware and software you are retiring. Obviously, you'll want to securely wipe all legacy devices that you are replacing, to ensure that there are no data remnants on the devices.

Proper Asset Disposal

Electronic devices can be hazardous because they typically have batteries and other components that could be harmful to people, animals, and the environment. If the organization is not going to reintroduce the device into its inventory, then it must decide what to do with its old devices. Some organizations choose to donate their old mobile devices to charitable organizations or schools, for example. If your organization decides to go this route, you should make sure that the devices are cleared of all sensitive data, and the operating systems are restored back to their factory configurations. It's also a good idea to donate all the applicable documentation and software licenses with the devices for the benefit of those who will receive them.

If the organization chooses not to donate its old devices to a charitable program, then the devices should be taken to an authorized recycle facility that deals in recyclable electronics. Some organizations may even choose to destroy their old devices to prevent them from being used improperly or accessed by unauthorized persons after disposal. The organization will need to make sure that the disposal method complies with the law and is environmentally sensitive.

 EXAM TIP   Understand that the key requirements in device disposal are device wiping, as well as legal and environmental considerations.

Configuring and Deploying Mobile Applications and Associated Technologies

In the previous chapter, I discussed Mobile Application Management (MAM) and how it is used in conjunction with MDM to manage the applications on a mobile device. Here, we're going to expand on that discussion, and go into a bit more depth on the various aspects of configuring and deploying mobile applications. I'll also cover the technologies behind the scenes that are used to manage mobile applications, including messaging standards and protocols, network configuration, and push notification services. Additionally, I'll go a bit more in depth into the types of apps that are used by mobile devices, as well as in-house app publishing.

Messaging Standards

Back in Chapter 1, I discussed general protocols and the ports associated with them, including those that are used in email messaging. For the most part, configuring email on a mobile device is not much different than configuring it on traditional desktop systems. Since email is an enterprise-wide service, it typically uses standardized protocols, ports, and configuration settings. We're going to revisit those particular protocols, and give you some additional information on them, as email is one of the primary apps that you'll manage, configure, and secure on a mobile device in the corporate world. Each of these protocols has a particular place in the messaging world, and each performs a particular function that must be configured appropriately. Here, we revisit some of the finer details of MAPI, IMAP, POP, and SMTP to ensure your preparedness for the real world and the exam.

MAPI

Microsoft's Messaging Application Programming Interface (MAPI) is not a networking protocol per se, as we also said in Chapter 1, but more of a programming interface that allows different clients to connect with email services for a variety of features. MAPI is useful for enabling programs that wouldn't otherwise be email-aware to connect to email services.

MAPI is fully integrated with Microsoft clients, Exchange servers, and Active Directory services, and allows them to use non-email content and services as well. It's also supported on a variety of non-Microsoft clients and services that need to connect to Microsoft services. As I also mentioned in Chapter 1, because MAPI isn't a network protocol, it doesn't use any specific network port, as do the true network email protocols I'll discuss in the next three sections. MAPI communicates over other network services, such as remote procedure calls, to remote servers and clients that also use MAPI.

IMAP

Internet Message Access Protocol (IMAP) is a networking protocol that handles the receipt and management of email messages. IMAP works on the client side of the connection, and is supported by most popular email clients. IMAP supports multiple connections to any mailbox at once, as well as by multiple clients simultaneously. IMAP also allows you to receive new messages automatically, almost in real time, without requiring you to reconnect to the server. The most current version, IMAP4, uses TCP port 143, but can also be used over a secure connection via SSL or TLS, and then uses TCP port 993. IMAP has multiple advantages over POP, the other popular client-side network email protocol, and is often the protocol of choice on mobile devices, providing the email infrastructure supports it.

POP

Post Office Protocol, or POP, is also a client-side networking email protocol that provides for receiving and managing email message from the centralized email server. The most current version is POP3, and is really considered a legacy protocol because the newer IMAP4 is much more feature-rich and enables more granular control over email. For example, POP3 doesn't automatically deliver email messages to the client; it requires periodic reconnection from the client to the server, usually on a timed or polling basis. POP3 also doesn't allow for multiple simultaneous clients or connections to an email account. Another important difference between POP3 and IMAP4 is that POP3 downloads emails from the server and then deletes them from the server inbox. IMAP4 simply downloads a copy of the email to the client and leaves the original email intact on the server. This difference alone can create issues on mobile devices if the user with POP3 configured as the email client protocol expects to be able to access her email later from a different device connecting to the server, and that email is missing or deleted. Like IMAP4, POP3 is supported by a wide variety of email clients and servers, so it's almost always a configuration option as well when configuring client email programs. POP3 uses TCP port 110, but can also use SSL or TLS over TCP port 995. POP3 is also supported by most mobile device email clients.

SMTP

Simple Mail Transfer Protocol (SMTP) is the server-side email protocol used to send email to other email domains that also run SMTP. SMTP uses TCP port 25 (and sometimes TCP port 587), and is supported on practically every email service in existence because it's one of the first and oldest major TCP protocols developed. It's also interoperable with almost every email service, including, of course, Microsoft Exchange, Unix's Sendmail, and others. SMTP can be secured by sending it over SSL as well (called SMTPS), using TCP port 465. Most mobile device email clients can be configured to communicate with SMTP or SMTPS over their standard ports.

All of these different email protocols and services are primarily concerned with the sending, receipt, and management of email and related services. None of these protocols are concerned with authentication and encryption of email, except for what's provided over their secure counterparts, which typically use SSL or TLS for those services. We discuss authentication and encryption in several later chapters in the book, as well as how to secure and troubleshoot email services.

 EXAM TIP Make sure you are familiar with the different messaging standards, protocols, and their ports for the exam.

Network Configuration

Network configuration is an important part of configuring apps to run correctly. The correct network settings are needed so the app will communicate to networked resources, look for updates, receive push notifications, and so on. In addition to the correct network settings, such as ports, protocols, and even authentication and encryption settings, there are other, larger considerations with the network configuration that are abstracted from the user, the apps, and even the device itself. Some of these are architectural design considerations, such as topology and security device implementation, while others may have more to do with cost effectiveness, tolerance for risk, and so on. In any event, I'll cover these considerations over the course of the next few sections, beyond the technical discussions we've already had on networking back in Chapters 1 and 2.

Vendor Proxy and Gateway Server Settings

There are some network configuration considerations with regards to implementing mobile device management technologies that you should be aware of and prepare for. If you use outsourced services that integrate into your enterprise infrastructure and mobile device management environment, then you should take into consideration some of the network settings you may have to push down to your devices. Some of these include proxy and gateway server settings for your devices that they will need to be able to access these third-party services. Some of these services could include vendor app stores, cloud-based storage services, or even cloud-based mobile device management services. These are but a few of the possible services that you may decide to outsource and provide to your mobile device users.

In some cases, your devices may be required to go through your own network infrastructure in order to communicate with these exterior services, especially if the devices reside within your own logical infrastructure boundaries. In many cases, however, because your devices are mobile, it's possible they'll need to access the services regardless of where they are connecting from or over any type of connectivity, be it cellular or Wi-Fi, whether it's from home, the office, or on the road. To make this happen efficiently, you may need to configure certain apps or network connections on the device with vendor-specific network settings. These may include default gateway settings, VPN server addresses, proxy server addresses, and so forth. It may also include configuring nonstandard port numbers for communications with vendor or cloud services. Along with vendor network settings, you may also have to configure encryption and authentication settings on the device in order to access the services securely. Examples may include trusted certificates from the vendor in order to establish a secure connection with the vendor's infrastructure.

Information Traffic Topology

Recall from Chapter 2 that a network topology defines how the network is laid out both logically and physically. It's closely related to the network architecture, which also adds devices and services into the topology discussion. While we're not going to rehash the discussion on topologies and architectures, the information traffic topology that will be used in your mobile device environment is important because you want to optimize the topology for bandwidth, speed, capacity, and efficiency. There is no one single correct topology for you to use; it's really based upon how your infrastructure is designed and built so you can implement the most effective way for traffic to flow on your network.

When integrating the mobile device management infrastructure into existing network, some of the considerations that you will need to examine include: placement of network access control (NAC) devices, upgrading existing devices to add more traffic capacity for your mobile devices, and adding additional equipment that is specific to mobile device implementation efforts, such as wireless LAN controllers, cellular repeaters, and so forth. Because you're going to increase network traffic with the addition of new mobile devices, as well as the overhead traffic used to manage them, capacity planning is an important factor here. While in-depth network design architecture is beyond the scope of this book, you should consider altering your existing design in order to accommodate the new mobile device infrastructure. Some of the design changes you may want to examine include addition of wireless access controllers, security devices, and even entry points into the network, such as VPN concentrators. Obviously, any new entry points will need to be secured with security devices and the appropriate technologies. Another consideration in managing the traffic flow for mobile devices is that it should be routed through specific devices, both to reduce latency and load-balance the traffic capacity throughout the network. In Chapter 10, I discuss some additional factors that you should consider, in both the network design and in troubleshooting network traffic and topology issues.

Hosting Solutions

Because we're discussing network architectures and design, with relation to how mobile device infrastructures are integrated into the enterprise, it may be good idea to discuss the different hosting solutions available for services that mobile devices rely on. Hosting services on the organization's premises is typically the traditional (and usually the desired) way to deploy services to the enterprise. This is largely an issue of control, in that the organization has a much larger degree of control over services and infrastructure that are located within its own physical and logical boundaries. Most traditional infrastructure models have dictated that both servers and the services they provide physically reside on the organizational network. Control is certainly an advantage to be gained from this model; the disadvantages include both the technical and managerial overhead involved with maintaining the services. On-premise hosting requires dedicated staff and resources to make sure that the services are kept running, secured, and available at all times.

Over the past several years, however, the traditional model has given way to other types of hosting. For example, some servers, services, and now even entire infrastructures can be hosted off-premise by a trusted third party that the organization has contracted with. This business model is viable simply because the organization, whether large or small, can take advantage of the fact that some third parties have built their entire businesses on hosting other organizations' infrastructures, so they can leverage economy of scale with those third-party providers. Now it's possible for even a small mom-and-pop type of business to have an enterprise-level infrastructure that is hosted purely by a third party, without having to dedicate their own physical space, facilities, or staff to maintaining those services. These hosted servers and services are located in and managed by third-party Network Operations Centers, or NOCs, which are responsible for the maintenance and security, as well as availability, of these outsourced infrastructure services. There are many "as-a-service" providers now that offer many of these services through cloud technologies, which are distributed across many geographical areas controlled by several different NOCs in some cases. Some of these cloud-based services include storage, infrastructure, and even mobile device management services.

While cost savings and efficiency are the advantages to third-party hosting, lack of control is sometimes the disadvantage. This issue can be resolved by establishing a solid SLA with a third-party provider that includes reassurances of proper security, backups, load balancing, redundancy, and so forth. These assurances can contribute to the organization's perception of having greater control over these outsourced assets.

 EXAM TIP   Understand that the service-level agreement (SLA) is the key to establishing the organization's requirements and control over issues such as security, access controls, protection of data, frequency of backups, and so forth for data stored with a cloud service provider or third-party host.

Types of Mobile Applications

From the user perspective, every function that can be performed on a mobile device is done through an app, which is basically just a program that runs on the device, similar to the programs you traditionally see running on desktops. But mobile apps have a bit of a different paradigm in terms of development, management, and security. Mobile apps usually have a different development focus, as they are designed with the mobility of the device taken into account. For example, in Chapter 6, I discussed the location-based services that mobile devices and their apps provide, so understandably many of these apps are developed with that service in mind. Apps are also developed to be more independent so that they don't need to rely as much on services, functions, and even data from other apps.

The real focus of our discussion, for the purposes of enterprise app management, is how apps are developed for the enterprise. Before we plunge headlong into a discussion on in-house enterprise app publishing, however, it's a good idea to sort out the different categories of apps that are present on mobile devices. There are three basic types of apps that the CompTIA Mobility+ exam objectives are concerned with, and I'll describe each one in turn. Understanding the different types of apps is helpful in understanding how each is developed, managed, and secured.

Native Apps

Native apps are those that are developed for a very specific hardware or device platform, or even for a specific mobile operating system. They are not really designed for porting over to a different platform, forcing developers to sometimes develop a completely different code base for a different platform if they want to have that same app available for other types of devices. Native apps are usually designed to take advantage of a particular piece of hardware and its lower-level device functions, as well as the API and other software hooks of a particular operating system. In other words, they are designed to fully take advantage of and be compatible with a specific OS and hardware device.

This was pretty much the development model early on when there were only one or two types of smart devices available. Earlier native apps were also developed with very specific programming languages and development platforms, as there may not have been a lot of choices for a developer to use when programming for a specific device. App developers quickly realized, as the mobile device consumer market exploded, that they would need to develop apps that could be ported across a wide range of devices, as consumers often changed platforms and expected the same apps and services on their new devices.

Web Apps

Web apps are developed from the perspective that a mobile device user will use specific, standardized protocols to access Web-based content in a browser. With that in mind, the developer does not have to develop for a specific operating system or hardware platform, as long as they use the accepted web development and communications protocols that most devices can understand. A web app is usually delivered through a user's browser, and browser differences notwithstanding (such as those found in Google's Chrome or Apple's Safari browsers, for example), the app experience will pretty much be the same and independent of the device's OS and hardware. Some web apps may be better suited for one particular platform than another, but the development goal is to deliver a uniform experience to the user, regardless of platform. Some web apps may occasionally still have some device-specific code that is rendered or delivered to the device on-the-fly to ensure that the content is properly displayed or accessible based upon the device, but this is usually kept to a minimum.

Hybrid Apps

A hybrid app offers the best of both the native and the web app worlds. It's an app that resides on the device platform itself, but it is written using web-based technologies that are common across most platforms and operating systems. This enables a common user experience, regardless of platform or device, for an app that actually is installed on the user's mobile device. It also allows for cross-platform support and permits developers to develop one main code base that can be used on different devices. The disadvantage of hybrid apps is that they may not fully take advantage of a specific operating system's APIs or device hardware features because they are written with web technologies that communicate to the device through an abstraction layer in the operating system. Examples of these common web technologies that may be used include HTML, Java, and so on.

EXAM TIP   Understand the characteristics of the three different types of mobile apps, including development models and how each is implemented on a mobile device.

In-House Application Requirements

I have discussed at length the development models and app stores of different vendors, especially Apple and Android platform vendors. However, I also discussed the need to develop in-house apps used for line-of-business applications specific to the organization. This may or may not be a necessary endeavor for the organization, depending upon what the business needs are. Many apps provided in vendor app stores may work just fine for a business; there are also third-party app developers out there that provide apps for specific business sectors and markets. However, disadvantages of using either vendor app stores or third-party apps include the fact that the organization must pay for these apps, as well as licensing them on a volume basis for its users. Depending upon the size of the MDM implementation in the organization, this may not be cost-effective in a very large organization, or in an organization that requires very specific requirements for their apps. For example, because security is a huge issue in mobile apps, the organization may want very specific security controls built into the apps that run on its mobile devices. If the organization can't find apps with this level of security from third-party providers, it may have to develop its own. As another example, the organization may have very unique business needs that third-party apps don't address. For whatever reason, the organization may decide that it is more cost-effective to create organization-specific in-house apps. That's the focus of the discussion in this section, and I cover the particulars of in-house app development in the next several paragraphs.

App Publishing

Several different factors contribute to the effectiveness of app publishing. First, the organization requires personnel that have the desired skill sets in several aspects of technology, including networking, security, and not the least of these, programming. Earlier I discussed three different types of apps that you can see on a mobile device: native, web, and hybrid. Depending on the type of app that the organization wants to produce, each of these may require different programming skills and knowledge of different programming languages. For example, producing a native app strictly for the iPhone or iPad requires a certain level of knowledge specific to programming languages used to create apps for these devices, as well as knowledge of the operating systems that run on these devices.

Second, the organization needs to have the infrastructure in place to deploy its own in-house development shop. In addition to personnel, there's equipment, software, and other materials that are required to support a development effort. These may include test servers and devices, development software, and so on.

Finally, the organization also has to have the infrastructure in place to deploy apps to the devices they are programmed for. This is where an MDM infrastructure comes in, as well as a MAM capability, coupled with a content management and distribution solution, as described earlier. There are some other considerations as well that I'll discuss in the next few sections.

Platforms

I mentioned in the previous section that you should have developers available with the appropriate skill sets, and these include skills in programming in the languages required to support your infrastructure. You may have only Apple devices in the infrastructure, so you'd obviously need someone with development skills to support iOS devices. If you have Android devices in the infrastructure, then of course you need someone to be able to program in the relevant languages and understand the operating systems necessary to support that platform. The same would apply to any other platform you have in the organization, such as BlackBerry or Microsoft. If it turns out that you must provide support for more than one platform, then you have probably just increased the number of people you need for your development shop, or you have to be lucky enough to find a few people that are skilled in all of those operating systems and corresponding programming languages. Remember that they also should have knowledge and skills in networking, security, and other technologies.

In addition to being able to support a particular operating system platform, such as iOS or Android, you may also have to look at supporting specific hardware devices. Not only does the programmer have to be familiar with the operating system and environment she is programming for, but also the hardware device she is developing around. In the case of Apple devices, this may not be a huge issue, but because both Android and Microsoft operating systems can be ported to a variety of devices, developers may need to have some hardware knowledge as well. This is probably more true on the Android side because each device vendor tends to customize the Android operating system a bit for its particular device. Microsoft devices, on the other hand, are likely to be easier to program for because they must meet stricter compatibility requirements to support Windows operating systems.

Vendor Requirements

Chapter 6 covered some of the requirements vendors levy on developers. With Android developers, Google doesn't impose as strict requirements as Apple does on its developers. That doesn't mean there aren't requirements; it just means that it may be a bit easier to develop for Android than it is for Apple devices.

Similar to operating system vendor requirements, there may also be requirements that hardware vendors impose on enterprise development shops. A hardware vendor, for example, could impose device certification requirements on a developer, requiring the developer to "certify" the app for security or compatibility with the device platform. In turn, the hardware developer may also have to provide the developer information on hardware hooks or other critical information the developer may need to build for that platform. This may only be the case if the organization intends to market their apps for that particular platform, but may also be required if organization requires lower-level proprietary information from the hardware vendor in order to write the apps for the device.

Certificates

You may be wondering why digital certificates are an issue in app development, and the reason is quite simple: trust in the app and its developers. As you will learn in Chapter 9, digital certificates help identify and authenticate an entity. In the case of software and apps, they can be used to verify that an app was developed and published by the owner of the digital certificate. The app can be digitally signed by the developer, which not only identifies the developer, but also proves to the user that the app comes from a trusted source.

In terms of enterprise app development, it's usually a good thing for the organization to acquire software signing digital certificates from a trusted source, usually from a third party that specializes in issuing digital identities. VeriSign and Thawte are just two of the many examples of widely known and respected, trusted third parties that do this. Having a digital certificate issued from one of the trusted third parties lends credibility to digitally signing the app because there is a rigorous process for obtaining a digital identity from one of these entities. Of course, the organization doesn't have to acquire a certificate from a third party. The organization can always issue its own self-signed digital certificate with which to sign software and apps. Unfortunately, this doesn't have a solid anchor (or *chain of trust*, as it is known in the business) to another trusted third party, so its value to anyone outside the organizational boundaries may be questioned. If the software or app is going to be used only by employees and users inside the enterprise, a self-signed certificate may be a valid option. If, however, the enterprise app store is going to distribute this app outside of the organization, obtaining a digital certificate from a trusted third party is probably a better way to go.

 EXAM TIP A digital certificate issued by a third party for the organization to use in signing its apps and software is a preferred method of proving identity and authentication for apps distributed outside the boundaries of the organization.

Data Communication

In-house programmers will have to take into account data communication, an element in app development. There are several aspects of development that involve network communications, as well as security. App developers must account for the mobile nature of devices when programming their apps so their apps should support a common occurrence of leaving one network and joining another dynamically. They must include provisions for the app to be able to connect to both wireless and cellular networks, and may need to include methods to connect to Bluetooth or other types of networks as well, depending upon the requirements of the app.

Generally, the app will use the existing network connection that the device uses, so the user doesn't have to worry about configuring connection properties such an IP address, default gateway, and so on. What a developer may have to watch for, however, is hard-coding any network connection requirements in the app that may not be easily or typically met by the device or the infrastructure. For example, if a developer hard-codes a specific port to be used by the app when communicating on the network, this may present problems if that port happens to be blocked on the firewall, or otherwise prevented from being used on the network. A better scenario may be to provide for a configuration method that the user or administrator could use to configure network settings dynamically, based upon how the network is set up. This might be as simple as a configuration screen in the app, or an XML file that an administrator can edit and send to the app for its configuration.

 CAUTION   Make sure you are aware of any hard-coded network configuration settings in the app such as static ports, for example. You may have to take these into account by changing the infrastructure, such as opening those specific ports on a firewall.

Security is also a programming issue in general, and programming for mobile devices is no different. Developers must make efforts to produce secure code by thoroughly testing it and taking into account specific vulnerabilities that may be associated with app programming. For example, developers should take into account input validation, as well as bounds checking, to ensure that any user input or action doesn't affect the app or its resources in a negative way. Faulty input validation has been known to cause issues such as injection attacks, and lack of bounds checking can result in application buffer overflows if an attacker can send malicious code into an app. Secure coding practices are outside the scope of this book, but the organization should definitely hire people who have a secure coding background or train them in these practices.

Other security requirements that should be considered when developing an app are the use of encryption algorithms and methods, as well as encryption strength. In Chapters 9 and 11, you'll see that incompatible encryption methods are a common issue that often prevents secure communications between network devices. If an app is not developed with compatible encryption methods in mind, this could be a serious issue affecting data confidentiality between the device and the network.

Authentication is another consideration in secure app development as well, for the same reasons described earlier regarding encryption. Authentication methods that are used by the app must be compatible with those used for the particular network service or resource the app is communicating with. If the service uses password or certificate-based authentication, the appropriate authentication mechanisms should be built into the app to support those methods.

Push Notification Technologies

Push notifications are small messages sent to a device from a central communications server. Notifications require that the device have a constant or always-on network connection, such as a Wi-Fi or cellular connection. Push notifications have been used since the early days of the first BlackBerry devices, although in these early days, they were primarily used to deliver email to the device. Recent advances in MDM technologies, however, allow push notifications to perform a wide variety of management functions. These functions could include notifications concerning updates, policy changes, and even configuration settings. Notifications can also be used to tell an app to schedule an update at a given time or under given circumstances, such as when connected to the corporate infrastructure. While notifications don't typically carry huge amounts of data, they are able to direct an app to perform a function or make a connection back to the central MDM services. From that perspective, they are more like control messages that are sent to devices. Unlike "pull messages," which was an older technique that used and involved apps that simply initiated connections to the server to check for updates, push notifications don't use up as much battery power as the older pull messages did, because they ran constantly in the background.

Note that push notifications aren't the only way to send control messages and management commands to devices; the Short Message Service (SMS) that is typically used over cellular services to send text messages can also be used. Because SMS doesn't require robust data services, it can be used to send messages to the device in the event that the infrastructure can't easily communicate with the device, which might be the case in the event that Wi-Fi data services, for example, are turned off, or a device has a very weak cellular connection. For this discussion, however, we're going to discuss three push notification services: the Apple Push Notification service that is unique to Apple devices, the Google Cloud Messaging services that are used by Android, and ActiveSync, which is a Microsoft protocol.

Apple Push Notification Service

The Apple Push Notification service (APNs) is used to push notifications down to iOS devices. These come from Apple's central notification servers for apps downloaded from iTunes, of course, but also from Apple even for enterprise-level app notifications. When an organization wishes to develop apps for its employees under the iOS Developer Enterprise Program, it also registers its services with Apple for the ability to send push notifications to the particular devices it develops for. These notifications must go through Apple's communications services and are forwarded on to the device. In earlier versions of iOS, APNs was very limited in what it could do in terms of messaging, managing, and controlling the behavior of apps. With the iOS 7 feature improvements, however, APNs allows a much wider variety of messages and the ability to control apps on a much more granular level. Push notifications are 256 bytes in size, and the protocol uses TCP port 2195. An additional protocol used by APNs is the APNs Feedback service, which uses TCP port 2196, and is used for failed notification delivery in the event the receiving device doesn't respond to the notifications.

Google Cloud Messaging

Android has its own type of push notification service, and it is called Google Cloud Messaging (GCM) services. GCM was previously known as the Cloud-to-Device Message (C2DM) service, but was revamped and re-released as GCM in 2012. GCM performs functions similar to APNs, but is able to carry larger messages to its devices. Organizations that wish to use GCM must also go through Google's servers to forward messages to their devices. GCM messages can be used to send simple notifications to the device, or even send up to 4KB (kilobytes) of data. Remember from my discussion in Chapter 1 that GCM uses TCP ports 5228–5230 for communications between devices and the Android Marketplace.

Exchange ActiveSync

Exchange ActiveSync (EAS) is a Microsoft protocol that has become widely used across a range of mobile operating system platforms and hardware vendors, including Apple and Android devices, as well as Microsoft mobile platforms. It was originally developed as a synchronization protocol for Microsoft Exchange corporate users, but has evolved over time to include more device control and management features. While not a full-fledged MDM solution, many organizations actually do use it to perform some device management functions over their mobile device population. EAS has the ability to set up and configure network connectivity and secure email options for clients that connect to Microsoft Exchange corporate servers, but it also has the ability to control the much wider range of functions. Some of these functions include the ability to set password policies, remotely wipe or lock a mobile device, and control some device settings. EAS can't, however, impose any management control on other apps or control the nature of secure network connections with most devices, so it's not a full MDM solution, but it could be used as a small part of one. EAS uses TCP port 2175 for its network communications.

 EXAM TIP   Make sure you know the characteristics of each of the three push notification services discussed in the preceding sections, as well as their network communications ports.

Chapter Review

This chapter focused on implementing an MDM solution in the enterprise. You need to consider several factors before the actual installation, which relate to planning and preparing the enterprise infrastructure and users for the MDM rollout. One of the primary tasks you'll need to accomplish before rollout is to make sure the infrastructure is ready for the installation. This requires careful coordination. You will also want to make sure your directory services architecture is prepared by creating the appropriate groups, organizational units, and even policy objects that will be pushed down to different devices. Certificates are another issue you must prepare for as well, in terms of enrolling users and devices and pushing those certificates down to the device. Other considerations that you need to plan on before installation include end-user license agreements, data containerization, and application sandboxing. Device and group profiles are also items you'll need to prepare because you will probably require different configuration and security settings pushed to different groups of devices and users.

Another important step in preparing for an MDM implementation is conducting a pilot test, in which you would test your rollout on smaller groups of users specifically chosen for the pilot. This would allow you to discover and resolve issues with the implementation before it is pushed out on a larger scale. You also need to carefully document every aspect of the implementation, including the MDM architecture, processes, and procedures used for the rollout. At that point, you're ready to launch your MDM implementation, keeping in mind that you should train users at the appropriate levels so that they can easily adapt to the new infrastructure. There are some key best practices you should consider for deployment as well, including scaling your deployment gradually from smaller groups to larger groups, based upon different characteristics of your user and device population. These characteristics could include business role or function, managerial level, and even device type.

This chapter also covered mobile device on-boarding and off-boarding, as well as the provisioning processes. In order to get a device successfully issued to a user, it must be adequately provisioned, which could include activation on cellular networks and getting configuration and security settings to the device. Profiles, which are used to get the settings to mobile devices, are basically simple XML files containing configuration settings relating to the network, security, and even apps. Provisioning can be done through several methods, including manually by the administrator, self-service by the user, through batch provisioning, and even remote provisioning. Off-boarding and de-provisioning were discussed as well, and include considerations such as employee terminations, removing and transferring sensitive data, wiping the device, and device disposal.

I also discussed day-to-day and long-term mobile device operations and management concepts. I covered topics such as centralized content management and distribution, as well as remotely managing devices. Life-cycle discussions in this section included considerations involved in the acquisition, development, implementation, sustainment, and eventual disposal of mobile devices and software. Life-cycle management also includes factors such as certificate expiration, application life cycle, and effecting orderly changes in the infrastructure through a formalized change management process.

Also discussed during this chapter was configuring and deploying mobile applications, as well as the associated technologies required for those applications. These technologies include messaging standards, such as MAPI, SMTP, POP, and IMAP. Network configuration is also critical to mobile devices and their apps functioning properly, and several aspects of networks in this context were discussed, including configuration settings needed for vendor and third-party services, traffic topology, and the different third-party and cloud-based network and service hosting solutions available for an organization to consider when implementing an MDM.

I also discussed the different types of apps that are available for mobile devices, including native apps, web apps, and hybrid apps, and how they are developed and implemented on different device platforms. And finally, I discussed push notification services, and covered the three major services used by the major vendors, the Apple Push Notification Service used by iOS devices, Google Cloud Messaging services used by Android devices, and Exchange ActiveSync, Microsoft's proprietary protocol, which is used to some degree by almost all the major OS and platform vendors.

Questions

1. Which of the following should be scheduled in order to integrate MDM with your existing infrastructure?

    A. Wireless LAN controller installation

    B. Downtime

    C. Mobile device synchronization

    D. Firewall changes

2. Which of the following file formats usually make up a profile?

    A. XML

    B. CSV

    C. X.509

    D. HTML

3. Your manager wants to fully roll out the new MDM infrastructure to all users at once. You would prefer to test the implementation first. Which of the following terms describes the type of test you should perform before a full implementation?

    A. Scaled

    B. Cross-sectional

    C. Incremental

    D. Pilot

4. Which of the following items are typically provisioned when initially issuing a device to a user? (Choose two.)

    A. VPN settings

    B. User data

    C. Digital certificates

    D. IMEI numbers

5. Which numerical identifier is used to identify the subscriber identity module in a device?

    A. IMEI

    B. ICCID

    C. IMSI

    D. MAC

6. Which of the following describes a draft security protocol published by the IETF and used to assist in securely enrolling mobile devices?

    A. SCEP

    B. SSH

    C. HTTPS

    D. XML

7. All of the following are valid reasons to off-board and de-provision a mobile device, EXCEPT:

    A. Employee termination

    B. Device disposal

    C. Device replacement

    D. OS upgrade

8. You have set up a content management and distribution solution that integrates with your environment. Which of the following are acceptable methods for securely distributing content to your mobile devices using the solution? (Choose two.)

    A. FTP server

    B. Web application

    C. SD card

    D. Shared folder

9. Which of the following existing security mechanisms and infrastructure can be integrated with a content management and distribution system? (Choose all that apply.)

    A. LDAP authentication

    B. HTTPS encryption

    C. File permissions

    D. Time-of-day restrictions

10. Which of the following is performed on a device to prevent unauthorized access to data in the event the device is irretrievably lost or stolen?

    A. Remote monitoring

    B. Remote lock

    C. Remote wipe

    D. Remote control

11. Which of the following terms describes the beginning-to-end process of planning, acquiring, developing, implementing, maintaining, and disposal of both MDM systems and related software in the enterprise?

    A. Pilot

    B. Process

    C. Life cycle

    D. Deployment

12. Which entity is usually in charge of implementing the formal change management process in an organization?

    A. Steering committee

    B. CCB

    C. Mobile device administrator

    D. Chief Executive Officer

13. All of the following are acceptable methods for proper asset disposal EXCEPT:

    A. Dumping them in a city landfill

    B. Donation to a charitable organization

    C. Transfer to a electronics recycle facility

    D. Environmentally safe destruction

14. Which of the following should be done at a device's end-of-life?

    A. Decrypt all of the sensitive data on it

    B. Donate the device to charity as is

    C. Reissue the device to a less important employee

    D. Securely wipe the device of all remnant data

15. Which of the following messaging standards is not a true networking protocol?

    A. IMAP

    B. EAS

    C. MAPI

    D. SMTP

16. You are configuring an app that will be pushed out to your organization's mobile devices. The app connects to a third-party cloud-based storage service that has been contracted by the organization. Which of the following should you make sure is configured properly in the app, in order for the mobile users to successfully connect to the third-party service?

    A. Proxy and gateway server settings

    B. User password

    C. Device IP address

    D. Organizational software signing certificate

17. All of the following are considerations in designing the information traffic topology you will use in the mobile device implementation EXCEPT:

    A. Latency

    B. Bandwidth

    C. Capacity

    D. Encryption

18. Which of the following are reasons to consider third-party service hosting? (Choose two.)

    A. Desire to maintain centralized control

    B. Desire to increase technical staff

    C. Desire to reduce costs

    D. Desire to reduce infrastructure

19. Which of the following types of apps are developed for a very specific hardware or device platform?

    A. Native

    B. Hybrid

    C. Web-based

    D. Third-party

20. Your company is developing an app that it will use in-house, as well as provide to its business partners. Which the following should the organization obtain in order to establish trust in both the app and the development process?

    A. Software signing certificate issued to a different developer

    B. Software signing certificate issued by the organization

    C. Software signing certificate issued by a third-party

    D. Software signing certificate issued by the business partner

Answers

1. **B.** Downtime should be scheduled so that you can take the time you need for the integration activities.

2. **A.** Profiles are typically text-based files, usually in an eXtensible Markup Language (XML) format.

3. **D.** A pilot program is how you would implement your MDM solution on a small-scale test basis to a select group of users and devices.

4. **A, C.** VPN connection profiles and digital certificates used for authentication to enterprise services can both be provisioned to the device. User data is normally not provisioned, although it may be restored from a backup at a later time. The IMEI number is not provisioned because it is already recorded with the device.

5. **B.** The ICCID number, which stands for Integrated Circuit Card Identifier, uniquely identifies a Subscriber Identity Module (SIM).

6. **A.** The Simple Certificate Enrollment Protocol (SCEP) was designed to deliver secure, encrypted settings to devices so they can securely communicate with the infrastructure.

7. **D.** Upgrading the operating system on a mobile device is not an adequate reason to de-provision it. All of the other choices are valid reasons.

8. **B, D.** Both web applications and shared folders are appropriate for securely distributing content to mobile devices, provided they have the appropriate access controls enabled on them.

9. **A, B, C, D.** All of these existing infrastructure security mechanisms can be used as part of a content management and distribution system.

10. **C.** Remotely wiping a device is done to prevent unauthorized access to data, should the device be irretrievably lost or stolen.

11. **C.** The life cycle refers to the beginning-to-end process of planning, acquiring, developing, implementing, maintaining, and disposal of both MDM systems and related software in the enterprise.

12. **B.** A change control board (CCB) is formally created and responsible for overseeing the change management process.

13. **A.** Dumping them in a city landfill is not an acceptable method of mobile device disposal.

14. **D.** Securely wipe all legacy devices, regardless of disposal method, to ensure that there are no data remnants on the devices.

15. **C.** MAPI is not a true networking protocol, but a programming interface that allows different clients to connect with email services for a variety of features.

16. **A.** Vendor proxy and gateway server settings should be configured in the app in order to access the third-party cloud storage services.

17. **D.** Encryption is a service that may be affected by one of the other characteristics of the network topology, but is not a characteristic of topology itself.

18. **C, D.** Reducing costs and infrastructure are two reasons to consider using third-party hosting solutions.

19. **A.** Native apps are those that are developed for a very specific hardware or device platform, or even for a specific mobile operating system.

20. **C.** The organization should acquire a digital software signing certificate from a trusted source, usually from a third party that specializes in issuing digital identities.

Mobile Security Risks

In this chapter, you will

- Explore risks, threats, and mitigation strategies inherent to mobile device infrastructures
- Learn about incident response

This chapter takes a look at the different security risks that mobile devices and their underlying infrastructures face. These include a wide variety of threats, including viruses, malicious apps, and even sometimes, the users themselves. Previous chapters touched on some of the different security issues that mobile devices, as well as other personal and network devices, are exposed to. This chapter, however, looks at some of the same threats from a mobile device perspective. You'll also take a look at the different mitigations that can be employed against some of these threats, from both a preventative and a remedial perspective. The CompTIA Mobility+ exam objectives covered in this chapter are 4.4, "Explain risks, threats and mitigation strategies affecting the mobile ecosystem," and 4.5, "Given a scenario, execute appropriate incident response and remediation steps." Covering these two objectives will set the stage for more in-depth discussions on mobile security technologies and troubleshooting security that are tackled in Chapters 9 and 11.

Mobile Infrastructure Risks

The risks that mobile devices and their supporting infrastructures face come from a wide variety of attack vectors. These attack vectors include the wireless media, the software that runs on the devices, device hardware, and even organizational risks. Organizations must take measures to mitigate these risks, but understand that rarely will a given type of risk be completely eliminated. The next few sections discuss these different types of risks in detail, and then cover some of the typical mitigations that organizations should employ in their infrastructures, on their devices, and even with their people.

Wireless Risks

The wireless media itself has several inherent risks. Chapter 5 covered the technical aspects of wireless, and also discussed some of security measures that can be taken to help secure wireless networks, particularly those that follow the IEEE 802.11 standards. You've learned about the different security protocols, such as WEP, WPA2, 801.X authentication, and some of the different security measures that can be taken on clients and access points. Even with some of these security measures, there are risks involved with wireless networks, both the wireless LAN type and a few specific to cellular technologies. We'll discuss these risks over the next few sections.

Rogue Access Points

A rogue access point is one that has been set up to attract unsuspecting users to it. Hackers may set up a rogue access point in order to entice users to connect to it so that the hackers can obtain user credentials and other sensitive data from the user. Often a rogue access point will be named similarly to another nearby access point that is legitimate. It may broadcast an SSID that is also similar to a legitimate wireless network. This would serve to trick users into attempting to connect to the fake access point. The attacker may use weak authentication methods in order to facilitate the users' connections to the access point. The AP may use open authentication, shared authentication, or even WEP. In any case, the user connects to the access point and authenticates, and then uses the access point as if it were the legitimate one. The attacker may also connect the rogue access point to the Internet in such a way that users may be tricked into using legitimate web sites while the attacker sniffs for passwords and other authentication credentials. The attacker may also use the opportunity to hack into the users' mobile devices in hopes of stealing data. Rogue access points are often found in corporate environments in an effort by employees to circumvent corporate security policies, or placed by hackers to use the above described techniques to acquire sensitive data.

Rogue access points are often seen in settings such as Internet cafés, bookstores, coffee shops, and restaurants that provide free wireless access to its customers. One form of rogue access point is called an *evil twin*, and duplicates the legitimate wireless network. The attacker may also offer a rogue access point that has a stronger signal than the legitimate one, which, of course, further entices the unsuspecting user to connect to their access point. Malicious rogue access points are also often used to carry out man-in-the-middle attacks. Figure 8-1 shows a wireless client that has found two different access points with the same SSIDs. One is an evil twin. Can you guess which one?

DoS

A denial-of-service (DoS) attack can be conducted against wireless clients using a variety of methods. One method, also used to attempt to crack WEP and WPA/WPA2 authentication credentials, includes sending deauthentication traffic to both the client and the access point it is connected to in an effort to cause them to have to reestablish contact and, in the process, re-authenticate. When this happens the attacker is sniffing the connection and hopes to capture the four-way handshake that WPA and WPA2 use to authenticate the client to the access point. When the deauthentication traffic is sent, it disrupts the connection between the client and access point. When this is done repeatedly, it prevents the client from communicating with wireless network. Figure 8-2 shows this very attack.

Currently connected to:
Unidentified network
No Internet access
Wireless Network Connection
thunderbird
Thunderbird
Information sent over this network might be visib
Connect autom
Name: Thunderbird
Signal Strength: Excellent
Security Type: Unsecured
Radio Type: 802.11n
SSID: Thunderbird
texaspride
VIZIO
Open Network and Sharing Center

Figure 8-1 A rogue access point masquerading as a legitimate one

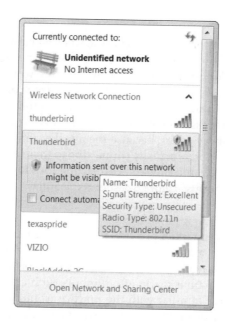

Another method of conducting a denial-of-service attack through a wireless network is to use a rogue access point that overpowers the legitimate one. The transmit power is set to a higher level than a legitimate access point and can cause issues with clients that attempt to connect to a wireless network. There are also devices and software that can be used to travel through the different wireless frequencies, sending disruptive traffic to any wireless network in the vicinity. This practice of disrupting signals between clients and wireless access points is called *jamming*. In addition to being disruptive, jamming is usually illegal in most countries.

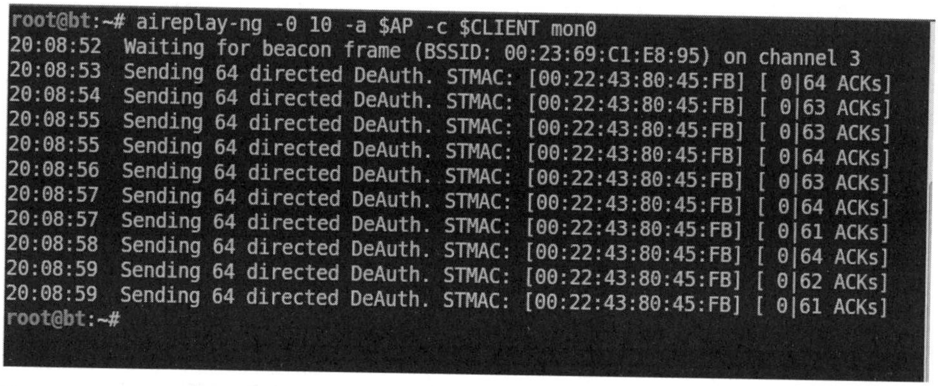

root@bt:~# aireplay-ng -0 10 -a $AP -c $CLIENT mon0
20:08:52 Waiting for beacon frame (BSSID: 00:23:69:C1:E8:95) on channel 3
20:08:53 Sending 64 directed DeAuth. STMAC: [00:22:43:80:45:FB] [0|64 ACKs]
20:08:54 Sending 64 directed DeAuth. STMAC: [00:22:43:80:45:FB] [0|63 ACKs]
20:08:55 Sending 64 directed DeAuth. STMAC: [00:22:43:80:45:FB] [0|63 ACKs]
20:08:55 Sending 64 directed DeAuth. STMAC: [00:22:43:80:45:FB] [0|64 ACKs]
20:08:56 Sending 64 directed DeAuth. STMAC: [00:22:43:80:45:FB] [0|63 ACKs]
20:08:57 Sending 64 directed DeAuth. STMAC: [00:22:43:80:45:FB] [0|64 ACKs]
20:08:57 Sending 64 directed DeAuth. STMAC: [00:22:43:80:45:FB] [0|61 ACKs]
20:08:58 Sending 64 directed DeAuth. STMAC: [00:22:43:80:45:FB] [0|64 ACKs]
20:08:59 Sending 64 directed DeAuth. STMAC: [00:22:43:80:45:FB] [0|62 ACKs]
20:08:59 Sending 64 directed DeAuth. STMAC: [00:22:43:80:45:FB] [0|61 ACKs]
root@bt:~#

Figure 8-2 A WPA2 deauthentication attack in progress

Wardriving

Wardriving is an older practice dating back to when wireless networks were first coming into mainstream use. In wardriving (and sometimes warwalking as well), a person would drive around looking for unprotected wireless networks, noting details about them, such as location, signal strength, channel, and what type of security they used (open, WEP, WPA, and so on). In some cases, the person doing the wardriving would also stop and connect to the wireless network if its security was easily broken. While the legality of wardriving isn't always clear, because the attacker may just be surveying the available wireless LANs in the area, actually connecting to them and using them usually is illegal. The laws regarding wardriving and connecting to WLANs vary from state to state, and from country to country. Figure 8-3 shows the author all set to go wardriving!

Warchalking

Warchalking is also an older practice, not really seen much anymore. It was used when there weren't really a lot of wireless networks proliferating the airwaves, as there are these days, so this is really more of a historical footnote than anything else. Warchalking involved placing peculiar marks and symbols, with chalk, near a wireless access point so others who were familiar with the practice would know that an access point (AP) was nearby, with a strong signal, and possible lack of security. Usually, they would be written on the sidewalk near the AP, or even on the side of a building, possibly obscured. Because wireless networks are almost everywhere nowadays, and most wireless devices

Figure 8-3 Preparing to go wardriving

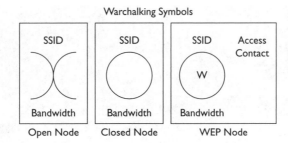

Figure 8-4
Warchalking symbols indicating different types of access points

can pick up the signal and information about the network anyway, this practice has more or less fallen into obscurity. Figure 8-4 shows an example of warchalking symbols.

EXAM TIP While you may see some mention of wardriving and warchalking on the exam, understand that these attack vectors are older and are not used very much anymore.

Wireless LAN Security Protocols

Wireless LAN security protocols can also have weaknesses that can be exploited. WEP, for example, uses very weak and repeating initialization vectors that can easily be intercepted and used to crack the WEP key. Because of this, the use of WEP is highly discouraged and should be discontinued as soon as possible. Although WPA and WPA2 are considered much stronger security protocols, and are the standard today in home and small business wireless infrastructures, they can also be exploited through the use of weak keys or easy-to-guess passphrases. Other wireless LAN security methods, such as MAC filtering and SSID hiding, are relatively ineffective against a determined attacker.

Weak Keys

As mentioned previously, weak encryption keys can help an attacker wage an interception attack on encrypted communications. Weak keys have the same effect as weak passwords; if they do not use large, complex character spaces and obscure characters for the password, they are easily broken. Weak keys can be created by uninformed users, or even by software that has not been implemented correctly. The algorithm used in encryption systems can also affect whether or not weak keys are used. For example, in WEP, the key is limited in length and also in implementation. While WEP uses RC4 as its streaming protocol, it does not use a large initialization vector, and repeats this initialization vector frequently. This makes it easy to intercept and crack the key(s). Even in WPA2 encryption methods used in wireless LANs, a weak or easy-to-guess password can be easily intercepted and cracked. Figure 8-5 shows a screenshot of the results of using the Aircrack-ng suite of tools to break a weak encryption key in WPA.

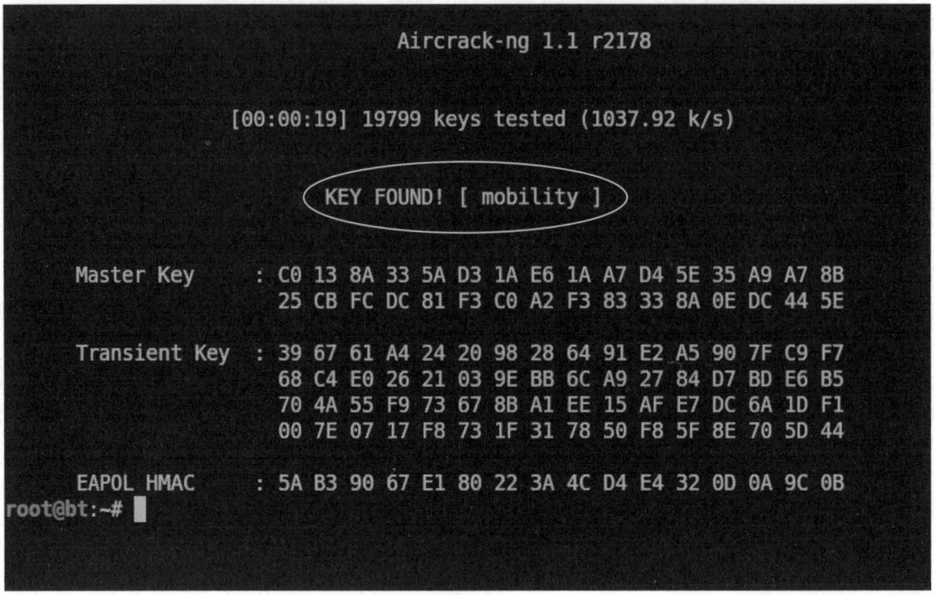

```
                          Aircrack-ng 1.1 r2178


             [00:00:19] 19799 keys tested (1037.92 k/s)


                      ╭─────────────────────────────╮
                      │  KEY FOUND! [ mobility ]     │
                      ╰─────────────────────────────╯


        Master Key     : C0 13 8A 33 5A D3 1A E6 1A A7 D4 5E 35 A9 A7 8B
                         25 CB FC DC 81 F3 C0 A2 F3 83 33 8A 0E DC 44 5E

     Transient Key     : 39 67 61 A4 24 20 98 28 64 91 E2 A5 90 7F C9 F7
                         68 C4 E0 26 21 03 9E BB 6C A9 27 84 D7 BD E6 B5
                         70 4A 55 F9 73 67 8B A1 EE 15 AF E7 DC 6A 1D F1
                         00 7E 07 17 F8 73 1F 31 78 50 F8 5F 8E 70 5D 44

        EAPOL HMAC     : 5A B3 90 67 E1 80 22 3A 4C D4 E4 32 0D 0A 9C 0B
    root@bt:~# ▮
```

Figure 8-5   Aircrack-ng breaking a weak key in WPA

Man-in-the-Middle

A man-in-the-middle (MITM) attack is so called because it involves an attacker that can intercept communications between the transmitter and receiver, and vice-versa. This type of attack can be used in several different ways. First, the attacker can intercept plaintext communications easily and either simply eavesdrop on the conversation or alter any intercepted data before retransmitting it to the unsuspecting receiver. In encrypted communications, this can be somewhat more difficult, although not impossible. Typically, attackers take advantage of weak encryption algorithms, weak keys, or a weakness in the implementation of encryption systems in order to break into the encrypted traffic.

Once the attacker has circumvented the authentication and encryption mechanisms, and entered the communications stream, they intercept traffic simply to eavesdrop on it, or to modify it in between the sender and receiver. On either end of the communication, each user believes that he is communicating directly with the other party, when in reality, they are both communicating with the attacker. Because there is another party in the middle of the conversation, delay in communications can be introduced, and data can be corrupted or modified. There are several variations to this type of attack, including shutting one of the parties completely out of the conversation or hijacking the session completely. MITM attacks may be difficult to detect, although the delay in communication and verifying data transmissions on both ends may indicate that this attack is being carried out against the users. Strong encryption methods that are implemented correctly, as well as using strong keys, can help prevent this type of attack.

Tower Spoofing

Tower spoofing is a technique that has recently come into the limelight over the past few years. It involves setting up equipment that can spoof a carrier's tower and infrastructure and cause a cellular device to use it instead of the normal tower equipment. It requires overpowering the nearest legitimate cell signal, causing the cellular device to lock onto it instead. Equipment used in tower spoofing can also be used to eavesdrop on any conversation, even if it is encrypted. In some cases, the equipment can be used to fool the device into turning off encryption completely.

The equipment required to spoof a cellular signal tower can be very expensive, although in 2010, demonstrations given by security professionals at different conferences showed how a determined hacker with some resources could purchase the equipment necessary to perform these types of attacks, for under US $1500. Just as hackers have been using this technique for a few years, law enforcement officials have been reportedly using it as well. Since 2010, there have been numerous court cases highlighted in the media questioning the admissibility of evidence obtained from cell signal interception. A device called a "Stingray" has been reported by the media as used by various federal, state, and local law enforcement agencies to intercept a suspect's cell traffic using tower spoofing equipment and techniques.

Software Risks

Now that I've discussed the wireless risks in mobile computing, let's spend some time discussing the software and hardware risks. I'll cover software risks first, and then move on to the hardware risks a bit later. Software risks include the use of untrusted or compromised applications on mobile devices. For the most part, depending upon the operating system and the vendor, software risks are kept to a minimum by controlling the source of apps, and through strict requirements on selling apps in the respective app stores from each vendor. That isn't to say, however, that software risks don't exist at all for these devices. As mobile devices proliferate the market space, new risks specific to mobile devices, as well as older risks that have always plagued computing, are appearing. Over the next few sections, we'll discuss some of the prominent risks with mobile software.

App Store Usage

For the most part, getting software from legitimate application stores run by the major vendors, such as Apple, Google, BlackBerry, Microsoft, and Amazon, is not only easy, but usually secure. Different vendors have different requirements for developers in order to get an app into the app store, and these include security requirements as well. Some of this stems from the development and support model used by the vendor. Apple and BlackBerry are very monolithic in their device and application structure, strictly controlling all aspects of both the device and the apps that run on it. Apple, for example, is extremely strict in terms of how developers must create an application that is sold in iTunes. Android, on the other hand, is based upon a multitier model, where the devices are developed separately from the applications, and even the operating systems that run on them. There are variations in the Android devices' operating system flavors that require developers to develop differently for each variation. What may run on devices sold by one vendor isn't necessarily guaranteed to run on another vendor's device, although they all use variations of the Android operating system. A prime example of this is Amazon's line of Kindle devices, which can only get apps from the Amazon Appstore. Additionally, Android apps aren't always subject to the same strict developer guidelines that Apple and BlackBerry apps are. That doesn't necessarily mean they are less secure, but this can cause issues for secure development, obviously.

The security weakness that exists with app stores is essentially getting apps from unapproved or unofficial sources. There are definitely legitimate app sources outside of iTunes or Google Play, for example, such as device manufactures, communications carriers, and even in-house corporate development sources. Some sources, however, are not so legitimate, and are usually unapproved by the vendors, manufacturers, and corporate customers for use. In some cases, you can get just the app, but getting it to run on the device may be problematic, as some of these apps require root-level access to the device. This is typically not allowed on most consumer devices unless the device is rooted (meaning that someone has obtained administrative or "root"-level access to the device) or hacked it.

When getting apps from questionable sources, problems include apps that contain malware, apps that steal personal data and transmit it to a third party, or apps that can even be used as hacking tools. Additionally, some apps require replacing the operating system with one that's not approved by the vendor, which not only invalidates the warranty on most devices, but also could cause the device to be unstable and not operate properly.

 CAUTION   Unofficial or unapproved app sources should not be allowed for enterprise-owned devices.

Malware

Malware is a short name given to malicious software, whose purpose is to infect the host and damage, destroy, or steal data. It can also be used to wage network attacks against one or groups of hosts. There are several different classes of malware, including viruses, Trojans, worms, spyware, keystroke loggers, and others. Malware can be used to steal authentication credentials, such as passwords, and send them back to the attacker, or it can be used to cause denial-of-service attacks against entire networks. I'll cover the different types of malware coming up the next few sections.

Virus

A virus is a type of malware that can be transmitted via files or executable software from device to device. We typically think of viruses as only living in the Windows-based PC world, but there have been instances of viruses specifically constructed for some mobile devices, such as laptops, tablets, and even smartphones. One important aspect of a virus is that it is not self-executable, nor is it self-replicating. A virus must be acted upon by a user or an external process in order to execute. It also must be copied in some form to another media or device in order to be replicated; it is incapable of replicating itself (unlike another form of malware, the worm).

Trojans

A Trojan is a very special piece of malware, whose mission is to infect a host masquerading as a useful program and then execute when the program is executed, performing all manner of malicious activities. Because Trojans hide in a useful looking piece of software, it actually gets its name from the legendary Trojan horse of Greek mythology. A Trojan can be used to steal data, such as authentication credentials, passwords, credit card numbers, and other sensitive data, and transmit it back to the attacker. Mobile devices, such as desktop PCs, have been hit by Trojans as well. Examples of mobile device Trojans include Gingermaster and DroidKungFu, two Trojans that work on various versions of Android devices. Both of these Trojans attempt to gain root access to the device and exfiltrate data from the device.

Worm

A worm is another particular piece of malware that can infect a host. However, unlike a Trojan or a virus, a worm has the ability to self-replicate across the network. This is why worms are very difficult to eliminate once they have infected the victim network. When one host is cleaned, it may be plugged back into the network, only to be re-infected when the worm spreads across the network from another infected host. Examples of famous desktop PC worms include the Morris Internet worm (the very first worm), MyDoom, and the Win32 Conficker worm. Mobile device worm examples include Cabir (for Symbian OS devices), and Ikee, which was the first identified worm affecting Apple's iOS devices. Note that Ikee, like most malware, requires that the device be jailbroken, a process explained a little later in this chapter.

Spyware

Spyware is a form of malware that, as evident by its name, is designed to spy on users and possibly steal data, such as passwords, credit card numbers, and so on. Spyware can be used to observe a user's actions on a host, such as surfing the Internet, typing a document, or any other activity. The activities that spyware records could be saved as pictures, video, or keystrokes. Spyware can also record activities, dump them to a list, and send them back to the attacker.

It's worth mentioning that some spyware has been used for somewhat legitimate purposes when used by employers, for example, to monitor the activities of employees. However, such activities should be undertaken with great care, as it may violate privacy laws in some states or countries. Employers who use such software should do so with care, and inform users that their activities are being monitored and to what degree, and what the information will be used for. Parents also use spyware to monitor activities of children when using the Internet. There are several pieces of legitimate spyware that perform these functions.

 EXAM TIP   Know the basic characteristics of the various classes of malware for the exam, including the virus, Trojan, worm, and spyware.

Jailbreaking

The term "jailbreaking" has only come about in recent years since the proliferation of smartphones and other smart devices, such as tablets. *Jailbreaking* means that the user will install a program on the device, typically not approved by the device manufacturer, which changes settings on the device that are not normally intended by the manufacturer to be changed. Jailbreaking allows a user to install software not normally allowed, such as apps that don't come from the manufacturer's legitimate app store, or applications that don't meet legal or quality requirements of the device manufacturer. Jailbreaking also allows a user to unlock functionality on the device. For example, some iPhones that use AT&T as a service provider can't be used to tether (which means to allow another device to use their Internet connection). Jailbreaking an iPhone can unlock that functionality and allow other devices to use the iPhone's connection to the Internet. Jailbreaking is normally not supported by the manufacturer at all; in fact, jailbreaking typically voids the warranty on a device. Additionally, the manufacturer or service provider, if they detect that jailbreaking has taken place on the device, can prevent the device from connecting to their services. In some cases, the jailbreaking process fails, and the device is rendered non-operational. Usually this can be fixed by restoring the device completely using a backup; however, this also removes the jailbreaking software. Rendering a device non-operational due to jailbreaking is popularly called "bricking" the device. In rare occasions, bricking a device can be permanent and can't be fixed by restoring from a clean backup. As you can see from Figure 8-6, there is no short supply of jailbreaking apps, for both Android and iOS.

 TIP Although you may hear or read the terms used interchangeably, technically, "jailbreaking" applies to an Apple iOS device, and "rooting" applies to an Android device.

Rooting

"Rooting" is a term that is similar to jailbreaking. When an Android device is rooted, it means that the user now has full administrative access to the lower-level functionality of the device. This is useful in that it allows the user to perform functions on the device that they would not normally be able to, and access functions that may be prohibited by the device manufacturer. Again, as in the case of jailbreaking, this is done to install software that could otherwise be used on the device, or to unlock functionality from a device. Although none of the popular device vendors condone rooting, in most cases, since the device belongs to the user, the vendors really have no recourse against this practice.

Key Logging

Key logging is a process by which a piece of software, or even a piece of hardware, is used to capture and record the keystrokes sent to a device from the keyboard. Most Trojan malware has keystroke logging capabilities. This type of software is very stealthy, and the user doesn't even know that it's running on the device. The key logger will typically

Figure 8-6

A short list of jailbreaking apps for Android and iOS

capture a finite amount of data before sending it to the attacker. It also may be designed to send this data in encrypted form, making it more difficult to detect.

Hardware keystroke loggers are usually plugged into a port on the device, such as a USB port, or even in-line with the keyboard itself. They are designed to be small and unobtrusive, making them less noticeable. When plugged into a wired keyboard, they simply look like an extension of the keyboard connector that plugs into the device's port. Hardware keystroke loggers may or may not have the capability to send data over the network to the attacker; they often simply sit and record keystrokes until the attacker is able to surreptitiously retrieve the device.

Unsupported OS

Earlier, I mentioned how some rogue applications downloaded from not-so-legitimate sources may not run on the stock operating system that is installed on the device. This is because a lot of applications that are obtained from these sources may have malware on them or other configuration settings that the operating system would disable or not allow. Unfortunately, the solution some determined users find for this is to install a replacement operating system on the device, one that does allow uncontrolled access to hardware and root level permissions on the device. These operating systems are obviously not supported or condoned by the respective manufacturers. These unsupported operating systems may have been developed from hacking into the legitimate operating systems' source code, altering the way it allows access to its kernel from users and applications.

For users' personal devices that are used only by them to process only their data, this might be okay because they are the ones taking the risk. However, from an enterprise perspective (which is the focus of this book), this should definitely not be allowed on devices that connect to the enterprise infrastructure or are used to access corporate data. The risks of malware or data loss from the device are too great. Additionally, from a practical perspective, unsupported operating systems could be problematic to manage with the Mobile Device Management (MDM) infrastructure.

Hardware Risks

Now we've covered software risks, let's talk about some of the hardware risks that exist in the mobile device world. Hardware risks include more common ones, such as device loss or theft, but also include some of the less common risks. These include device cloning, and even replaced firmware that acts as malware or allows unauthorized root level access to the device. I cover these in the next few sections.

Device Cloning

Device cloning is a method that hackers use to basically create a duplicate of a smartphone, including its firmware configuration. This enables a hacker to impersonate the legitimate owner of the device, to include making and receiving phone calls, text messages, and accessing their data. Cloning the device also enables the hacker to eavesdrop on the legitimate user's communications. Typically, to clone a device, a hacker has to have almost the exact same model and specifications as the target device, as well as access to a device backup file, or access to sophisticated phone configuration and programming software.

Another aspect of cloning a device may be legitimate, in that sometimes an enterprise can maintain software-based clones of devices, in the form of virtual machines or devices, for testing applications and configuration changes. In some security infrastructures, these virtual devices can actually be linked to the actual physical ones, receiving configuration updates and synchronization from the physical device and vice versa. These virtual devices should be secured in a controlled area so they are not accessed by unauthorized persons.

Device Theft

Theft of mobile devices is an obvious threat against hardware. Theft is typically a low tech threat, carried out during a moment of opportunity, or sometimes planned. When a mobile device is stolen, it is subject to being used and examined by the thief, at least for short periods of time, until the theft has been detected, and the device is either reported as stolen, or the services turned off. One of the largest threats against a mobile device in the event of theft is unauthorized access of data. Unless the device is protected with a personal identification number or other authentication method and encrypted, a thief can retrieve any of the data stored on the device by the user, including contacts, phone numbers, and so on. As critical data such as usernames and passwords, credit card numbers, and other personal data is being increasingly stored on these mobile devices, this is a huge risk for the user. Frequently, users have email accounts or even bank accounts set up on the devices so they can remotely access these accounts without having to log in. This may allow the thief to easily access these accounts and use them for malicious purposes. The thief can also use stolen devices to gain information about the user, including where the user lives, the user's family members and friends, and the user's work. Sensitive data may even be posted to social media sites. In the event that a corporate-owned device is stolen, the thief can access sensitive or proprietary information, or even gain access to the corporate network through any network connections or email connections that are set up on the device.

Device Loss

Along the same lines as device theft, the simple act of losing a device can cause the same issues. Although not a malicious act like theft, losing a device can also mean losing data. Anyone finding the device may be tempted to go through it to view the data stored on it, and this can result in unauthorized access. Even if the device is not found by anyone, just its loss could mean a loss of data if critical data was stored on the device and not backed up anywhere.

EXAM TIP   In order to prevent data loss from device theft or loss, implement device encryption and remote wipe capabilities as a preventative step. Device location through GPS services, as well as remote lock and remote alerts can also be used to prevent data loss from devices that are stolen or lost.

Organizational Risks

In addition to software and hardware risks, there are also risks associated with the organization, usually from an operational and managerial perspective. Even these risks, when leveraged with other technical or physical vulnerabilities, can become attack vectors for malicious entities targeting a mobile device infrastructure, or at least, be impediments to the organization effectively managing and securing its infrastructure. These risks are those that the organization may incur due to policy, planning, or even implementation of the mobile device infrastructure, and include a variety of items such as BYOD implementation, personal device security, use of removable media, security, control and wiping of personal data, and the infiltration of unknown and unmanaged devices into the network. These are only a sampling of organizational risks, of course, and we will discuss those throughout the next several sections.

BYOD Ramifications

Any organization that makes the decision to allow users to bring their own personal devices into the network is doing so at some risk. The organization must balance the security requirements of the network infrastructure against the users' personal privacy and right of ownership over the device. The problem occurs when the device is used to access corporate data, which then may be stored or transmitted from the device. The company still owns that data, but the user owns the device. So the question is how much control the organization has over the users' personal device.

In some organizations, users must consent to allowing the organization access to their device and the ability to wipe the device remotely if it is lost or stolen. Users may also have to consent to the organization being able to control what type of activities the user is allowed to do on the device or what resources they are allowed access. For example, an organization that allows a user to bring their own device into work and connect to the corporate infrastructure may require the user to connect to the corporate proxy server and firewall, so the organization can control what content on the Internet the user accesses. The normal acceptable user agreements that employees are required to sign in order to use corporate computing assets may also be enforced on the users' personal device. Additionally, issues such as device security, antivirus, forensics, patching, and upgrades should be considered when implementing BYOD. This all brings up the real issue with BYOD, which is how much control the organization and the user each has over the device and the data that it stores and processes. Organizations must determine how much risk is involved with such an effort, and what mitigating factors can be applied to those risks. Users, likewise, must determine if being able to use corporate resources over their personal devices is worth the potential invasion of privacy and giving up control over their device to the organization.

 EXAM TIP The most important issues with BYOD are level of organizational versus personal user control, and privacy.

Securing Personal Devices

Device security is a responsibility of both the organization that owns a device and the users that use it every day in their work. Securing a mobile device is not dissimilar to securing other computing assets, such as PCs, for example. Laptops, tablets, and even smartphones can be similarly secured using some of the same practices and techniques used in traditional devices. As discussed, this includes mitigating risks, such as malware, device theft, wireless risks, and software risks. Obviously, some of these techniques have to be adapted to the nature of a mobile device. This means that the device is not always connected to the infrastructure in order to monitor it and update it, and the organization doesn't always have full control over the device, especially when the user has it in their possession away from the office. Additionally, some of these techniques require a mobile device infrastructure implemented in the organization that can be used to control mobile devices. Some also require specialized software and management devices on the network.

Use of Removable Media

The use of removable media has several different ramifications for any organization's mobile device security posture. On one hand, removable media can be used to secure a mobile device, in that corporate data can be stored only on removable media, which can then be removed and stored securely when the device leaves corporate property. This can help to ensure that corporate data never remains with the device. On the other hand, the organization must balance the use of removable media with the security risks involved as well. It is, of course, possible to steal corporate data using unapproved removable media. In addition to establishing a policy on removable media, the organization should also establish technical controls, including the data loss prevention controls discussed later in the chapter. The organization should also look at removable media encryption as way to protect corporate data. In addition to encryption, the organization can enforce only approved, corporate-issued media to be used in the device. As a last resort, the organization could also ban removable media so that the only place that corporate data resides is on the device itself, which can be remotely encrypted or wiped.

Wiping Personal Data

One security solution, and a potential security issue, is wiping personal data from mobile devices. If the device is corporate-owned, only a minimum of personal data should be stored on it. However, if it is a user-owned device, and is allowed to connect to the corporate network, there may be issues with wiping a user's personal property. The user may not have backed up her personal data, and wiping it would result in a significant data loss. On the other hand, if the device is lost, the organization would definitely want to remotely wipe the device in order to ensure that there was no unauthorized access to corporate data.

This is an issue that has to be closely examined before it is implemented, and it should be formally stated in the security policy for both users and the organization to adhere to. Users should be educated on the policy and cautioned about the importance of backing of personal data, as well as the risk of losing data in the event a remote wipe becomes necessary. Users should carefully consider the risks involved in using personal devices on the corporate network because ultimately, the organization has a responsibility to protect its information assets. One solution to this particular issue may be in using data containerization techniques, discussed later in the chapter.

Unknown Devices

One security risk that occurs frequently in corporate networks is the connection of unknown or unapproved personal devices to the corporate infrastructure. In most cases, this will be employees bringing a device from home and connecting it to the network. In a few instances, however, it may be a malicious user, or attacker connecting a rogue device to the network with the intent of stealing data or disrupting operations. Unknown devices should be carefully controlled using any of several techniques.

One such technique is to use a network access control (NAC) device that doesn't allow any device to connect to the corporate network unless it has gone through a quarantine process, which serves to identify and authenticate the device, determine what software is running on it, and limit its connectivity until the user obtains the proper permission to connect it to the corporate network. Another technique is the use of virtual LANs, which involves forcing unknown devices to connect to a quarantined VLAN, with no connectivity to the main network. Once the device is identified, it can be assigned to a less restrictive VLAN within the organization. Mutual authentication requirements are yet another technique that can be used to minimize unauthorized devices on the network. Authorized devices may have to use the corporate-issued PKI certificate, for example, to identify themselves and authenticate to the network or authentication server. These and other techniques can reduce the number of unknown and unauthorized devices that attempt to connect to the corporate infrastructure. In addition to technical measures, policies should be developed regarding the authorization process for introducing new devices into the network.

 EXAM TIP Understand the different options an organization has at its disposal to deal with unknown devices that attempt to connect to the corporate infrastructure. These include NAC, restricted VLANs, and certificate-based authentication.

Mitigation Strategies

Now that we've covered the different security risks that are inherent to not only mobile but also traditional networks, it's time to discuss some of the things you can do within your organizations to reduce these risks. Keep in mind that most of these risks can't be completely eliminated; that is the nature of risk, but the mitigations discussed in the upcoming sections can be used to reduce these risks to an acceptable, manageable level. The next several sections discuss technical and organizational controls, such as antivirus solutions, firewalls, policies, and procedures to help mitigate these risks.

Antivirus

Antivirus and anti-malware solutions are probably the best solutions an organization can implement in order to prevent malicious software from infiltrating and infecting the network, and the devices that connect to it. Typically, at the enterprise-level, centralized anti-malware infrastructure exists that provides protection to the entire network in the form of anti-malware servers that automatically update all devices with the latest anti-malware software and signature files. Obviously, this type of centrally managed solution would be ideal to cover all the different mobile devices and infrastructure devices in the enterprise network. However, because of the mobile nature of some of these devices, this isn't always practical to implement. In some cases, a hybrid solution of centrally managed anti-malware services, and decentralized antivirus solutions, used on a limited individual basis, may be the answer. Figure 8-7 shows an example of user-level antivirus software for an Android device.

 NOTE "Malware" is the general term used to describe many types of malicious code, and includes viruses, spyware, adware, Trojans, and worms. Most antivirus and anti-malware solutions include protections against most of these types of malware.

Figure 8-7
Antivirus app for Android

Any anti-malware solutions used should cover the widest possible range of malware threats, and should, to the best degree possible, cover the widest range of mobile devices used in the corporate network. In a heterogeneous infrastructure, because there may be different varieties of mobile devices using different operating systems and coming from different vendors, a one-size-fits-all anti-malware solution may not work. Different solutions may be necessary for the different devices present on a network, or, in some cases, different modules that cover specific types of devices may be available from the vendor to integrate into an enterprise-level anti-malware solution.

In any case, the important part of the enterprise-level anti-malware solution would be to deliver timely updates to the devices on a routine basis. Network access control solutions can ensure that when a device attempts to connect to the network, it is checked for the latest anti-malware signatures and updated as necessary before being allowed to connect to the network. In the case of user-managed solutions, when necessary, policy, network access control, and other technical solutions may be needed to ensure users are updating their own devices in a timely manner.

Software Firewalls

Software firewalls are typically installed on individual hosts, and as such, are normally used to protect the host itself from network-based threats. Now, while a network-based firewall also serves the same function, it's essentially used to protect the entire network from a wide variety of threats. Because different hosts may be running different applications or processing different types of data at varying sensitivity levels, the network firewall may not catch threats specific to what the host requires. A software firewall is typically installed into an existing operating system. There are firewalls that are integrated with the Windows operating system, for example, but there are also ones that exist as third-party products that can be installed. Linux also has firewall packages built into its operating system, as do most of the other OSes available. Unfortunately, there is a surprising lack of software firewalls for mobile devices in general, but mobile devices often rely on other security measures to compensate for this shortfall. One example of a software firewall for Android is shown in Figure 8-8.

Figure 8-8
An Android
firewall app

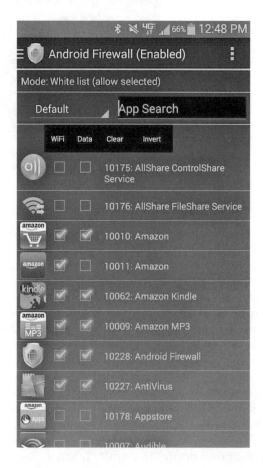

Most of the software firewall packages include basic rule elements that can be used to construct rules to filter specific traffic coming into the host. Many of these packages also include anti-malware solutions, and basic intrusion detection solutions. Some of these software firewall solutions are standalone and have to be configured and managed by the user, whereas some are enterprise-level solutions and can be centrally configured, updated, and managed by the systems administrator. Keep in mind that the software firewall packages work at the very basic level and can't possibly keep out every single network threat that the host is exposed to. But they do serve as a second line of defense for the host, and are part of any good layered defense-in-depth security design.

Access Levels

Access levels refer to the different types of access available to the user or a process. One of the first things an organization should do is determine the sensitivity levels for each type of data it processes. Based on the sensitivity levels, access can be assigned to different users or groups of users based upon their need to access data, the equipment that processes it, and the areas in which it resides. Frequently, several factors are used by an organization to determine access levels to sensitive data, equipment, or even work areas. These factors could be things such as security clearance, proper identification of the individual, and the need-to-know the individual may have. If an individual doesn't possess any one of these things, then the organization may deny him access to the area or computing asset. Policies regarding access should be formally developed and implemented by the organization. Procedures for granting access to sensitive areas and data should also be implemented and followed. Access levels for individuals should be documented, especially for highly sensitive data. Remember that an individual should have only the level of access she needs to perform her job, and no more.

Permissions

Permissions typically refer to technical access controls that are used to determine what an individual is authorized to do on a system or with data. Permissions typically apply to data objects, such as files, folders, printers, and so on. Permissions may be applied in different ways, depending upon the data itself, how it is accessed either on the local machine or across the network, and how the operating system itself implements access control. For example, in a Linux or Unix-based system, the basic permissions for a file or directory include read, write, and execute. In a Windows system, however, a wide variety of granular permissions can be used for both folders and files. Some of these permissions are specific to accessing the data locally while sitting at the host, and other sets of permissions are geared toward accessing the data across the network. Some Windows permissions include full control, modify, read, write, delete, and so on.

Permissions are assigned based upon the access control model in effect on the system. In a mandatory access control model, the system assigns permissions based upon the individual's security clearance and labels assigned to the data. In a discretionary access control model (the most common), the creator or owner of the asset has the discretionary power to assign permissions to anyone he or she chooses. In a role-based access model, permissions are assigned to specific roles rather than individuals. The access control model in use determines both who can assign permissions and how they can be assigned.

Permissions can be explicitly assigned, which ensures that an individual or group will be able to access a data object and perform the tasks they need. Permissions can also be denied explicitly, so that an individual will never get permission to the data object, regardless of other permissions they may have. Permissions should be documented and reviewed on a periodic basis to ensure that they are still required for an individual to perform their job.

Host-Based and Network-Based IDS/IPS

An intrusion detection system (IDS) is a device used to detect malicious traffic or unauthorized access. A host-based intrusion detection system is typically a software package that is installed on a device that protects the host itself. It can be installed as part of a combination package that includes a host-based firewall and anti-malware software as well. In the case of a network-based intrusion detection system, it is a device that is strategically placed at various points on the network in order to detect malicious network traffic and unauthorized network access. It can be used to detect hacking attempts or denial-of-service attacks. Some of these devices are also intrusion prevention systems (IPS), and when detecting malicious traffic or attacks, they reroute traffic or dynamically block traffic based upon port, protocol, or IP address. Most intrusion systems perform both functions in terms of detecting and preventing attacks.

IDS/IPS systems work in a couple of different ways. Some systems are signature-based, meaning that they are loaded with known attack signatures or rules describing known bad network traffic. The signature-based systems have to be periodically updated with current signatures and rule sets. Other systems are known as anomaly-based systems, and must learn the unique traffic patterns inherent to the network they are protecting. Usually an administrator will allow them to run in a "learning mode" for a while, where they do not actively detect or prevent malicious traffic; rather, they learn what kind of traffic is typical for the network. When actively detecting and preventing malicious traffic, they will take action or alert based upon traffic that does not meet known traffic patterns. For example, if there is an unusual amount of traffic at 3 a.m. on the network, where there normally isn't any at that time, the IDS/IPS will perform actions based upon its rule set, which may include blocking the traffic, rerouting it, or sending an alert on the traffic. Anomaly-based IDS/IPS systems usually must be adjusted or fine-tuned whenever changes to the normal network traffic patterns occur, such as installation of new network devices or services that may generate new traffic unknown to the IDS/IPS.

EXAM TIP Be familiar with the different types of IDS and IPS for the exam, to include host-based and network-based systems, as well as signature (rule-based) and anomalous (pattern-based) systems.

Application Sandboxing

Sandboxing is a method whereby applications are run in a restricted memory space and not allowed to interact with any other application or even certain hardware on the device. The operating system may have an abstraction layer that interacts with the application itself and transfers data to the device components as needed, serving as a sort of firewall that insulates the operating system and applications from each other. In addition to protecting the data that is exclusive to one application, including its files and configuration information, this arrangement also prevents potentially harmful actions from one application from interfering with another application or even the device itself. In a sandbox, all applications are firewalled from each other and typically can't share data or files unless the user or device allows them to. Most of the current operating systems prompt for the user's permission to allow app interaction with the device or other apps. All modern mobile device operating systems implement some form of sandboxing. Another form of sandboxing—called virtual sandboxing—can be used to create sandboxes from existing memory and storage space for apps or processes that the user chooses. In effect, this is user-level sandboxing, and uses third-party software instead of the operating system to create these virtual sandboxes.

Of course, apps that are not controlled using sandboxing may be dangerous by virtue of the fact that they often run with the full access permissions and rights of the user, so there is a potential for damage to data, other apps, or even the operating system when using them. Note that application sandboxing isn't a panacea for mobile device security; there are other measures both the enterprise and the user should take to ensure security between and with applications. For example, downloading a malicious application from an unapproved source can still cause data loss or device issues because of the actions the application may take and how it interacts with both the user and the device.

Data Containers

Containerization of data is a technique used to separate one class of data from another. There may be sensitive data that should not intermingle with other types of data so this data is separated into its own restricted area and cannot be accessed except by certain applications, users, and processes. The most commonly seen implementation of data containerization in mobile device infrastructures is the separation of corporate data from personal data. This is usually seen in an organization that allows BYOD in the mobile infrastructure. This helps solve problems related to corporate ownership versus private ownership of the device and the data it contains.

Data containers can be created to separate corporate and private data so that each can be managed separately in terms of access control, restrictions, encryption, and so on. This can be implemented on the device level, as well as in the larger infrastructure, such as mass storage areas that contain device backups. Containers are not only used to separate data, but also corporate and private applications as well, including their interactions with the separate data. Policies used in data containerization can also control and enforce restrictions on unauthorized applications and data use, to include copying and pasting between applications and email of corporate data to personal email accounts, for example.

One advantage to data containerization is the ability to selectively remove corporate data from the device at will while not harming the employee's personal data. Other advantages include the ability of the employee to do everything on one device without having to carry, maintain, and care for a corporate-owned device as well as their own device; choice of devices beyond what the enterprise offers; and most importantly, privacy from the corporate IT personnel and infrastructure. An MDM infrastructure used to secure BYOD implementations is required to effectively establish and maintain data containerization.

 EXAM TIP Remember to distinguish between sandboxing and data containerization. Although the two may seem similar, sandboxing is for applications, and data containerization separates two discrete types of data, usually personal and corporate.

Trusted Platform Modules

A trusted platform module (TPM) is a hardware device, typically implemented as a firmware chip on the mobile device board itself, which provides security functions. These functions are usually focused on cryptographic functions, which enable the user to encrypt the entire device itself, its storage, and even removable media. A TPM can also control boot-level authentication into the device prior to even loading the operating system. TPMs have been used for a few years now on laptops and have also made their way into other mobile devices such as tablets and smartphones. If a TPM's encryption and authentication functions are enabled, this can provide extra layers of security in the event the device is lost or stolen.

Unlike software security controls, it would be very difficult for an attacker to alter or compromise a particular TPM on a device. It usually requires very sophisticated equipment and physical possession of the device before it enters the market in order to compromise a TPM. This makes it a very secure alternative to software-based authentication and encryption measures. Additionally, the TPM or device can be controlled through the use of software on the device that communicates directly with the enterprise infrastructure. This enables the MDM infrastructure to load and change certificates, encryption keys, and so on remotely for the device.

Content Filtering

Content filtering is a method of restricting the types of data and files that can either enter into or exit from a device or network. Typically, content filtering is used to prevent users from downloading certain content, such as music files, video content, or even executable files that can be used to spread malware throughout the organization. The filtering is usually performed by a security gateway device, such as a firewall or proxy server, although it can also be restricted on the device level.

Content filtering can operate on a wide variety of parameters, including file extensions, sizes, application types, and so on. Content can also be filtered based upon the end device or even the user account involved. Additionally, filters can be placed upon the source of the content, effectively blocking certain domains, sites, and

IP addresses from delivering content to a device or network. As you will see later on in the discussion of data loss prevention, content filtering can also be used to prevent sensitive data from leaving the organization.

DLP

Data Loss Prevention, or DLP, as it is known in the industry, encompasses policies, techniques, and technologies used to prevent loss of an organization's sensitive data. This could include loss through data breaches, hacking, and even unauthorized exfiltration of data by insiders, whether their motivation is malicious or not. The techniques used by DLP are based upon monitoring and detecting potential data loss while the data is in use, being transmitted, or even in storage. DLP is normally a formalized program within large organizations.

DLP is a multilayer security effort. The techniques used in DLP include those typically used in normal information security, which include policy, user training, data sensitivity categorization, and so on. Other normal security measures, such as authentication, access control, and encryption are also used in DLP. There are other techniques used outside the normal security controls that include content filtering for traffic leaving the infrastructure, as well as detection algorithms designed to detect massive information removal from databases and other electronic stores. Metadata marking can also be used in DLP, as certain software and devices can recognize information included in a file's metadata, for example, which may indicate the file should not be passed beyond the infrastructure. Content filtering can be applied in DLP to make use of keyword searches, or filtering on data sensitivity labels, if they are applied in the file's metadata. Content filtering can also be used at the basic level to prevent certain files from exiting the infrastructure, such as encrypted or compressed files, or files exceeding a certain size.

DLP highly depends upon the ability of the organization to classify data in terms of sensitivity and identify that data in some way. This might be accomplished using sensitivity labels (as used in a mandatory access control model) and through the use of file metadata, as well as other access controls, such as additional authentication methods and encryption. Data use and handling controls should also be used for data deemed as protected under the DLP program. This could include the storage of sensitive data in restricted folders, databases, or media, as well as additional operational procedures for accessing and using this data. DLP can also make use of physical controls, in that highly sensitive data may only be stored and processed in certain areas and on certain systems that are not connected to the network, or have no external media connections enabled (such as optical media drives removed and USB ports disabled).

Device Hardening

One of the basic tenets of information security is the concept of defense-in-depth. Defense-in-depth essentially prescribes that the security of the system should not rely too much on any one measure or control. This is because, should that one control fail, the entire infrastructure may be at risk. In defense-in-depth, there are several layers of security at various points in the infrastructure, including at the perimeter, the device, hardware, software, and even user controls. With this in mind, device hardening is yet another layer of security that can be used to prevent malicious action or data loss.

Device hardening involves securing the device through configuration and software control. Mobile device hardening is actually no different from hardening standard desktop PCs or a server, as far as the concept goes. Like traditional devices, hardening an enterprise mobile device also involves locking down its configuration, controlling unauthorized access, allowing the user to perform only those actions that are necessary for them to do their jobs, and software control. How it is implemented may be slightly different because of the operating system, apps, and multiple functions involved.

Each of these functions should be looked at for hardening, including cellular telephone use, communications and Internet access, software installation, and user permissions. Unnecessary ports, protocols, and services should be removed or disabled, as should unneeded software. Vendor-approved security patches should also be installed after proper testing in the environment to ensure that the latest security vulnerabilities are mitigated. Antivirus software should also be kept up-to-date. Device hardening should be based upon security policy, which, of course, varies with each organization, as well as the degree of freedom allowed the user. There also may be different device hardening policies based upon whether it is an organizationally owned device versus a user-owned device. In any case, device hardening is one layer of the defense-in-depth strategy that should be used.

Some of the more common device-hardening techniques for mobile devices include requiring a strong passcode, setting the "autolock" feature on the device, and requiring encryption for the device and removable media. Additionally, keeping the device updated with the most current OS versions and patches is also strongly recommended. It should go without saying that rooting or jailbreaking shouldn't be done on the device. Specific to the different operating systems on the wide variety of available devices are hardening techniques published by vendors such as Apple and Google as well as the device vendors themselves. There are also various hardening techniques and checklists published by security researchers that can be helpful to both end users and the enterprise.

Physical Port Disabling

One other mitigation strategy, from a physical device perspective, is turning off physical ports on the device that should not be used. This might include infrared ports, USB ports, SD card slots, Bluetooth, and cameras as well. Disabling the physical use of these particular pieces of hardware and ports can help keep the device secure, especially if they are not needed or present security risks in the environment in which they are used. For example, while smartphones may be permitted in certain secure areas, their cameras and their ability to take pictures may not be. Disabling the camera itself might prevent data exfiltration from malicious insiders who may photograph sensitive equipment, areas, or documents. Disabling unused communication ports, such as infrared ports, for example, helps support the defense-in-depth concept by minimizing the attack surface, in conjunction with other security measures. Bluetooth should also be turned off if not in use to minimize the possibility of an attack. If external media is not permitted or is restricted, disabling USB ports and media slots should be considered.

Incident Response

All organizations suffer from adverse incidents from time to time. Sometimes we think about incidents as disasters or issues that affect our business in terms of ability to provide products and services. Some of these incidents are covered under business continuity and disaster preparedness programs, and those are not the focus of this discussion. For our purposes here, I'm going to be discussing security incidents. Security incidents include things such as hacking, data theft, unauthorized information access or modification, malware, and even equipment theft and sabotage.

Incident response is used to help an organization prepare for, respond to, and recover quickly from events that can harm the infrastructures, equipment, data, personnel, and business. Not only does incident response make sense from a sound business and security perspective, but it is also required as part of legal governance. Incident response is also considered necessary to fulfill the due diligence and due care responsibilities for the organization and help prevent, or at least reduce, legal liability.

The next few sections will discuss how an organization can prepare for incidents by developing policy, plans, and procedures, and preparing a team to respond to a variety of security incidents. While most of the things I'll discuss aren't particular to mobile infrastructures, there are some items that can directly apply to mobile device management.

Preparation

If an organization does not prepare for an incident, it likely will not be able to respond to it with any effectiveness. No one will know what to do in the event of an incident; no one will take the correct actions, and the actions that are taken may possibly even make the incident worse. That's why preparation is so important. Preparation involves establishing an incident response policy that directs risk management in terms of sudden, unexpected events that may occur, as well as the formation of an incident response team. Preparation also involves establishing an incident response plan and then testing that plan periodically to ensure that it works.

 TIP Inadequate preparation can cause a very poor incident response. Make sure the organization, or at least your part of it in the mobile device infrastructure, is adequately prepared and ready.

Policy

Policy is the first step in preparation. Most security professionals understand that unless policy is created and promulgated down to the organization, there's no structure in terms of determining whether something is acceptable or not, or there is no direction to perform any action or adhere to any governance. There are typically security policies that dictate what users can and can't do on the network, how backups are performed, how data is managed in terms of sensitivity, and how access and authorization work in the organization. Incident response is no different, in that a policy should be created that dictates what incident response is for the organization, how important it is, who is responsible for the incident response program, and so on. As with all security policies, and incident response policy does not tell you how something will be done, just that it *must* be done, why, who is responsible for it, and what requirements it must meet. Incident response policies are typically created and approved at top levels of management with consultation from lower-level supervision and even technical personnel. Once the policy is created and approved, it applies to the entire organization, and must be carried out.

Incident response policy should, at a minimum, include the types of events that the organization will respond to, who is in charge of the incident response program, and the priority and criticality of response, and it should direct that the incident response plan and team be formed. The policy will typically not contain specific procedures, team member names, and so on. Those are usually contained in the incident response plan itself.

EXAM TIP The organization's incident response policy drives the entire incident response program, and is directive in nature. It's at the very top in terms of governance in requirements. Everything else in the incident response program serves to support the policy.

Incident Response Plans

Incident response plans are created based upon the organization's commitment of resources to potential threats. The response plan should include procedures and actions for a wide variety of threats, including malware infections, critical service outage, data loss, malicious insider actions, as well as hacking attempts. Some of these threats will have common actions and procedures, such as notification and escalation to key personnel, enacting different security protocols, and so on. Others may require very specific actions to be performed, depending upon the threat or incident.

An incident response plan should dictate who the key incident response team members are, as well as their roles during an incident. This could include the incident team leader, area supervisors, technicians, and even administrative personnel. Each role should have defined duties and actions, as well as procedures for each that the individuals must perform from the outset of the incident. These duties could include notification of the chain of command, first responder and triage duties, containment duties, and coordination between internal organizational responders and external ones, if necessary. There are even duties that may be required to interface with law enforcement personnel as well as media outlets if the incident warrants it due to severity or scale.

Let's say your organization is under attack from a hacker. Obviously, the technicians who discover the attack by monitoring the intrusion detection systems or during routine connectivity or outage problem troubleshooting must notify the chain of command and initiate incident response procedures. They would normally notify a designated person, such as the security manager or incident response team leader, who would activate the incident response team. While the team is assembling, the same technicians may have the responsibility to either stop the attack by disconnecting systems from the network and the outside world, or shutting down critical servers to prevent data theft or destruction. When the incident response team arrives, the incident team lead must make sure everyone immediately has the resources they need to contain the incident and respond to it, and begin to do so according to the procedures laid out in the plan. The same person may have to notify law enforcement in some cases, and coordinate with media in the event the breach becomes public knowledge. As you can see, there are a great many details in the incident response plan that relate to duties and responsibilities.

 EXAM TIP   Remember that, while policy drives the incident response program in an organization, the incident response plan supports the policy. Policy dictates that something will be done, who is responsible for it, and why it will be done, and the plan contains the procedures that dictate what will be done, how it will be done, who will do it, and when.

Testing Incident Response Plans

Incident response plans shouldn't simply sit on the shelf waiting for the day they will be needed. Once you have created the plan and have buy-in from management, as well as resources dedicated to it, people have to be trained on it. The incident response plan should be promulgated down to everyone that has any part of it, regardless of organizational level or responsibility. Periodic training is required to make sure everyone stays familiar with the plan and what their responsibilities are during an incident. Some personnel may require specific training if their duties are very critical or highly technical. Others may need to be trained if their incident response duties are not part of their normal day-to-day duties.

Beyond training, the plan has to be tested occasionally. Unless people periodically perform some of the actions they may be required to do during an incident response, they will forget, training will get stale, and they will not be prepared when an actual incident occurs. There are several ways you can test incident response planning in your organization. These range from simple document reviews to full-blown simulations and exercises. For a simple document review, the major players get together periodically to review the plan and update it for changes. Responsibilities, actions, timelines, and key personnel should be reviewed to ensure that everyone understands what they're required to do, how they must do it, and when they have to do it.

Another type of test may be a walk-through type of test. In this test, key personnel actually walk through the plan step-by-step, going over key responsibilities and actions that have to take place during an incident. There may be a simulated incident scenario for the response team to react to. During this type of test, each person in turn may discuss what they're going to do as the incident response team leader offers different incident response scenarios or questions. Usually, during this type of test, personnel don't actually take the actions they would in an actual response, because of the time and possible expense involved.

An exercise or simulation is designed to test the response plan under actual conditions that the team may encounter during an incident. Because everything is simulated, no actual data or systems are affected, but test data and devices used for training may be used to simulate outages, malware infections, hacking attempts, and so on. An exercise can be very elaborate or very simplistic, depending upon the organization's resources and commitment, as well as the scale of the incident being exercised. The response team would actually have to perform the actions they've been trained to accomplish during an incident, to include triage, incident containment, network and service monitoring and restoration, notification procedures, forensics, and so on. This type of test can show deficiencies in training, resources, and procedures because the actions are actually performed instead of being simply discussed.

 EXAM TIP   Understand the types of testing that can be performed on the incident response plan, to include documentation review, walk-through, and simulations and exercises.

Incident response plans should be reviewed, updated, and exercised periodically. You should aim to do this at least once a year to start with, but changes in the organization and its infrastructure may warrant doing this more often. For example, if a new system is brought online, this should be reflected in the incident response plan. Likewise, if a key person on the incident response team leaves the organization, someone new should be appointed, trained, and documented in the plan. When any changes or exercises happen, they should be documented, to include changes made in the plan, results of exercises, and areas of improvement.

 EXAM TIP   The incident response policy, plans, and procedures should be reviewed and updated at least annually, but more often if the organization, governance, and its environment requires it.

Outsourcing Considerations

One other item you should be aware of in incident response is how you will prepare for it when using outside service providers. Organizations frequently use outside providers for a wide variety of services, including cloud storage services, infrastructure services, and so on. It is critical that any outsourced services be included in your incident response planning. You should ensure that service level agreements (SLAs) are written such that incident response is included in them, and the third-party organization knows what its responsibilities are with regards to incident reporting and response. The SLAs should specify how incidents are dealt with, how your organization is notified of incidents, and what steps the provider will take to both prevent and respond to them. Incidents you may be concerned about include service interruptions and outages, and especially security incidents.

Response

Now that we've discussed the planning part of incident response in-depth, it's time to look at the actual response and how it's carried out. If the organization has a good, thorough plan in place, the response should go fairly well. Understand, however, that it may be impossible to plan for every single small thing that can go wrong during an incident. This means that the plan must be solid in terms of procedures, yet flexible enough to make changes based upon unexpected conditions when they occur. If the team has also been adequately trained, and the plan periodically exercised, these unexpected events can be kept to a minimum. The next few sections describe the basic steps of incident response, and can be used to develop the organizational incident response plan and any checklists or work aids. Pay special attention to these steps, which include identifying the incident, determining and performing policy-based response, and reporting the incident, which may include, escalating the incident, documenting it, and capturing logs.

Incident Identification

The first part of the response is the identification of the incident. This may occur several ways, from a customer or user that notices something unusual on the network, or through an alert from a security device, such as an intrusion prevention system. It can also come from outside the organization, such as the upstream service provider, or even a law enforcement agency in the event of something like a large-scale cyber-attack. You can even discover an incident through normal troubleshooting of what seems to be at first an innocuous network problem. In any case, once the incident has been identified as a security-related incident, the incident response process should be activated. The incident response plan should identify who gets notified based upon different events that occur on the network and whether the incident response team should be activated or not. It should also identify immediate actions that should be taken as part of triage to determine the nature and scope of the incident, as well as any first response actions that should occur.

Determine and Perform Policy-Based Response

The incident response plan, as part of the policy it supports, should drive the response effort. When an incident is identified, the key actions should be notification, triage, first response, and sometimes even containment. Containing an incident may require some time in the involvement of the entire response team, but the initial actions taken to contain possible security incidents are critical in terms of data loss prevention or compromise.

Personnel from the incident response team, or any other required area, must be on hand to review logs, monitor intrusion detection systems and firewalls, be ready to shut down services or disconnect them from the outside world if necessary, contain malware, or backup/restore data immediately if necessary. While many security actions will be common to any type of threat, the specific incident that's occurring will dictate what actions are taken. For example, if the incident results from an insider threat, such as a violation of policy, data exfiltration, introduction of malware, or equipment theft or destruction, the team should respond accordingly by isolating the segment of the networking equipment affected by a malicious action. If the incident is caused by an outsider threat, such as an external hacker, the team should consider whether or not to shut off services from the outside world and coordinate extensively with the upstream service provider to stop the attack, preserve evidence, and investigate further.

From a mobile infrastructure perspective, incident response actions may mean remotely wiping a user's device if it has been lost or stolen, or at least deactivating it if an unauthorized person is using it. It could also mean locking the account of the user to prevent an unauthorized person from attempting to infiltrate the network through the mobile device, and blocking the device completely from connecting to the network. Wireless access points may also have to be shut down or reconfigured in the event they have been compromised through attack or infiltration. Each incident will have its own particular measures of response that must be accomplished as quickly as possible.

Reporting Incidents

Once the incident has been responded to and contained, the team should assess the damage to the infrastructure and data, and report it. This may be an ongoing effort while the response is still occurring, or after the incident has concluded. In any case, the response effort may have to be escalated as well if the team cannot handle it due to technical ability or scope of authority, or if it extends to outside the company's infrastructure, such as with an Internet service provider. The incident also may have to be escalated to law enforcement officials if it affects breaches of privacy, violations of the law, or certain protected systems, such as banking, medical, and interstate commerce systems. Escalation should be a determination that is made by the incident response team lead, as well as upper management. The incident response plan should include thresholds or specific scenarios leading to incident escalation.

 CAUTION Understand when you should escalate an incident, and under what circumstances. Failing to escalate or report a serious incident, especially to other affected organizations or law enforcement officials, can result in legal action against the organization!

It's important that the incident and all the facts surrounding it be documented as quickly and as accurately as possible. This would include information about how the incident was discovered, the actions the first responders took, the actions of the incident response team, and what the ongoing status of the incident is if it has not been resolved. Documentation will be of critical assistance in the event the incident is escalated, and may be required by governance or law.

Documentation surrounding the event could include artifacts such as security and system log files, network traffic captures, and even configuration files or details from various systems. It also may include statements from customers, users, and incident response team members regarding the incident, what was observed, and what actions were performed by each. In the event of a physical incident, such as equipment theft or destruction, even security camera video footage can be included as documentation.

In any case, anything that helps to add to an understanding of the incident should be included.

 EXAM TIP   Remember the incident response steps and their order, as you may need to know these for the exam.

Chapter Review

This chapter discussed security threats and risks to mobile device infrastructures, as well as possible solutions and mitigations for those threats. We categorized those risks in terms of wireless risks, software risks, hardware risks, and organizational risks. Wireless risks are those inherent to the physically open nature of the medium that wireless uses. These include threats from attackers against wireless infrastructure, such as rogue access points and encryption mechanisms. It also includes attacks using spoofed cellular infrastructure equipment. Wireless risks include denial-of-service attacks as well as man-in-the-middle attacks. This chapter also discussed wardriving, which is an older practice used to drive through an area looking for unsecure wireless access points, as well as warchalking, another older practice used to identify wireless access points to others.

Software risks include threats stemming from using apps from unapproved sources that may not have been vetted from a security standpoint. These apps could contain malware, or even attempt to gain root level access to the user's device. Rogue apps could also be used to steal a user's private data from the device and send it to a third party. I discussed the different types of malware that can be used in software attacks, including Trojans, viruses, spyware, and worms. The chapter also looked at unapproved user actions that can be taken against the device, such as jailbreaking and rooting. Usually, these actions are taken in order to run unauthorized software on the device, or get root level access to it. You also looked at other threats, such as key logging, which involves recording a user's keystrokes and device actions in the hopes of capturing user credentials or other sensitive data. Finally, the chapter discussed the ramifications of unsupported operating systems on mobile devices, which can lead a manufacturer to void its warranty, and allow uncontrolled root access to the device.

In terms of hardware risks, the chapter covered the common ones, of course, such as device loss or theft, and some of the actions that can be taken by the thief to compromise the device and retrieve information from it. I also discussed more advanced hardware threats, such as device cloning and unauthorized firmware. Device cloning can allow an attacker to not only intercept the user's communications, but also to actually use the device and masquerade as the user.

The last group of risks I discussed are those taken by the organization itself. Some of these risks are from a policy or infrastructure perspective. For example, the BYOD initiative that many organizations are beginning is a risk to the organization in terms of balancing the security of the enterprise against the needs of the user. This opens the debate on how to secure these personal devices in order to protect the organization's data while affording the user the privacy and control that comes with device ownership. Some of the security controls available to the organization include controlling removable media, as well as wiping personal data in the event the device is lost or stolen. Both of these are touchy subjects that the organization has to examine in formulating BYOD policy.

Another organizational risk comes from unknown devices that attempt to connect to the organization's network infrastructure. These can be innocent devices brought in by employee, or even a device that's introduced into the network by a malicious entity. In either case, they pose a threat to the organization because of the possibility of introducing malware into the network, or exfiltration of data outside the organization. There are several ways to control unknown devices, including the use of certificate-based authentication, VLANs, and network access services.

After discussing the various threats and risks to the mobile infrastructure, this chapter examined various mitigation strategies that can be employed to counter some of these threats. From a software perspective, we looked at antivirus and anti-malware solutions and how they can be employed at the enterprise-level. The chapter also discussed application sandboxing, which can help protect the device application software from other apps. At the host level, I discussed both software firewalls and host-based intrusion detection systems. TPMs are also used at the host level to provide hardware-based authentication and encryption mechanisms. They can protect devices from misuse, data loss, or data theft by encrypting the devices and requiring authentication prior to loading the OS. I also discussed other methods used to secure a device, including disabling physical device ports, such as infrared and USB ports, disabling hardware devices such as cameras, and hardening the device through configuration and software controls.

At the infrastructure level, I discussed network-based intrusion detection and prevention systems, and other controls that can help protect the enterprise. These included techniques such as content filtering, data loss prevention, and data containerization. Additionally, the chapter covered two essentials in infrastructure security: access levels and permissions.

The chapter also covered incident response from a preparation and execution perspective. Incident response starts with policy, which dictates what must be done in terms of incident response, who is responsible for it, and why. Incident response plans support the policy and contain the procedures and actions that must be carried out during the incident. The plan also should contain information about the incident response team members' roles and responsibilities. The plan should be as detailed as possible to account for a wide variety of security incidents, including malware infection, hacking, malicious insiders, equipment theft or destruction, and so on. Additionally, the chapter also covered testing the incident response plan, and covered the different types of tests you can perform on the plan to ensure it works as intended.

Executing the incident response plan involves quickly identifying an incident and immediately notifying the responsible organizational and incident response team leadership, who will make a determination about how to scale the response and activate the incident response team. Team members must perform triage to determine the scale and scope of the incident, and begin containment as soon as possible to minimize the damage to organizational data and equipment. Team members will have specific actions they must perform based upon the type of incident they are experiencing.

If the scope and scale of the incident is beyond the knowledge, experience, expertise, and ability of the incident response team, it may have to be escalated to higher levels and involve more of the organization's personnel. Additionally, if the incident extends beyond the organization's infrastructure, it will have to be escalated as well to the appropriate external personnel. This could be a service provider's organization or other external entities. It also may have to be escalated to law enforcement agencies due to the nature of the incident, especially if it involves a breach of privacy, violation of law or attack on a legally protected computer system.

I also discussed the importance of documentation and how it is critical to document and report the incident. Documentation includes reports from customers, users, and any others involved in the incident. Observations and actions, especially from those on the incident response team, are important in helping understand the nature of the incident, how it occurred, the damage it caused, and how to investigate it. Documentation also includes security camera video footage, configuration files, and system and security logs.

Questions

1. What type of attack involves the use of sophisticated equipment to intercept cellular phone signals?

    A. Tower spoofing

    B. Wardriving

    C. Warchalking

    D. Rogue access point

2. You suspect that someone has broken into encrypted communications between a remote user and the corporate network for the purposes of intercepting sensitive data. What type of attack is this called?

    A. Cracking

    B. Man-in-the-middle

    C. Spoofing

    D. DoS

3. Which of the following vendors use a single-tier, monolithic model for the device, operating system, and app development? (Choose two.)

    A. Apple

    B. Microsoft

    C. BlackBerry

    D. Google

4. All of the following can result from jailbreaking a mobile device, EXCEPT:

   A. "Bricking" the device

   B. Root-level access

   C. Unauthorized functionality

   D. Replacement device from the manufacturer after it is rendered unusable

5. Which of the following types of malware can self-replicate across a network?

   A. Virus

   B. Trojan

   C. Worm

   D. Spyware

6. A hacker is able to make phone calls and send text messages from a company cellular phone number, seemingly at will. The legitimate device user insists that she did not make the calls or send the text messages, and she has her smartphone on her at all times. What is one likely explanation for this scenario?

   A. The device has been rooted.

   B. The device has been cloned.

   C. The device has been jailbroken.

   D. The device has a virus.

7. Which of the following is the key issue in the BYOD paradigm?

   A. Users making unauthorized phone calls

   B. User control and privacy

   C. Use of non-approved apps

   D. Bill sharing between the company and the user

8. Which of the following techniques could be used to deal with unknown user devices that connect to the corporate infrastructure? (Choose two.)

   A. Device registration

   B. Encryption

   C. NAC quarantine

   D. Restricted VLANs for unknown devices

9. Which type of security measure can be installed on a host's operating system to filter potentially malicious traffic?

   A. Hardware firewall

   B. Intrusion detection system

   C. Proxy server

   D. Software firewall

10. What options are available to an organization to deal with the possibility of data loss on a lost or stolen device? (Choose two.)

    A. Remote device wiping

    B. Reporting it to the police

    C. Offering a reward

    D. Device encryption

11. Which of the following is a security mitigation used to contain applications in their own restricted memory space?

    A. Data containerization

    B. Sandboxing

    C. Virtual device

    D. Jailbreaking

12. Which of the following can be used to encrypt a device and require authentication to access the device even before the operating system loads?

    A. Data containers

    B. Sandbox

    C. Certificate-based authentication

    D. Trusted Platform Module

13. Which of the following does the concept of having separate data containers on a mobile device apply to? (Choose two.)

    A. Malware

    B. Trusted software

    C. Personal data

    D. Corporate data

14. You have been directed by corporate management to prevent certain types of files from entering the network, including mobile devices. Which of the following would help accomplish this?

    A. DLP

    B. Content filtering

    C. Port blocking

    D. Sandboxing

15. You have been directed to implement a Data Loss Prevention (DLP) program in the infrastructure. All of the following are elements of DLP you should implement, EXCEPT:

    A. Anonymous access

    B. Content filtering

    C. Information removal detection algorithms

    D. Metadata marking

16. All of the following are typical device hardening best practices, EXCEPT:

    A. Updating antivirus software

    B. Permitting users to be administrators on the device

    C. Locking down the device configuration

    D. Restrictive permissions

17. Which of the following hardware devices should be disabled on a device if not used? (Choose two.)

    A. Audio jack

    B. Bluetooth

    C. Camera

    D. TPM

18. Which of the following types of Intrusion Detection Systems must learn the network traffic patterns?

    A. Network-based IPS

    B. Host-based IDS

    C. Anomaly-based IDS

    D. Signature-based IPS

19. Your supervisor would like for you to investigate an incident involving unauthorized access of an organizationally-owned workstation in a secure area. You wish to collect all of the relevant incident-related documentation for your investigation. All of the following are considered valid incident documentation, EXCEPT:

    A. Security video

    B. System logs

    C. User statements

    D. Security advisories

20. Your organization has not previously developed an incident response capability, and must now do so in order to meet compliance requirements with government regulations. Your company's CEO has directed that you should immediately write incident response procedures, but you state to her that corporate-level governance must be developed first. Which of the following is the primary source for incident response governance in an organization?

    A. Incident response plan

    B. Incident response policy

    C. Security logs

    D. Incident response procedures

Answers

1. **A.** Tower spoofing involves the use of sophisticated equipment to intercept cellular signals, causing a cellular device to be fooled into connecting to the hacker's equipment instead of the carrier's.

2. **B.** A man-in-the-middle attack can be used to break encryption schemes and intercept communications between two parties.

3. **A, C.** Both Apple and BlackBerry use this model in that they control all aspects of the device, including the operating system and app development. Microsoft and Google both use multi-tiered models.

4. **D.** Jailbreaking normally violates a manufacturer's warranty so manufacturers typically will not replace the device if it is rendered unusable due to jailbreaking.

5. **C.** Once they have infected the network, worms can self-replicate across the network without any user intervention, but the other forms of malware listed cannot.

6. **B.** The device has been cloned, and the hacker is using a cloned phone to make calls and send text messages.

7. **B.** The extent of a user's control over their own device, as well as privacy considerations, is a key issue to consider when implementing BYOD programs.

8. **C, D.** Using either a network access control device to quarantine an unknown device, or using restricted VLANs are two ways of dealing with unknown devices that may try to connect to the corporate network.

9. **D.** A software firewall is an effective host-based security measure to prevent potentially malicious traffic from entering the host.

10. **A, D.** Two preventative measures to help avoid data loss on a lost or stolen device are to enable remote wiping software and enabling device encryption.

11. **B.** Sandboxing is used to keep apps running in a separate memory space so they do not interfere with each other's data or configuration.

12. **D.** A Trusted Platform Module (TPM) is a hardware device or firmware chip that can encrypt an entire device and require authentication to access the device even before the OS loads.

13. **C, D.** Data containerization is used to separate personal data from corporate data in a BYOD scenario.

14. **B.** Content filtering can restrict files entering the network by file type, size, extension, and other parameters.

15. **A.** Anonymous access is not an element of DLP, and would not help in preventing unauthorized data exfiltration from the organization.

16. **B.** Permitting users to be administrators on the device is typically not a desired practice when hardening a device.

17. **B, C.** Bluetooth and device cameras should be disabled if not in use, especially in secure areas.

18. **C.** An anomaly-based IDS must establish a traffic pattern baseline by "learning" the normal traffic patterns for the network.

19. **D.** Security advisories may have helped to prevent an incident, but are usually not relevant documentation concerning the actual incident itself.

20. **B.** The incident response policy is the primary source of governance within an organization for the incident response program and its execution.

Mobile Security Technologies

In this chapter, you will

- Learn about encryption methods used in mobile devices to secure data in-transit and at-rest
- Explore access control methods used on mobile devices

The last chapter discussed some of the security risks inherent to enterprise mobile device infrastructures. This chapter covers some of the different security technologies that you can use to help mitigate these risks. Many of the technologies explored here involve using encryption in various ways. This includes encrypting data both at rest (in storage) and in transit (during transmission or reception). You'll be examining various encryption algorithms used to protect data, as well as the applications of encryption, such as full disk encryption, file encryption, folder encryption, and even encrypting removable media.

You will also take a look at various access control concepts, including authentication, and using PKI. The CompTIA Mobility+ exam objectives covered in this chapter are 4.1, "Identify various encryption methods for securing mobile environments," and 4.2, "Configure access control on the mobile device using best practices." Covering these two objectives will help solidify your knowledge of mobile device security, and help prepare you for further discussion in Chapter 11 when I discuss monitoring and troubleshooting security issues.

Encryption

I have touched on encryption earlier in various parts of the book, but it helps to have a little bit of a quick review before moving into some of the more complex encryption techniques and algorithms. Recall that encryption basically allows you to convert a readable text message or data into an encrypted form, which is generally undecipherable unless the process is reversed. Before we begin to discuss the specifics of the mechanics of encryption methods, it's helpful to go over a brief refresher of encryption basics.

Encryption Basics

The goal of encryption is to protect data confidentiality, as well as (in some cases) data integrity. Encryption involves two important parts, the algorithm and the key. The *algorithm* is the mathematical construct or theory that is used to manipulate the human readable data (called *plaintext*) into something that is unreadable (called *ciphertext*). These mathematical algorithms are generally well known and have been tested for strengths and weaknesses. Algorithms are also publicly known to ensure compatibility between applications and systems. The *key* is the portion of the encryption process that is kept secret or unknown. The key (also called the *cryptovariable*) is the unknown or variable piece that introduces uniqueness and secrecy into the process of encryption that the algorithm performs. Combining the algorithm and the key together, along with the plaintext data, results in converting plaintext to ciphertext, called *encryption*. The reverse of this operation is called *decryption*.

Encryption keys, which can come in the form of PINs, passwords, passphrases, or even electronic keys stored in digital certificates, should be constructed so as to be cryptographically strong. This means that, in general, keys should be complex and be of sufficient length as to resist attempts to crack them. *Key space* contributes to complexity in that the more possibilities exist for characters of the key, the stronger the key is. For example, a key space that includes numerals 0 through 9 is not really strong, simply because there are only ten possibilities for a particular character. Even if the key is eight characters long, this is still *only* 100 million possibilities! This may sound like a lot of possible keys, but most modern computers and key cracking programs could go through this key space in very little time. A complex key would use a large key space, such as numerals 0 through 9, lowercase alphabetic characters *a* through *z*, uppercase alphabetic characters *A* through *Z*, and special characters, such as !, %, #, &, (,), and so on. Large key space, combined with key length, ensures a cryptographically strong key.

Algorithms can be broken up into two major categories, *symmetric* algorithms and *asymmetric* algorithms. Symmetric algorithms involve the use of only one key, which must be shared between any party that wishes to encrypt and decrypt communications. One problem with symmetric cryptography is that this shared key, sometimes called a *secret key*, must be shared securely and not intercepted by any party that shouldn't participate in the communications session. Another problem in symmetric cryptography is that of scalability. The more parties you wish to securely communicate with separately, the more keys you must maintain in order to keep your communications with each and every one of them separate from the others.

In asymmetric cryptography (also called *public key cryptography*), a pair of two keys is used for the encryption and decryption process. One key is a public key that anyone can have access to, and the other key in the pair is kept private or confidential by the individual that owns the key pair. Because anyone can have the public key, key exchange is typically not a problem. I discuss asymmetric keys and their application in cryptography later on in the chapter in the discussion of public key infrastructure.

One other way to classify encryption algorithms is to categorize them as *block* or *streaming* algorithms. Block algorithms encrypt plaintext in defined chunks or sizes. For example, an algorithm may encrypt plaintext in 64-bit or 128-bit block sizes. A block algorithm encrypts a specified chunk of text, and then the next chunk, and so on until it encrypts the entire message. A streaming algorithm encrypts only one bit at a time of plaintext. The advantage to this is that it is faster than block ciphers, but a disadvantage is that streaming ciphers are typically easier to crack. Most of the algorithms that you will see in everyday cryptography are block algorithms, and in fact, only one streaming algorithm is covered in this chapter. I'll point out the streaming algorithm when we come to it in the upcoming sections.

Encryption can be used for a wide variety of functions in information security; it is primarily used to ensure data confidentiality and data integrity. The application of encryption is found in processes such as hashing, drive and folder encryption, certificate-based authentication, and other applications. The next sections discuss the application of encryption in protecting data, both at rest and in transit.

 EXAM TIP Know the definitions of the basic encryption terms for the exam, such as *algorithm*, *key*, *symmetric*, and *asymmetric cryptography*.

Data-at-Rest

The term "data-at-rest" refers to data that is stored on some type of media, versus being transmitted, received, or in the midst of being processed by a device and residing in system RAM. Data in storage requires protections that are often provided by encryption technologies, as well as other types of access control. This section looks at some of the different encryption algorithms that are used to protect the confidentiality and integrity of data in storage.

DES

The Data Encryption Standard (DES) is an older block cipher encryption algorithm (implemented in 1977) that has, by and large, been replaced by other more secure algorithms. DES is based upon the Lucifer encryption algorithm developed by Harst Feistal. While Lucifer originally had a 128-bit key size, in its implementation as DES, it uses a 64-bit key; however, eight bits are used for computational overhead (as parity bits), resulting in a 56-bit key.

DES operates in five *modes*. These modes determine how plaintext is input and manipulated to produce the resulting ciphertext. You don't necessarily have to know how each mode works for the exam, but it's useful to know what they are called. Additionally, other encryption algorithms use these and other similar modes for their operations, so it's helpful to understand what they are. These modes operate on plaintext by manipulating it in specified size blocks of data (in the case of block ciphers), and introduce keys (variables) into the process to encrypt the data. The mode may also involve performing certain other mathematical functions on the data to more thoroughly encrypt it, sometimes repeating operations over several *rounds* (iterations). The modes also may manipulate the data and keys in different ways to produce stronger encryption.

The first mode is Electronic Codebook mode (ECB). This mode operates on plaintext blocks of 64 bits or less and is the simplest of the five DES modes. In this mode, identical 64-bit (or less) blocks of plaintext will always produce the same ciphertext. Cipher Block Chaining mode (CBC) is more complex than ECB, and stronger because each block of plaintext that is input will produce a different piece of ciphertext, even if the input blocks are the same. CBC introduces an initialization vector (IV) as well as the XOR (exclusive or) mathematical function into the process. When the first block of plaintext is converted to ciphertext, that ciphertext is XOR'd and used as an IV input to the second block of plaintext. Each resulting block of ciphertext is chained with the next block of plaintext. When decrypting, this process is reversed, and the recipient needs only the original initialization vector used to encrypt the first block of plaintext.

In Cipher Feedback mode (CFB), the plaintext input is divided into different bit-sized segments (1, 8, 64, or 128-bit sizes) and works similarly to CBC mode in that the resulting ciphertext (also of those corresponding bit sizes) is fed back into the encryption process as an IV. Because of the variable way CFB mode works, it seeks to emulate a streaming cipher operation, although DES is a block cipher. Output Feedback mode (OFB) is also very similar to CFB mode, except that it uses 64-bit IVs that are fed back into the encryption process for each subsequent block of plaintext to be encrypted. The last mode I'll discuss is Counter mode (CTR). Counter mode is typically used when DES requires faster operations. In Counter mode, a 64-bit random block of data is used as the first initialization vector. It is different for every subsequent block of plaintext, and is usually incremented by a predefined number, or counter (usually 1) for each new block of plaintext. Understand that I could spend several chapters on how some of these modes and encryption algorithms work, but anything beyond a definition-level of understanding is beyond the scope of this book and the exam.

 EXAM TIP You do not have to know how each of these modes works for the exam, but you should be familiar at least with their names and basic functions.

3DES

Triple DES (unofficially abbreviated as 3DES, but also seen in official documents as TDEA) is another version of DES that uses larger key sizes. It was originally created in attempt to mitigate some of the weaknesses of the original DES specification. It doesn't really have any newer or more complex features; it simple repeats DES's encryption three times for every block of plaintext, respective of mode, albeit in a sometimes different order. Because it essentially uses three 56-bit DES keys, it is commonly considered to have a 168-bit key strength. Although largely replaced by other more secure algorithms, 3DES can still be found occasionally in some technologies. Curiously enough, some technologies still use 3DES for encryption, including those used to encrypt copyrighted DVDs.

Twofish

Twofish was developed by famed cryptographer Bruce Schneier, and was a new version of his Blowfish encryption algorithm. It operates using variable key sizes (128-, 192-, and 256-bit keys) on 128-bit blocks. It was one of the five finalists for the Advanced Encryption Standard competition sponsored by the National Institute for Standards and Technology (NIST). This competition was designed to develop and establish a new national encryption standard. Although it did not win the competition, Twofish is a very strong algorithm that is still used in some software and hardware implementations.

AES

The Advanced Encryption Standard (AES) is the algorithm that was selected from among five finalist entries to become the national encryption algorithm standard for the United States, based upon a competition sponsored by NIST in 2001. The algorithm that won the competition and was selected to become AES was the Rijndael (pronounced as *rain-doll*) cipher, and it was selected from among four other competitors, including Twofish, Serpent, RC6, and MARS. Rijndael uses block size of 128-bits, but can use three different key sizes (128-, 192-, and 256-bit).

AES is a symmetric block cipher, and uses 10 rounds (iterations or repetitions) for 128-bit keys, and 12 and 14 rounds for 192- and 256-bit keys, respectively. AES has been certified for use in all U.S. Government applications and confidentiality levels, through Top Secret, with certain stipulations. AES is also used, with the CCMP mode described later on, in the IEEE 802.11i (WPA2) standard wireless encryption protocol. While some theoretical direct attacks on AES have been proposed, most effective attacks are likely to be *side-channel* attacks, which don't directly attack the algorithm itself; they attack the underlying implementation of the algorithm in the system.

ECC

Elliptic Curve Cryptography (ECC) is popularly used on mobile devices, smartcards, and tokens, because it requires less computing power than typical encryption algorithms. It requires shorter keys because of the way its discrete logarithmic algorithms work, so this allows for faster operation using less memory and CPU power. ECC algorithms are also known for their speed and encryption strength.

 EXAM TIP You may not be asked to know in-depth information regarding these and other algorithms, but you should be familiar with their names and basic information about them, including block and key sizes, and where they are implemented.

Hashing

One interesting use of cryptography is technically not encryption all, and is known as *hashing*. Hashing isn't really considered encryption because there is no decryption process. Hashing uses complex mathematical algorithms to create a fixed-length numerical representation of variable length data. In plain English, this means that you could take a letter, a word, a string of characters, or even entire document or file and perform hashing against it and produce a fixed-length numerical string (usually represented by hexadecimal numbers) called a *hash* or a *message digest*. This hash doesn't contain the data itself; it's really a very unique metadata representation of the data. One key characteristic of hashing is that it is extraordinarily difficult to find two different pieces of text that, when hashed, produce the same digest (difficult, but not impossible, and when you do, it's called a *collision*). Because of this, if any single bit (a 1 or a 0) changes in all in the variable-length text, the resulting hash would be completely different. The application of this hashing concept is used to provide for integrity of a file or a piece of text. Hashing can be used to determine whether a piece of data has been altered or not, thus assuring its integrity.

Hashing uses cryptographic algorithms to produce these fixed-length message digests. There are several hashing algorithms available, which produce different size digests. For example, the popular MD5 (which stands for Message Digest version 5) produces a 128-bit (32-character) hash, versus the 160-bit (40-character) hash that the Secure Hashing Algorithm version 1 (SHA-1) produces. From a mathematical perspective, obviously, the longer the hash value, the stronger the algorithm. The two algorithms just mentioned have actually both been determined to be mathematically unsecure, and have already been replaced by stronger ones, although both are still in common use.

Hashing is used in a wide variety of applications, including secure authentication (hashes of passwords are sent across the network instead of the actual plaintext password), message integrity, and digital signatures. As hashing isn't actually intended to be an encryption–decryption process; it is typically considered to be irreversible. That is, you can't take a hash and reverse it to produce the original data, as it really doesn't actually have the original data contained in it—it's only representation of the original data.

 TIP Hashing is an important concept to understand before attempting to learn how digital signatures work, since hashes are a critical part of digital signatures.

Data-in-Transit

"Data-in-transit" refers to the state of data as it is being transmitted or received over a communications session. Data-in-transit requires protection from interception and modification by unauthorized parties. There are several ways to accomplish this. One way might be to encrypt the data as a static file that is sent over an unprotected session. The data could still be intercepted, but could not be read. It also couldn't be modified without the sending and receiving parties knowing about it. Another way is to encrypt the communications session itself, so that any data sent over the session is encrypted as it is sent, and decrypted as it is received. The encryption algorithms and technologies covered in the next few sections demonstrate how to accomplish both of these methods.

SSL

Secure Sockets Layer, or SSL as we have been referring to it, is used to encrypt sessions between two hosts. SSL can offer server authentication, client authentication, or mutual authentication, which requires hosts on both ends of the communications session to authenticate with each other. SSL uses digital certificates to provide for authentication and encryption. SSL is most commonly seen in typical Internet-based transactions, such as buying something or checking a bank account online

SSL authentication and encryption work as follows: First, when the client accesses the server, the server identifies itself using its own digital certificate. Once the identity of the destination server is verified, the client extracts the server's public key from the certificate. The client then creates a secure session (symmetric) key in order to encrypt sensitive data. The client then uses the server's public key to encrypt the session key, so it can be securely sent to the server. This is one way of getting around the key exchange problem that plagues symmetric cryptography. The client then sends this encrypted session key back to the server, and the server uses its private key to decrypt the session key. At this point, both the client and the server have the session key in their possession and can begin to encrypt data sent back and forth between the two hosts using the session key. The session key is configured to only last for the duration of the session, and in fact is only valid for a short period of time. If the communications last longer than the validity period of the session key, or a new session is initiated between the client and server, a new session key must be renegotiated. Figure 9-1 demonstrates the SSL authentication and key exchange process.

TLS

Transport Layer Security (TLS) is considered the stronger replacement for SSL. It's actually fairly transparent to the end user to use, as most of the time the client is configured to use either SSL or TLS, depending upon the requirements it can negotiate with the destination host. TLS also uses TCP port 443 for compatibility with applications and technologies that use SSL. Figure 9-2 shows a mobile device email configuration setting for both SSL and TLS.

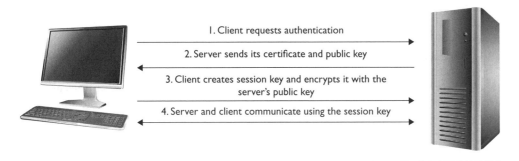

1. Client requests authentication

2. Server sends its certificate and public key

3. Client creates session key and encrypts it with the server's public key

4. Server and client communicate using the session key

Figure 9-1   The SSL connection process

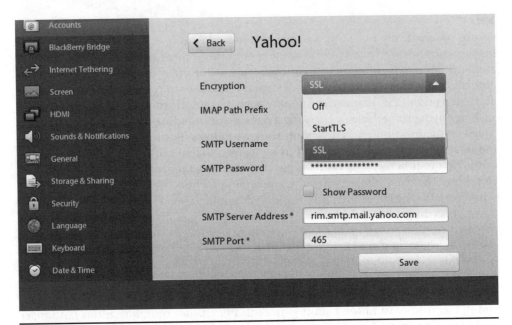

Figure 9-2 Mobile device email configuration for SSL and TLS

HTTPS

I discussed HTTPS way back in Chapter 1, and as you'll recall, it is a secure version of the ubiquitous web protocol, HTTP. HTTPS is essentially HTTP over SSL/TLS. HTTPS typically uses the client browser as its application to connect to the server and transfer data, but this is not necessarily required. At a minimum, HTTPS requires a server certificate from the destination server so that it can be authenticated. Often the client device is not authenticated using HTTPS, but can be if required by the server. As it uses SSL, HTTPS also uses TCP port 443.

SSH

Secure Shell (SSH) is a protocol that originated in the Unix-Linux world, and is used to secure communications between two hosts, typically on the same network. It provides for encryption and authentication between the two hosts, for both the user, if required, and both devices. SSH replaces older remote communications protocols, such as telnet, rlogin, and so on. It can also replace FTP, as it has built-in file transfer and copy utilities. Not only is SSH a protocol, but is also implemented as a suite of utilities as well. These include SCP, which is a secure copy utility, and SFTP, which provides for secure file transfer between remote hosts. SCP is normally used in an ad-hoc manner between two single hosts to transfer a few files, while SFTP is a better more permanent stable solution to replace an FTP server that is accessed on a more frequent basis from various hosts. SSH, as well as all of its utilities, uses TCP port 22. Figure 9-3 shows an example of using SCP to copy files between two Linux hosts.

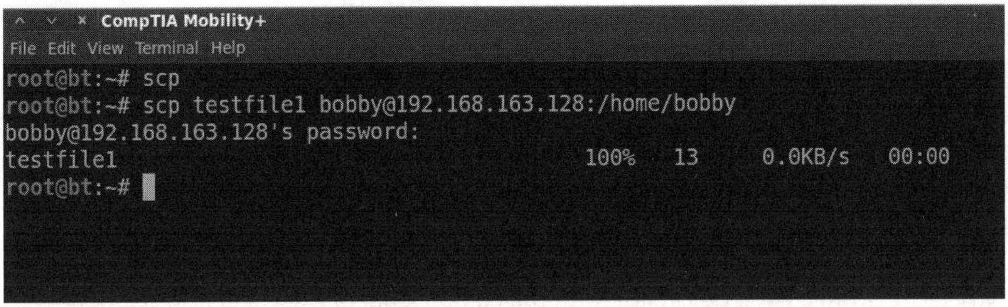

```
^  ∨  ×  CompTIA Mobility+
File  Edit  View  Terminal  Help
root@bt:~# scp
root@bt:~# scp testfile1 bobby@192.168.163.128:/home/bobby
bobby@192.168.163.128's password:
testfile1                              100%   13     0.0KB/s    00:00
root@bt:~# █
```

Figure 9-3 Copying files between Linux hosts using SCP

IPsec

Internet Protocol security (IPsec) is a protocol suite that allows for encrypted communications between two hosts using a wide variety of methods and algorithms. It is not an algorithm itself, but allows the sender and receiver to agree upon the algorithms and methods that will be used to secure communications between hosts. IPsec has the ability to encrypt and authenticate every single IP packet transmitted from a host. IPsec is composed of several different components, and resides at the network layer of the OSI model. IPsec is an open standard (non-proprietary), so it can be used across many different systems and OSes. As it works at the network layer, it can be used to protect data at any layer above it, as opposed to application-layer protocols (for example, SSH, SSL) that can only protect data at that layer.

The first component of IPsec you should be aware of is the Internet Security Association and Key Management Protocol (ISAKMP), which is used to negotiate the *security association* (SA) between hosts. The SA is the possible combination of authentication and encryption algorithms and that are supported between two hosts. The negotiated SA is the most secure standard that the two hosts mutually support. The Internet Key Exchange (IKE) is the protocol that is used to negotiate encryption and keys between two hosts. UDP port 500 is used for the IKE protocol and must be opened on security devices for IPsec to work properly through them.

IPsec also uses two important protocols, the Authentication Header (AH) protocol and the Encapsulating Security Payload (ESP) protocol. The AH protocol is used to provide integrity and authentication assurance to IP traffic. It works on the entire IP packet, including the header information. ESP provides for confidentiality through encryption (and in some configurations, integrity and authentication) for the data portion of the IP packet between two hosts, and includes the headers as well when used in tunnel mode, described shortly.

IPsec works in two distinct modes of operation. The first, and simplest, mode is called *transport* mode, and is commonly used between hosts on the same network. In this mode, the IP header information, containing the host and destination IP addresses, is not encrypted because it must be read by the different hosts and network devices that process the packet in order to ensure it is delivered to the correct host. The data, however, is encrypted and cannot be read if intercepted. The second mode is called *tunnel* mode. Tunnel mode is used when IPsec is implemented in Virtual Private Networking (VPN) technologies. In this mode, both the IP header and the data are encrypted (by ESP), and then encapsulated in another protocol, called a *tunneling protocol*, and sent across an untrusted network, usually a wide area network, such as the Internet. The tunneling protocol is used to get the IPsec packet from one network to another through an untrusted network. Figure 9-4 shows the IPsec configuration for mobile devices.

 EXAM TIP   Be familiar with the components of IPsec and how its modes are used.

VPN

Recall from Chapters 1 and 2 that VPNs are used to establish secure connections between a mobile client and the corporate infrastructure, or two different sites, such as a branch office and the corporate office. VPNs are typically implemented using tunneling methods through an unsecure network, such as the Internet. In a client VPN set up, the host has client VPN software specially configured to match the corporate VPN server or concentrator configuration. This would include encryption method and strength, as well as authentication methods, such as a username and password combination, or even digital certificates. A site-to-site VPN scenario uses VPN devices on both ends of

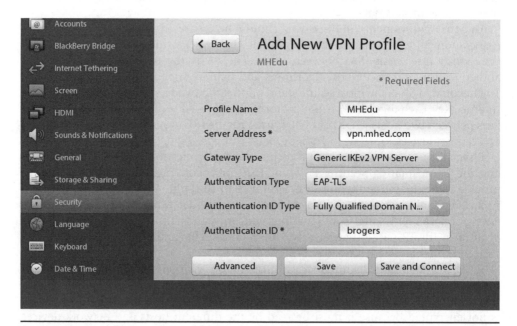

Figure 9-4   Configuring the IKE settings for an IPsec VPN on a BlackBerry PlayBook

the connection, which are configured to communicate only with each other, while the hosts on both ends use their respective VPN concentrators as a gateway. This arrangement is usually transparent to the users at both sites; hosts at the other site appear as if they are directly connected to the user's network.

VPNs can be created using a variety of technologies and protocols. The most popular ways to create a VPN is to use either a combination of the Layer 2 Tunneling Protocol (L2TP) and IPsec, or Secure Sockets Layer. When using the L2TP/IPsec method, UDP port 1701 is used and must be opened on packet filtering devices. In this form of VPN, the user connects to the corporate network and can use all of their typical applications, such as their email client, and can map shares and drives as they would if they were actually connected on-site to the corporate infrastructure. The client essentially becomes a part of the corporate network completely, even though it is not physically located at the corporate site. Figure 9-5 shows how IPsec is encapsulated and sent over L2TP.

An older VPN protocol combination uses the Point-to-Point Tunneling Protocol (PPTP) as the tunneling protocol, along with Microsoft's Point-to-Point Encryption (MPPE) protocol. PPTP is a method of tunneling the older Point-to-Point Protocol (PPP), used in the early days of dial-up networking, and tunnels data encrypted with MPPE. PPTP/MPPE uses TCP port 1723, and is rarely seen in modern VPNs.

An SSL-based VPN, on the other hand, uses the standard SSL port, TCP 443, and is typically used through a client's web browser. It doesn't normally require any special software or configuration on the client itself. However, SSL-based VPNs can be somewhat limited in functionality, and present the corporate resources through the client browser in the form of an access portal.

 EXAM TIP Remember that, as far as VPN protocols go, IPsec is paired with L2TP, and MPPE is paired with PPTP. SSL VPNs use only the native security capabilities of SSL.

WEP

I discussed wireless security protocols at length in Chapter 5, but this section will serve as a reminder for the key protocols used to secure wireless networks. Wired Equivalent Privacy (WEP) is an early security protocol used in IEEE 802.11b wireless networks. It was developed to provide some semblance of security and data protection, comparable to that of a wired network. It provides for data encryption and limited authentication between two devices.

Figure 9-5
Encapsulating
IPsec into L2TP

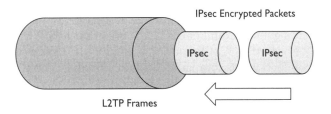

IPsec Encrypted Packets

IPsec   IPsec

L2TP Frames

WEP is very unsecure and was compromised early on in its implementation. It uses the RC4 symmetric streaming protocol, and very small (24-bit) initialization vectors (IVs), which repeat periodically. The combination of the very small IVs and the frequent repetition in higher traffic scenarios makes it very susceptible to cracking. It has largely been replaced by WPA and WPA2, except in the case of older devices and operating systems that may not support these two newer protocols. WEP should be avoided at all costs, and any legacy devices that cannot use the newer, more secure, protocols should be upgraded or taken off the network at once.

RC4

RC4 is a symmetric streaming protocol used in a variety of implementations, with the most popular one being WEP. Remember that a streaming protocol is one that encrypts one bit at a time, as opposed to a block protocol, which encrypts entire blocks of bits. This makes it very fast and very efficient, but, unfortunately, not as secure as block ciphers. While RC4 isn't necessarily by itself an unsecure protocol, it was very poorly implemented in WEP. This is one reason why WEP is so unsecure and easily cracked. It is also used more securely in some TLS implementations, and is known for its simplicity and speed. It has key sizes that range from 40 to 2,048 bits. It was developed by Ron Rivest of RSA Security, and is officially known as "Rivest Cipher 4."

 EXAM TIP Understand the characteristics and weaknesses of WEP and RC4 for the exam.

WPA/TKIP

Wi-Fi Protected Access, or WPA, was implemented in 2003 because of the recognized weaknesses of WEP. While the IEEE was involved in the long process of drafting and ratifying a standard for better wireless security, the different technology manufacturers and vendors came together and developed a stopgap security standard in order to quickly enhance the security of wireless networks. WPA was their solution, and involves the use of more advanced encryption methods. One of the methods used is the Temporal Key Integrity Protocol, or TKIP, as its method of creating and transmitting secure keys. While TKIP was also eventually implemented for WEP, the use of 40-bit or 104-bit static keys made WEP easy to crack. The implementation of TKIP included in WPA allows for a 128-bit key for each packet, so there are none of the repetition problems that plagued WEP.

WPA2

Eventually, the IEEE officially ratified their new security standard, 802.11i. This standard became unofficially known as WPA2, and except for some details, was very similar to WPA. One significant difference was the use of the AES as its primary encryption protocol. WPA2 uses AES-CCMP mode, discussed in the next section, in order to encrypt. It is, however, backwards compatible with devices that cannot use AES-CCMP, and will default to using TKIP in those cases.

Like WPA, WPA2 has two different implementations, and usage is based upon the type of security required. WPA2-Personal is primarily used on small business and home networks. In this implementation, a 256-bit pre-shared key (PSK) is used to encrypt network traffic between devices. This key can be based on a passphrase that is either a string of 64 hexadecimal digits, or as an ASCII passphrase consisting of 8 to 63 characters. It is worth noting that despite secure implementation of WPA and WPA2, the use of weak passphrases makes cracking either one quite easy.

The second type of implementation of both WPA and WPA2 is the enterprise version, which uses the IEEE 802.1X authentication standard. This standard allows for multiple types of security protocols to authenticate both users and devices to the enterprise. In addition to standard username and password combinations, 802.1X can use certificate-based authentication through the use of the Extensible Authentication Protocol (EAP) framework, which is covered a bit later in the chapter.

CCMP

Counter-mode Cipher Block Chaining Message Authentication Code Protocol (or shorter, Counter-mode CBC-MAC Protocol), or CCMP, is an encryption protocol implemented in the IEEE 802.11i standard, also known as WPA2, which ensures data confidentiality and integrity, as well as authentication. CCMP is based upon, and is used by, the Advanced Encryption Standard. It uses a 128-bit key with a 128-bit block size, and 48-bit IVs to help prevent replay attacks. CCMP uses the combination of CTR and CBC-MAC modes to provide for different components of assurance. It uses CTR mode to assure data confidentiality and CBC-MAC to create a message integrity code (MIC) that assures data authentication and integrity.

SRTP

Secure Real-time Transport Protocol (SRTP) is a secure version of the Real-Time Transport Protocol (RTP) associated with Voice-over-IP (VoIP) technologies. SRTP was developed by engineers from Cisco and Ericsson to provide confidentiality and integrity services for VoIP traffic. RTP is an application-layer protocol, and was developed to create a standard packet format, used to deliver both video and audio over IP. RTP is managed by the Real-Time Control Protocol (RTCP), which also has a secure version called Secure RTCP (SRTCP). Like SRTP, SRTCP provides confidentiality and integrity services for RTCP. SRTP uses AES as its encryption protocol.

RSA

RSA is an encryption algorithm used extensively in asymmetric cryptography, and stands for the last names of its inventors, Rivest, Shamir, and Adleman. Recall that in asymmetric cryptography, there are actually two keys that exist as a pair and are issued to the user. The public key can be known by anyone, but the other key in the pair, the private key, is kept secret and only used by the owner. RSA, as the most popular public key algorithm, works on the mathematical basis of factoring extremely large prime numbers. It is primarily used to develop the public and private keys used in digital certificates employed for authentication and encryption services. I discuss public key cryptography a bit later on in the chapter in the "PKI Concepts" section.

EAP

The Extensible Authentication Protocol (EAP) is more of an authentication framework than an actual protocol, although there are protocols that fit into the EAP framework. EAP was developed to provide for a wide variety of authentication and encryption methods, including the use of certificate-based authentication. EAP is also used in the IEEE 802.1X standard as the primary method of authentication. Both EAP and 802.1X are used in the WPA/WPA2-Enterprise authentication, which allows for multiple possible authentication methods, including PKI authentication, mutual authentication, directory-based authentication services, and many others. EAP, along with 802.1X authentication, makes it possible to not only authenticate devices to each other, but also individual user entities to the network. There are several different forms of EAP. EAP-TLS, Protected EAP (PEAP), EAP-MD5, and EAP MS-CHAPv2 are a few of the variants of EAP.

 EXAM TIP   Know the different characteristics of the protocols discussed in this section of the exam.

Implementing Encryption

Now that you've learned the different encryption algorithms available to protect data both at rest and in transit, let's discuss some of the ways you can actually apply these encryption technologies to protect data. Some of these methods are built into the operating systems of devices, while others require the use of third-party software or utilities. The next few sections discuss full disk encryption, block-level encryption, file and folder encryption, and encryption of removable media.

Full Disk Encryption

Full disk encryption can be a scary prospect because the disk is encrypted when the device is powered off, and decrypted when the device is powered back up, after the user is properly authenticated to the device, and usually before the operating system even loads. The scary part is wondering if the disk will actually decrypt before the operating system loads, which can occur because of problems with the authentication method or keys. Over the past several years, this technology has improved drastically, and it isn't as worrisome as it used to be.

With full disk encryption, the user must authenticate to the device before the disk is decrypted. There are several ways this can happen. First, if a Trusted Platform Module (TPM) is built into the device, it can be used to encrypt and decrypt the drive, as well as provide for an authentication utility to access the drive. Many laptops use TPMs and protect their drives, in the event that they are lost or stolen, by using full disk encryption. If the drive is removed from the laptop, the TPM can no longer be used to decrypt the drive. This means that an unapproved user cannot connect the drive to another system and attempt to decrypt it.

Another method is to use removable media, such as a USB removable drive, that can provide the keys used to decrypt the disk. This method is primarily used for older devices that don't use TPMs. When setting up full disk encryption, the decryption key is stored on the USB device. It should also be backed up to protect it from becoming corrupt. When the device is booted, a software utility on the drive looks for the decryption key on the removable media. If it's found, it decrypts the drive. If not, the drive remains encrypted and unusable.

Windows BitLocker Drive Encryption is an example of software built into an operating system that can be used to encrypt the entire drive or even a partition or volume on the drive. Figure 9-6 shows the configuration screen for configuring disk encryption using BitLocker on a Windows 8 tablet.

Block-Level Encryption

Block-level encryption does not rely on the particular file system used. It encrypts blocks of data, regardless of the file system structure. This means it is perfect for encrypting an entire disk. File and folder encryption are typically functions of the operating system itself, and vary between operating systems and even different versions of the same operating system. Because block-level encryption is completely independent of the file system, it doesn't matter what type of file system resides on the host drive. Typically, block-level encryption is performed by an independent piece of software or even firmware. Trusted Platform Modules can use block-level encryption to encrypt entire disks.

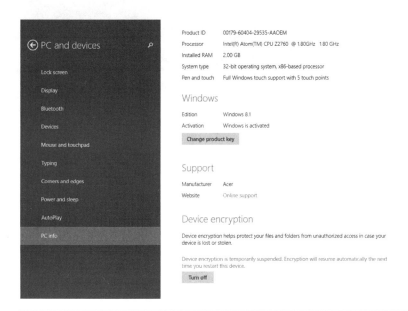

Figure 9-6   Configuring disk encryption on a Windows 8 tablet

File-Level Encryption

File-level encryption has become popular for end users over the past few years, as ordinary (non-security) people become increasingly aware of security threats to data. It can be used to encrypt a particular file or set of files, or even a folder that contains certain files. Because file-level encryption is typically a function of the operating system itself, most operating systems, such as Windows and Linux, have file encryption capabilities built-in. Because they are operating system dependent, they are usually incompatible between different OSes and file systems. In Windows, the file encryption key is derived from the user's login credentials. When decrypted, a file can be used by its owner as they would any other file, but to another user, the encrypted file can't be accessed, to include opening, copying, moving, or even deleting it. In Windows, files encrypted with EFS also are shown in the graphical interface with a different font color to visually distinguish them from non-encrypted files. Figure 9-7 displays how simple it is to enable Windows Encrypting File System (EFS) on a Windows 8 laptop.

Figure 9-7
Windows EFS

Folder-Level Encryption

In addition to file-level encryption, entire folders can also be encrypted. Folder-level encryption can be used to encrypt an existing folder with files already stored in it, or to create an empty encrypted folder that automatically encrypts any file placed in it. Windows EFS can perform folder-level encryption, as can several third-party utilities. Keep in mind that in reality, at the operating system level, the folder itself is not actually encrypted by EFS; it only serves as a logical construct in the file system that designates all files within as encrypted. If the folder itself has been encrypted, then it is actually considered an encrypted container. The same restrictions that apply to files encrypted with EFS also apply to EFS-encrypted folders.

Removable Media Encryption

In addition to encrypting a device's permanent storage, removable media can also be encrypted. There are various ways to do this as well, including using utilities built-in to the device's operating system, as well as using third-party utilities. With removable media encryption, the media is inserted into the device, and the user typically receives a prompt for them to enter the correct passphrase in order to decrypt the media. If the removable media is used in a device that doesn't have the correct utility, the media will simply appear to be blank and unformatted. The user may get prompted to prepare the media for use, which will effectively destroy all the data that is encrypted on it.

Removable media encryption is useful in scenarios where sensitive data may be stored on a mobile device, and can be restricted to the removable media in the device only, instead of on the permanent storage of the device. The organization could require that all removable media used in this manner be removed from the device and stored securely at the end of the business day or whenever the user takes the device away from the business premises. Even if the removable media is not removed, the fact that it is encrypted can help prevent data loss in the event the device is lost or stolen. Figure 9-8 shows an example of a USB storage device being encrypted by Windows BitLocker.

Figure 9-8
Encrypting
removable media
with Windows
BitLocker

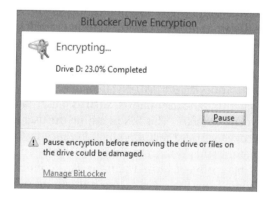

Access Control

Access control is the use of various security measures to ensure that only entities that have a legitimate need and permission to access data and systems can do so. Access control involves several components, including identification, authentication, and authorization. Each of these components will be described in the sections that follow. Access control can apply to technical measures, operational measures, and even physical measures. Technical measures can include privileges and permissions assigned to user. Operational measures can include written policy and data sensitivity controls. Physicals access controls include, of course, secured server rooms that require badge access, for example. For the purposes of our discussion, however, we are primarily going to look at technical or logical access controls, starting with authentication concepts.

Authentication Concepts

There are several components to authentication. The first is identification. Before a user can be authenticated, they must present some form of identification and assert that the identification applies to them. This can be in the form of a username and password combination, or even a smartcard and personal identification number (PIN). The form of identification that the user presents is called the *credentials*. Credentials assert and seek to prove an identity. Credentials can include smartcard, a username and password, or even a photo ID. Regardless of the method of identification, credentials submitted to the system are checked against a database of credentials, which is used to authenticate the user. If the credentials the user submitted match those in the database, then this essentially means that the user's identity has been confirmed, or authenticated. The next step after authentication is authorization. Even if a valid user has been properly identified and authenticated, it still doesn't mean that they have the ability or authority to access data or systems. This is where authorization comes in.

Authorization can be in the form of privileges, rights, and permissions to data objects. Once an individual is properly identified, authenticated, and authorized, he or she can access data and systems. There are several ways to identify and authenticate an individual, which I discuss in the upcoming sections.

Single-Factor Authentication

The first form of authentication that I'll discuss is the simplest, single factor authentication. In order to understand single factor authentication, it helps to understand what the authentication factors actually are. First, there is the knowledge factor. This is attributed to something the user knows, such as a username and password. The second factor is the ownership factor. This is something the user has in her possession, such as a smartcard or token. A third factor is the inherence factor. This is something the user either is or something they do. An example of an inherence factor is a biometric identifier, such as a fingerprint or retinal pattern. You commonly hear these three factors referred to as something the user knows, something they have, and something they are. There are other authentication factors as well that are not as commonly considered in security authentication, but exist nonetheless. For example, something you do (an inherence factor) could be used in an authentication scheme whereby the user draws a certain pattern on the screen of a mobile device. Another authentication factor is the location factor, or somewhere you are. This can be used if the individual's location can be pinpointed via GPS or some other method. The individual may be required to be at a certain location in order to log in to the system, for example. Yet another factor is the temporal factor. This factor is time based, and may require logon at a certain time of day, for example, within a narrowly defined time period, or even within so many seconds or minutes of another event.

Now that you know what the various authentication factors are, let's get back to the simplest form of authentication, the single factor authentication. In single factor authentication, the user has to use only one of the factors in order to authenticate. You most commonly see this in authentication methods that require only a username and password combination. Although it seems that these are two different elements, it is still only one factor being used, the knowledge factor.

 EXAM TIP   Don't confuse the username and password combination with multifactor authentication. Remember that only one factor is being used here, the knowledge factor. This makes the username and password combination a form of single factor authentication.

Multifactor Authentication

It makes sense that since single factor authentication uses only one factor, multifactor authentication uses more than one factor. In the past, the trend was to refer to it as two-factor authentication, which meant that the authentication scheme required more than one factor. Over the years, however, authentication methods have developed that require even more than two factors, so it's become more correct to say single factor and multifactor authentication. Multifactor authentication can use a variety of methods, as long as it uses more than one. There are many examples of multifactor authentication that you may see in your everyday life. For example, when you use a bank's ATM, you're using multifactor authentication because of something you possess (the ATM card) and something you know (the correct PIN). To access data and systems, multifactor authentication can use that method as well, and others that I will discuss in the next few paragraphs.

Biometric

Biometric elements are included as inherence factors. As mentioned previously, this can include physical features, such as fingerprints, retinal patterns, voice patterns, and even DNA. Combined with other authentication factors, biometric elements can provide a very secure authentication mechanism. An example of biometric authentication might be presenting a smartcard to a proximity badge reader and then having your fingerprint scanned on a fingerprint reader before being granted access to a secure area.

Tokens

A token is a small electronic device that the user carries on them. It usually displays a series of rapidly changing numbers or characters. When the user attempts to authenticate to a system, the user is prompted to input the characters shown on the token. If they match what the system expects as input, the user is authenticated and allowed access to the system. Most tokens are time-based in that the characters displayed on the screen change according to a predetermined time increment. In addition to using the token, another factor is usually required, such as a password or PIN, or even a biometric element, such as a thumbprint. Figure 9-9 shows a typical security token, an RSA SecurID.

PIN

A personal identification number (PIN) is a knowledge factor element that is often combined with something you physically possess, such as a token or smartcard. In some cases, however, a PIN alone could be used as single factor authentication. Think

Figure 9-9

A security token

of a smartphone, for example, that uses a PIN, also known as a passcode, to unlock the device. Although single factor authentication methods are not as secure as multifactor ones, it is still better than not using any authentication at all. Obviously, the longer in length the PIN, the more secure it is.

Device Access

Employing some of the encryption and access control techniques covered so far is necessary to ensure secure access to mobile devices. For example, there are many methods that can be used to unlock a device, such as the use of a PIN, or even using a finger pattern on the screen. Mobile devices can also make use of the different encryption techniques discussed, such as full disk encryption, file encryption, and even removable media encryption. All of these techniques can be used to control access to the device.

Wireless Networks

The use of encryption technologies has been discussed extensively during this chapter and others. Encryption is found in wireless networks, of course, in the use of secure communications and authentication protocols, such as WEP, WPA, WPA2, IEEE 802.1X authentication, and EAP. The combination of all of these encryption and authentication protocols goes a long way in securing wireless networks, but keep in mind that you are not restricted by only these protocols. You can use other encryption technologies as well to secure data over wireless networks. For example, there's no reason you can't encrypt a file and send it securely over WPA. You could also use IPsec, if you like, in combination with another protocol, even WEP, to provide for additional security. Because most wireless technologies work at the first two layers of the OSI model, other encryption technologies, such as SSH and SSL, can be used at the higher layers of the OSI model to protect data, even when sent over a wireless network. In any case, the use of multiple encryption protocols and technologies can be used to provide defense-in-depth for wireless networks, compensating for weaknesses that might be found in other protocols, and adding extra layers of security in the event another layer is compromised.

Application Access

Applications can also make use of encryption technologies, both for authentication and data encryption. The application of encryption in authentication methods allows client and server applications to require secure authentication, while protecting the confidentiality and integrity of data. These methods could include username and password authentication, of course, but can also include more robust methods of authentication, including smartcards and certificate-based authentication for application access.

Additionally, applications that transmit data over the network can make use of encryption to protect that data. The application itself may have cryptographic modules built-in that provide encryption services for the data, or external encryption methods, such as SSL or IPsec, could be used. These can be configured on the network itself, or in some cases, in the operating system on the device.

PKI Concepts

Public Key Infrastructure, or PKI, is the collection of tools, techniques, and processes used to manage formally issued and controlled digital certificates. PKI is usually a formal structure, composed of different components that interact with each other to manage the digital certificate lifecycle. It's important not to confuse the PKI structure itself with public and private key cryptography, although they are often thought of as one and the same. For example, PGP encryption and techniques with a *web of trust* model, discussed later in the chapter, use public and private key technologies, without the formal infrastructure that PKI brings.

Basics of Public Key Cryptography

Before we get heavy into the PKI infrastructure itself, let's quickly review the basics of public key cryptography. Remember that public key cryptography, also known as *asymmetric key cryptography*, uses two separate, but mathematically related, keys in a pair. This is unlike *symmetric key cryptography*, which uses only one key that must be shared between any users that needs to encrypt or decrypt data. One of the keys in the pair is designated as the public key, and one of them is designated as the private key. The public key, obviously, is one that will be shared with anyone who requires it. The private key is kept secret and protected by the user. Figure 9-10 shows an example of the distribution of public and private keys.

One important thing that you need to know about public key cryptography is that what one key in the pair encrypts, only the other key in the pair can decrypt. This means that you can't use the same key to both encrypt and decrypt data. This relationship is very important in understanding how asymmetric encryption works. So, if someone encrypts something with your public key, confidentiality is ensured because only *you* can decrypt it with the corresponding private key. In order to encrypt something and send it back to them, you would need to possess *their* public key so that you could encrypt something that only they could decrypt with their private key. In order for a conversation of sorts to occur back and forth between two parties, each party must have a key pair, and each party must also have the public key for the other party. Figure 9-11 illustrates how this process works.

Figure 9-10
Distribution of public and private keys

Private Keys Kept Secret

Mary's Private Key

Meghan's Private Key

Mary's Public Key   Public Keys Exchanged   Meghan's Public Key

Figure 9-11
Exchanging encrypted data using public and private keys

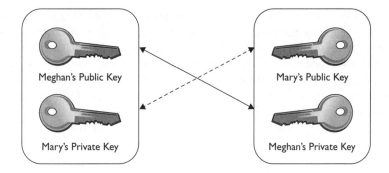

One interesting twist to this relationship between keys is the use of the private key to encrypt something that is then sent out. Obviously, if the private key encrypts the data, then only the corresponding public key can decrypt it. However, because *anyone* can possess the corresponding public key, this doesn't assure confidentiality at all. What it does do, however, is provide for a measure of authenticity in that only the person with the private key could have encrypted the data in the first place. This process assures authentication of a message from a party and fills the requirement of non-repudiation that a PKI requires.

Certificate Management

Before I discuss how the PKI works and how its components interoperate with each other, you should have a good foundation of knowledge concerning the certificate life-cycle. The certificate lifecycle refers to the entire life of the certificate, from request and issuance through renewal, to revocation and retirement. The different elements of PKI that are covered in this chapter are used to manage the certificate lifecycle.

Digital Certificates

Digital certificates are nothing more than electronic files. These electronic files come in very specific formats and can be stored on a wide variety of media, or in central storage within an organization, or even a publicly accessible server on the Internet. These electronic files are used to verify the identity of an entity, such as an individual person or a host (a client or server device). The certificate is digitally signed by the issuer. This proves that the issuer has confirmed the identity of the owner of the certificate through a stringent process, and verifies that the user of the certificate is whom they purport to be. This digital signature also verifies the public key is assigned to the identity of the entity listed on the certificate.

The certificate file contains public keys that are issued to the individual entity. Usually, the public key belonging to the entity is embedded in the file, and as the digital certificate is distributed to anyone who desires it, the public key goes with it. The private key, as I have mentioned before, is kept separate and secret by its owner.

Certificates are also used for specific purposes, such as digitally signing software or email. When used in this manner, they are used for authentication purposes. Certificates can also be used to encrypt items, such as files or email. The user could have several certificates, each used for very specific purposes, or only one certificate that allows the user to use it for all of the purposes mentioned. This depends upon the policy of the issuer and the purpose for which the certificate was issued. Different vendors may define different classes of certificates, which designate their intended purpose. For example, a Class 1 certificate may be intended for individual email use, a Class 2 for authentication purposes, and a Class 3 may be used for server identification or software signing.

Digital certificates typically meet the International Telecommunication Union Telecommunication Standardization Sector (ITU-T) X.509 standard, which dictates how digital certificates are constructed. In addition to the public keys of the user and the digital signature of the issuer, digital certificates also contain other elements, such as the certificate serial number, the algorithm used to create the digital signature of the issuer, identifying information for both the user and the issuer, validity dates of the certificate, the purpose for which the certificate is intended to be used, and the thumbprint of the certificate (a hash of the certificate itself).

The X.509 standard also defines some of the infrastructure involved in requesting, processing, and issuing a certificate. For example, a certificate can have a file extension of .pem or .cer, depending upon how it is encoded. RSA Security, Inc., also published a series of file types and formats, called Public Key Cryptography Standards (PKCS), which are used throughout the industry, although considered proprietary by RSA. For example, a PKCS#7 file is a certificate used to sign or encrypt digital messages, such as email. A PKCS#12 file contains both public and private keys, and is encrypted via asymmetric algorithm and used for secure key storage. Additionally, individual vendors, such as Microsoft, Apple, VeriSign, and so on, each have their own proprietary way of managing certificates, although for the most part they are X.509 compliant.

Digital certificates can typically be stored and accessed on a mobile device through an application such as a browser, or an app that is been designed to access and use digital certificates. Certificates can be downloaded from centralized storage on the Internet or from an organization's web site, sent via email and downloaded, or transferred via removable media. Figure 9-12 shows an example of a digital certificate as viewed through a mobile device configuration screen.

Certificate Lifecycle

Now that I have discussed what digital certificates are, we need to go over the process of how a digital certificate is requested, created, issued, and managed by an organization. The process for creating a digital certificate and issuing it, as well as managing it through its life, is called the certificate lifecycle. This process can be rather complex, depending upon several factors, including the size of the organization, the validity period of the certificate itself, and how the organization manages its certificates, among other things.

The certificate lifecycle starts with the request from the user for a digital certificate. If the certificate will only be used within the organization and not out in public, or between organizations, the lifecycle process can actually be a little less complex. For

Figure 9-12 A digital certificate

certificates used only within an organization, the organization controls all aspects of the request, issuance, and certificate disposal. For organizations that issue certificates used outside the organization, such as with communications with the public or with another organization, the lifecycle can be a little bit more complex. In some cases, a user request will not go to the organization, but may go to a third-party certificate provider, such as VeriSign, Thawte, or the other numerous providers in the commercial world.

In any case, when the user requests a digital certificate, it typically accompanies proof of identity that the user submits to assert her identity. Different organizations have different requirements, but this proof could be in the form of a birth certificate, passport, driver's license, fingerprints, and so on, and may even include a background check. When the issuing organization receives the request and the proof of identity, it must validate the identity against the user to ensure that it is issuing a certificate that is bound to that particular verified identity. The organization usually has its own internal processes for this. In the case of a commercial certificate-issuing organization, the organization will also likely charge a fee for the service.

Once the organization has verified the requester's identity and issued a certificate, the certificate is valid only for specific purposes and for a particular time period. A certificate may be valid for three years, for example, and only good for sending and receiving secure email. When the certificate validity period is almost expired, the user may have the option of renewing the certificate for an additional time period, depending upon the renewal policy of the issuing organization. If the user does not renew the certificate, it will become invalid on its expiration date. After this time, it can't be used to authenticate, encrypt, sign, or whatever other purpose for which it was issued. At this point, it's considered an expired certificate.

There are some other events that could happen before the certificate expires, other than the renewal process. An organization could suspend a certificate temporarily from an employee or individual if he is on extended leave, for example, or under investigation. The organization can also suspend the certificate in the event the individual violates the agreements under which he received the certificate. A suspended certificate typically can't be used temporarily, but it can be reinstated when the conditions of the suspension are no longer valid; that decision is based upon the issuing organization's policy. Another event that can occur is the revocation of a certificate. A certificate can be revoked by its issuing organization under similar circumstances as a suspension. For example, if an employee is fired or even voluntarily leaves the organization for another position, the certificate is normally revoked. A revoked certificate, unlike a suspended one, usually can't be reinstated; a revocation is usually permanent. If the circumstances of the revocation are no longer valid (maybe the employee later returns to the organization), at the organization's discretion, it can reissue another new certificate, but not the revoked one.

Once a certificate is expired or revoked, that usually signals the end of its lifecycle. Any attempt to use the certificate will result in error messages regarding the validity of the certificate, and the inability to perform any related functions with the certificate. When a certificate has been revoked and is no longer valid, this action is normally published on a centralized list, called a Certificate Revocation List (CRL). This list is published through various means, such as from the Internet or from a centralized server. The list is typically downloaded during normal software updates from the vendor, or through the operating system. It may also be pushed by the organization to all of its host devices.

 EXAM TIP Understand the different characteristics of the certificate lifecycle, to include certificate issuance, renewal, expiration, suspension, and revocation.

Certificate Management Infrastructure

The certificate management infrastructure, usually prescribed by the organization's PKI, consists of the different components used to manage the certificate from its request all the way through its disposal. Several components of the PKI are used to manage certificates; PKI is not limited to just the digital certificates and algorithms themselves. It is the policies and infrastructure, including hardware, software, processes, and procedures, that are needed to manage all aspects of digital certificates. The first major component of the PKI is the Certificate Authority, or CA. The CA is the main component of the PKI responsible for issuing and validating digital certificates. The term "certificate authority" is often used rather contextually. The CA can be the organization that issues a certificate, such as the organization itself or a third-party issuer, like VeriSign or Thawte, for example, and it can also be the server that actually creates and issues the electronic file that is the digital certificate. It also could be an individual or organizational element that is responsible for approval decisions on digital certificates that are issued.

So you may hear the term "certificate authority" used several different ways in conversation and in different literature.

In any case, the CA must first confirm the identity of the requestor before it issues and digitally signs the certificate and public key that is bound to a particular user entity. In smaller organizations, this may all be done by one organizational element. In larger ones, there may be a function separate from the CA, called the Registration Authority (RA). The Registration Authority may have the job of verifying the identity of the user applying for a digital certificate before it passes the request on to the CA. If the certificate is to be used outside the organization, another component may be present. This is the third-party Validation Authority (VA), and is usually needed to either verify the individual's identity from a trusted third-party, or even to issue the root certificate to the organization, so any certificates it issues will be trusted outside of the organizational boundaries.

The root CA is the initial server set up to issue certificates. It does so using a *root certificate*. The root certificate could be a self-signed certificate, in the case of organizations that will only issue certificates to be used within the organization itself. An organization that desires for its certificates to be trusted outside of organizational boundaries, such as with another organization or the general public, will typically apply for and receive a root certificate from a trusted third party. The root certificate and root CA are typically used to issue certificates to intermediate or subordinate CAs. These CAs are used in larger organizations to manage the daily workload of issuing out the actual user entity certificates. For security purposes, the root CA is usually taken off-line and not used to issue out certificates to users. It is normally taken off-line to protect it because, if the root CA is ever compromised, the entire chain of certificates that are issued from it, including subordinate CA certificates as well as any user certificates issued, are rendered untrustworthy. Figure 9-13 demonstrates a possible arrangement of root and subordinate CAs. Note that subordinate CAs can also be subordinates of other CAs, and also be given specific authority to issue only certain types of certificates, or any type, depending upon the needs and size of the organization.

Figure 9-13
Root and subordinate CAs

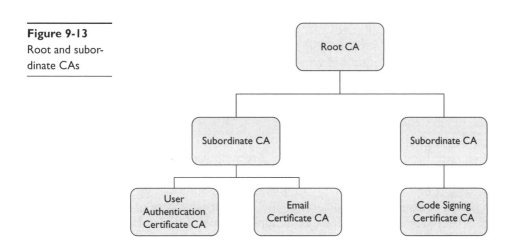

There are other elements of PKI that are beyond the scope of both the exam and this book, but you should be aware of them in order to understand PKI better. One of these elements is the concept of trust. A trust model is used between elements of the Public Key Infrastructure to describe how each element is validated and trusted in terms of creating, issuing, and using digital certificates. You've already learned about the hierarchy of trust model, which was discussed earlier , and described the root and subordinate CA infrastructure, as well as the identity verification methods used. This is the most common model used in PKI. However, there also other trust models, such as the *cross-certification* trust and the *web of trust* model. In the cross-certification trust model, one organization that issues its own certificates establishes a trust relationship with another organization. In this model, each organization may issue certificates to the other that allow each of their own issued certificates to be trusted between them both. This model can actually be incorporated between organizations that use the typical hierarchy of trust. In the web of trust model, there is no centralized PKI infrastructure utilizing CAs or RAs; certificates are usually self-validated and issued by individuals to only those people they themselves trust. This would work fine between any two individuals, but can become complex and increasingly untrustworthy as the number of certificates exchanged between groups and individuals increase. One popular example of the web of trust model is the use of Pretty Good Privacy (PGP) to issue and manage digital certificates.

Software-Based Container Access and Data Segregation

There many different reasons for separating applications and data from each other. Data, in particular, is often separated because it has different access requirements. For example, a user or application may have read permissions to data that another user or application does not. Segregating data allows a unique set of permissions to be assigned to an entity, such as a user or application, so that only that entity can interact with that data. Segregating data ensures confidentiality by allowing only those authorized users to access the data. It assures integrity by allowing only authorized users and applications to modify or change data. From another perspective, separation prevents one set of data from co-mingling with another set, which could result in potential contamination or modification if allowed.

This section covers some of the different methods for segregating data in order to protect it from unauthorized access or modification. I discuss permissions, access controls, and containerization. I also discuss a key aspect of that separation, which is policy, of course, because policy will drive what measures we need to take to ensure data is segregated appropriately.

Data Segregation Policies

As discussed throughout this book, policy drives almost everything you do in the corporate security infrastructure. Data segregation policies, like other policies I have discussed, must be carefully thought out, planned, and approved by management. Data segregation policies basically determine the levels of sensitivity of different classes of data, as well as what protections should apply to those different classes. For example, public data, as a sensitivity class, would be protected somewhat differently than say, employee financial or health data, or company proprietary data. These classes of data should be determined and assigned different levels of protection. Data segregation policies may state who, in general, would be authorized to access those classes of data and who would approve access, along with handling procedures for each class of data. Access control procedures for segregated data could include technical controls, such as encryption, object permissions, and sensitivity labels, and operational controls, such as duty requirements, security clearance, and need-to-know. Physical controls should also be employed when necessary. While the policy should state what has to be done and why, procedures have to be developed to support that policy. The procedures have to detail how data separation and control should be implemented throughout the enterprise on both traditional and mobile devices. These procedures may be implemented on a per device basis or as part of enterprise policy that is pushed down through the MDM infrastructure.

Data Access Controls

I've discussed different controls throughout the book that could be employed to maintain access control over the different classes of data the organization wants to protect, but they're worth mentioning again in this section. The primary methods of access control for data segregation are usually authorization controls, such as data object permissions (permissions to individual files and folders), encryption, and even strong authentication methods. Let's briefly cover each of these.

Object permissions are typically assigned based upon a user account or even a group membership in a traditional network. For mobile devices, however, there are a couple of different ways this can be implemented. First, for the device itself, as it is unlikely that multiple users would be using the device, object permissions would likely be set for the primary user account on the device. There could also be different object permissions set for particular apps, so that apps could only access certain data or device functions. This may be implemented by device-specific software, but could also be implemented as part of policy that is pushed to the device from the MDM infrastructure. Object permissions could apply to both files and folders, or even individual data elements in some cases. These individual data elements could be, for example, items such as a person's name and address, as well as any identifying information such as a driver's license number or social security number. It also may include data such as credit card information stored on the device. Permissions could be very specific and allow only certain actions such as read-only or write, or could even be expanded to include the ability to modify, delete, or add data. Rarely, an individual user or app may need some type of full-control permissions over a dataset so she can perform all of the aforementioned actions, as well as assign permissions themselves to other entities.

I've also mentioned encryption several times during this chapter as a method to segregate data. Encryption could be used to protect the data from unauthorized access as well as maintain its integrity. Encryption can be implemented on a per-file or per-folder basis, or can even be implemented as an encrypted container. This particular implementation will be described a bit more in the next section. There's a wide variety of both enterprise-level and device-specific encryption software that could be utilized for data protection.

I've also discussed strong authentication in various chapters in the book, and I've talked about how strong authentication is the key to data protection, simply because authentication can prevent an unauthorized person from accessing data or systems he shouldn't. Strong authentication methods in the enterprise will typically be multifactor systems, such as the ones I described earlier, to include certificate-based authentication. Many data protection schemes integrate with strong authentication methods to allow data access only when an individual has been authenticated, and may even use the same keys to decrypt data for the user.

Containerization

I discussed containerization a bit in Chapter 8. Remember that data containerization is a way you can separate one particular type or class of data from another. This could be data that should not be mixed with other types of data, or accessed by particular users or apps, so it is separated into its own protected storage area. It normally can't be accessed except by certain applications, users, and processes. You normally see data containerization in the form of separation of corporate from private data, especially in a BYOD environment. However, data containerization can have much more broad uses. Data containerization can be used on a commonly used device, such as a tablet, to separate individual user data, or to separate different sensitivity levels of data. It also could be used to segregate data from different applications or processes on the device itself.

Data containerization works like this: You can create secure "containers" on the file system of the device that contain files, folders, and other data objects, all secure inside one single logical container. Typically, this logical container is in the form of a file itself, and, when not in use, is usually encrypted and may have other access controls implemented on it. In some cases, the data container may not be visible to unauthorized users in the file structure, and even when it is visible, it may not be possible to copy or move the file from the media, unless by an authorized user after decryption. When decrypted, it may appear as a mounted folder, volume, or drive letter on the device, and would allow access to whatever data is stored in it. In many cases, executable programs stored in the container may even be run from the container itself, or data files such as documents and media may be accessed and used while the container is in its unencrypted state.

There are several different types of applications, both device-specific, as well as enterprise-level, that can create and manage data containers. TrueCrypt is one such example of software that can be used to create secure containers in which to store files. The container appears as a large file itself until it is decrypted, and is then mounted as a logical drive letter in the operating system. As with all other security controls, it is best to manage this on a policy basis, meaning that there should be a stated policy about data containerization and that it should be implemented using centralized MDM policy elements and procedures in the infrastructure. Figure 9-14 shows an example of a data containerization app for Android devices.

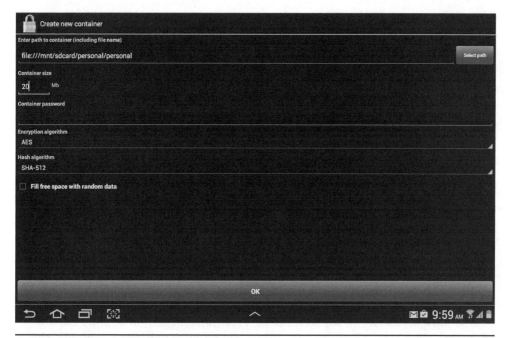

Figure 9-14 Data containerization on a mobile device

Chapter Review

This chapter covered mobile security technologies that use encryption and other access control methods to ensure the security of mobile devices and the data they contain. I discussed encryption methods used to secure data both during transmission (in-transit) and in storage (at-rest). I reviewed several key components of encryption, including algorithms and keys, and discussed the different ways that algorithms are categorized. Algorithms can be broken up into symmetric and asymmetric, as well as block or streaming. I discussed the basics of the encryption process, in terms of how plaintext is combined with an encryption algorithm and a key, and how this process produces ciphertext. I also discussed the importance of key space and key length in producing a key that is cryptographically strong.

Having reviewed the basics of encryption, I then discussed some of the methods used to protect data at rest. I covered several different protocols that are used to encrypt data while at rest, including some older protocols such as DES and 3DES. I discussed the five modes that DES operates in because they provide the basis for later, more advanced encryption algorithms as well. Of the advanced, more modern encryption algorithms, I discussed Twofish, which was a finalist for the AES competition, as well as Rijndael, which actually won the competition and became the Advanced Encryption Standard that we know today. I discussed AES at length, describing its block and key sizes, as well as its implementations. I also discussed elliptic curve cryptography and how it is popularly used on mobile devices because of its high-speed and low computational requirements. I also described the process of hashing and its application in providing data integrity.

The chapter then delved into how to protect data-in-transit, which is data that is neither in storage nor currently being processed. The primary method of protecting data while in transit is to encrypt the communications channel used to transmit and receive the data. There are several methods and algorithms used to protect data during transmission, including Secure Sockets Layer (SSL), which uses a hybrid of both public key and symmetric key cryptography to authenticate hosts to each other, and to produce and exchange a secure session key in order to encrypt data passed between hosts. I also briefly discussed Transport Layer Security, which is the stronger replacement for SSL. HTTPS is able to use either SSL or TLS in order to secure unencrypted HTTP traffic. Secure Shell is also a protocol used to protect data sent across an unsecure communications session, usually between Unix or Linux hosts. SSH consists of both the protocol and utilities used to manage the secure connections, including secure copy utilities and secure file transfer utilities. SSH replaces older unsecure protocols, such as telnet and rlogin.

I also discussed IPsec at length, which is a complex method using different protocols to provide for authentication and encryption between two hosts on the same network (using transport mode) or across a secure VPN implementation (tunnel mode). We know that VPNs are used to establish secure communications between mobile clients and the corporate infrastructure, or between a corporate site and a branch office site, for example. VPNs can use tunneling protocols, such as L2TP or PPTP, which "tunnel" or encapsulate data encrypted by either IPsec or MPPE, respectively. VPNs can also use SSL to establish a limited access portal type of VPN.

I also revisited some of the wireless security protocols that I discussed in Chapter 5. I discussed WEP and why it is extremely unsecure, as well as its underlying symmetric streaming protocol, RC4. I also discussed more modern, secure protocols, such as WPA and WPA2, as well as their underlying security methods, TKIP and AES-CCMP. Keep in mind that while the WPA2 is backwards compatible with WPA, it must default to using TKIP for older devices that don't support AES. I also briefly discussed SRTP, a secure protocol that provides confidentiality and integrity services, and is usually seen in VoIP solutions. I finished the discussion of encryption protocols with the RSA encryption algorithm used in public key cryptography, as well as EAP, which is frequently seen as the primary authentication framework used in 802.1X implementations.

I then covered implementing encryption in various ways to protect data. I examined different applications of cryptography to data protection, including full disk encryption, block-level encryption, file and folder encryption, and even removable media encryption. Some of these methods use technologies that are OS dependent, while others, such as block-level encryption, are independent of the underlying operating system and file system.

The next section looked at access control methods and reviewed the concepts involved with controlling access to data and systems. I discussed authentication concepts, including identification, authentication, and authorization. I reviewed single factor authentication concepts, as well as multifactor authentication. In our discussions, I examined different authentication methods, such as biometrics, tokens, username/ password combinations, and PINs. I discussed how to apply these methods to device, network, and application access control.

I then went in-depth on Public Key Infrastructure, describing PKI and the basics of public key cryptography. In the discussions on certificate management, I described digital certificates, the data they contain, and what they are used for. I also discussed the lifecycle of digital certificates, beginning with the certificate request process, through issuance, renewal, suspension, and revocation. The management infrastructure controlling certificates during their lifecycle included components such as the certification authority (CA), the registration authority (RA), and even sometimes a third party known as the validation authority (VA). I discussed the functions of each of these, as well as their roles, responsibilities, and elements. I also briefly discussed the different trust models used in and related to PKI, such as hierarchy of trust, cross-certification, and web of trust.

Finally, I discussed data segregation and containerization topics. Major points included data segregation policies, data access controls, and data containerization. All of these items are designed to protect data while on the mobile device, from data sensitivity and access control policies implemented by the organization, to data access controls such as permissions and encryption, and finally, to containerization, where data is segregated by access necessity.

Questions

1. Three encryption algorithms are listed here, in no particular order. From the choices provided, select the one that best describes the order of strongest to weakest algorithm:

    1. AES

    2. DES

    3. 3DES

    A. 1, 3, 2

    B. 1, 2, 3

    C. 3, 2, 1

    D. 2, 1, 3

2. Which of the following contributes to a cryptographically strong key? (Choose two.)

    A. Increased key length

    B. Unknown algorithm

    C. Publicly tested algorithm

    D. Larger key space

3. Which type of algorithm encrypts only one bit at a time?

   A. Block

   B. Streaming

   C. Symmetric

   D. Asymmetric

4. Which of the following algorithms uses a 168-bit key?

   A. DES

   B. AES

   C. 3DES

   D. ECC

5. In which mode is a 64-bit random block of data used as the first initialization vector and then incremented with each subsequent block of plaintext, usually by a value of 1?

   A. Output Feedback mode

   B. Counter mode

   C. Cipher Feedback mode

   D. Cipher Block Chaining mode

6. In a Secure Sockets Layer transaction, what type of key is generated by the client and securely sent to the server?

   A. Session

   B. Private

   C. Public

   D. Asymmetric

7. You are setting up a secure communications channel between two hosts on your network. Which of the following would be the best solution for this as a permanent secure channel?

   A. SSH

   B. IPsec in transport mode

   C. An SSL VPN

   D. IPsec in tunnel mode

8. You have set up an L2TP/IPsec VPN solution, and need to ensure that the proper port is set up on the network security devices to allow VPN traffic to pass

through the devices. Which of the following ports would you need to open? (Choose two.)

    A. TCP port 1723

    B. UDP port 1701

    C. TCP port 22

    D. UDP port 500

  9. Which of the following protocols use TKIP and AES encryption methods? (Choose two.)

    A. WPA

    B. WPA2

    C. SSL

    D. SSH

10. Which of the following is *not* a characteristic of AES-CCMP?

    A. 128-bit key

    B. Uses a MIC to ensure confidentiality

    C. Uses CTR mode to ensure confidentiality

    D. 128-bit block size

11. Which of the following encryption algorithms is based upon factorization of extremely large prime numbers?

    A. RSA

    B. ECC

    C. AES

    D. Twofish

12. EAP is used in what major type of authentication?

    A. Single factor

    B. Multifactor

    C. 802.1X

    D. 802.11i

13. All of the following could be used as multifactor authentication EXCEPT:

    A. Username and password

    B. Smartcard and PIN

    C. Smartcard and thumbprint

    D. Token and PIN

14. In asymmetric key cryptography, which key can decrypt something encrypted with the user's private key?

    A. Another user's public key

    B. A session key

    C. Another user's private key

    D. The same user's public key

15. What open standard covers the format of a digital certificate?

    A. Class 1, 2, and 3

    B. PKCS

    C. X.500

    D. X.509

16. Which of the following entities is responsible for issuing digital certificates?

    A. Registration Authority

    B. Certification Authority

    C. Validation Authority

    D. Verification Authority

17. All of the following actions in the certificate lifecycle could possibly result in the use of a certificate EXCEPT:

    A. Renewal

    B. Issuance

    C. Revocation

    D. Suspension

18. All of the following are reasons to implement data containerization, EXCEPT:

    A. Access control

    B. Data integrity

    C. Data sensitivity

    D. Collaboration

19. Which use of data encryption would be file system and operating system agnostic?

    A. Block-level encryption

    B. File encryption

    C. Folder encryption

    D. Container encryption

20. You are trying to secure legacy wireless devices on a network that use WEP. Replacing or upgrading the devices is not an option. Which of the following could help mitigate the security risks with using WEP?

    A. Use DES to help protect the data-in-transit.

    B. Use IPsec in transport mode to encrypt the data between hosts.

    C. Use only the AH protocol in IPsec tunnel mode.

    D. Encrypt each file separately that is sent over the wireless network.

Answers

1. **A.** The correct order is 1, 3, 2. AES is stronger than 3DES, which is stronger than DES.

2. **A, D.** A longer key length and large key space both contribute to cryptographically strong keys.

3. **B.** A streaming algorithm encrypts only one bit at a time.

4. **C.** 3DES uses a 168-bit key.

5. **B.** In Counter mode (CTR), a 64-bit random block of data is used as the first initialization vector, and then is incremented with each subsequent block of plaintext, usually by a value of 1.

6. **A.** A session key is generated by the client, encrypted with the server's public key, and sent to the server during an SSL transaction.

7. **B.** IPsec in transport mode would likely be the best scenario, as a SSL VPN is not needed, tunnel mode is for VPN connections across WANs, and SSH is usually for short-term secure connections.

8. **B, D.** UDP ports 1701 (for L2TP) and 500 (for IPsec's Internet Key Exchange protocol) should be opened on the security devices.

9. **A, B.** WPA and WPA2 use TKIP and AES as their encryption methods, respectively.

10. **B.** AES-CCMP uses a message integrity code (MIC) to ensure data authentication and integrity, not confidentiality.

11. **A.** RSA, a popular public key algorithm, works on the mathematical basis of factoring extremely large prime numbers.

12. **C.** EAP is used extensively in IEEE 802.1X authentication.

13. **A.** A username and password combination is a single factor authentication method because they are both knowledge-based.

14. **D.** In a key pair, what one key encrypts, the other decrypts. In this case, the same user's public key is the only key that could decrypt something encrypted with the user's private key.

15. **D.** The ITU-T X.509 standard prescribes the format of digital certificates.

16. **B.** The Certification Authority (CA) is responsible for issuing digital certificates.

17. **C.** Revocation would permanently terminate the use of a certificate. Suspension would only be a temporary loss and might eventually be lifted, resulting in continued certificate use.

18. **D.** Collaboration is not a reason to implement data containerization; in fact, containerization would severely limit collaboration, as it restricts access to data through the use of secure containers.

19. **A.** Block-level encryption does not rely on operating or file systems to function.

20. **B.** Using IPsec in transport mode to encrypt the data between hosts would help mitigate the risks of using WEP. Even if the WEP key is broken and the wireless traffic is decrypted, the IPsec traffic would still be secure.

Troubleshooting Network Issues

In this chapter, you will

- Learn how to implement a troubleshooting methodology in various scenarios
- Examine how to troubleshoot common over-the-air problems in various scenarios

The first five chapters of this book covered a great deal about different mobile technologies that provide the infrastructure for mobile devices. Then I spent several chapters talking about mobile devices in the corporate infrastructure and how they are provisioned, managed, and secured. All the knowledge that you acquired throughout the book up until this point will be applied in our discussion of troubleshooting. As we all know, technology doesn't always work the way it's supposed to, for a variety of reasons. Sometimes we blame the technology itself, but often the problem occurs because it was implemented incorrectly, or is incompatible with another piece of technology, or sometimes even the user of the technology can cause issues. In any event, when the technology goes wrong, you troubleshoot.

This chapter covers the first two parts of troubleshooting. I discuss troubleshooting methodologies that you can use in various scenarios, and I talk about particular troubleshooting steps you can take with network connectivity issues. Other troubleshooting topics, such as troubleshooting security and client devices, are covered in the last two chapters of this book. The CompTIA Mobility+ exam objectives covered in this chapter are 5.1, "Given a scenario, implement the following troubleshooting methodology," and 5.4, "Given a scenario, troubleshoot common over-the-air connectivity problems." Coverage of these two objectives will introduce you to troubleshooting in general, and give you the knowledge needed to troubleshoot over-the-air and network connectivity issues.

Troubleshooting Methodology

When your car doesn't start, you usually follow a series of steps in order to determine what the problem is and how you can fix it. Although you may not realize it at the time, you follow a methodology of sorts. That is, you follow a series of logical steps to identify the problem, gather information, evaluate the causes, come up with a solution, and then implement the solution in the form of a fix that hopefully will cause the car to start. Troubleshooting for almost any other piece of equipment is actually very similar, in terms of following a logical process and methodology.

This section examines those logical process steps involved in general troubleshooting. Although I specifically talk about networking later on in this chapter, and other areas in subsequent chapters, you could apply this troubleshooting methodology to a wide range of equipment problems and issues. You can use this troubleshooting methodology in your normal day-to-day work processes as part of your troubleshooting toolkit, so to speak.

Troubleshooting Steps

The troubleshooting methodology that I describe coming up in the next few sections is actually very common to CompTIA examinations. Similar troubleshooting steps are required for the CompTIA A+, Network+, and Security+ certifications. So if you have taken any of those examinations, the troubleshooting steps described here should seem very familiar to you.

The troubleshooting methodology described here should be followed in a logical, orderly manner for best results. This actually makes sense because it's hard to fix a problem unless you are able to identify its cause or evaluate possible solutions, for example. This methodology is actually easily adaptable to any work center checklist as well, so learning it will serve you not only for the exam, but also in your real life work as a CompTIA Mobility+ certified professional. The troubleshooting methodology, as derived from the exam objectives, is shown in Figure 10-1, but I'm going to cover each step in detail throughout the first part of this chapter.

There are two important items you should know about the troubleshooting process before you go on any further: First, when troubleshooting something, change only one thing at a time. If you change multiple things at once, and the problem gets fixed, you don't know which thing that you changed is the one that fixed it. Conversely, if the problem gets worse, you don't know which thing you changed might have made it worse. Second, try not to make things any worse! If you change something and the problem gets resolved, that's great. If, however, it gets worse, or doesn't get any better, then it's probably a good idea to put it back the way it was. These two important observations probably won't be on the exam; these are just two things from experience that you can apply in the practical world, and hopefully save yourself some heartache.

 CAUTION Remember to only change one element at a time when trouble-shooting a problem. Changing multiple elements, such as swapping cables and changing the client configuration at the same time, won't tell you which fixed the problem—or made it worse!

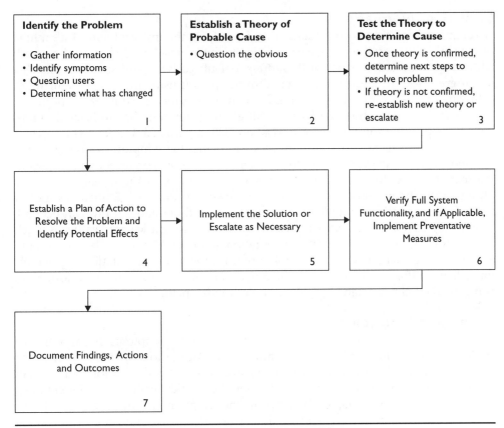

Figure 10-1   The CompTIA Mobility+ troubleshooting methodology

Identify the Problem

Although this seems to be common sense, identifying the problem is the first major step you should take in this troubleshooting methodology. It would also seem to be the easiest step in the methodology, but this is because technicians often tend to jump into a problem and mistakenly identify the symptoms of a problem as the problem itself, without taking the time to discover what the actual problem is. For example, if a user calls you on the phone and says that the Internet is down, to them that may be the actual problem, but you as the technician know that the Internet itself is not down. Likely the actual problem is a bad or loose cable, a problem with the host, or even a network device that's down. But you won't know what the actual problem is until you investigate and gather some information. Identifying the problem consists of a few steps, namely, gathering information, identifying symptoms, questioning users, and determining if anything changed with the host or the network. I explain each of the steps in the sections that follow.

Gathering Information

Gathering information is your first step in identifying the true problem. If it's a host or network issue, you'll of course want to know the location of the host and network addressing information, such as IP address, default gateway, and subnet mask. You might even need to know the username of the person using the device. Sometimes gathering this bit of information can help point you in the direction of the problem, such as an incorrect subnet mask or default gateway, but don't be too hasty in jumping to one possible solution too quickly. You also might need to know which device the host is plugged into, or communicates with, what the symptoms of the problem are, and what exactly the user was doing when the symptoms occurred. If you're working remotely away from the user, you may gather some of this information over the phone or via the network, using remote administration software. Sometimes it's not the user that identifies a problem; rather you may find out about it through network monitoring software that alerts you to an issue. In any case, the key is to gather as much information as you can about the configuration and how the host or device is set up. As you'll find out a bit later in the troubleshooting process, documentation is extremely important, so this is a good time in the process to start documenting the information you gather to help you start along the troubleshooting path.

Identifying Symptoms

Sometimes you're only alerted to a problem because of the symptoms that appear. This could be network users calling sporadically to complain of loss of connectivity to the Internet, for example, or several alerts coming from your network monitoring devices. In any case, the symptoms are your first indication the problem exists, but as I mentioned earlier, don't fall into the trap of thinking that the symptoms are the problem itself.

As pointed out in the previous section, you should take the time to document the symptoms as you discover them in order to help you diagnose the problem, as well as keep a record of the issue in case it occurs again in the future. It may help to categorize symptoms in terms of network connectivity, symptoms affecting the host, symptoms that affect one user versus many users, symptoms that may affect particular operating systems only, and so on. As you collect data on the symptoms, you are closer to determining what the problem really is. However, you also need to question the users and find out if anything on the network or the host changed.

Questioning Users

Questioning users helps you to further gather information regarding what the problem may be. Obviously you'll get the user view of what the problem is, and usually this is from their perspective of what they see the system doing (or not doing) while they are interacting with it. You'll want to ask the users what they were doing when the symptoms occurred, and find out if the issues were sporadic or continuous, whether it started all of a sudden or gradually happened, and so on. Information you can get from questioning users includes which applications were in use when the issues occurred, the timeframe of when the problem occurred, and how long it lasted. You also may ask them what they were doing before the problem occurred that may have contributed to it, and what they did in terms of trying to fix the problem.

These last two items are important because you may need to know what actions the users were taking when the problem happened because they may have contributed to the problem in some small way, and any actions they took when the problem occurred in order to correct it may have changed it or worsened it in some way. And once again, you'll want to document a summary of what information the users were able to give you to help you identify the problem. It also goes without saying (but I will anyway) that talking to the users about the problem shows them that you are concerned with their issues, and it gives them a sense that you are all about providing good customer service.

TIP Some organizations have a "script" that customer service and help desk technicians use when questioning the user, in order to ensure a standardized, thorough way to elicit the right information from them. This can be very helpful, but be aware that sometimes it might not cover every single scenario, and you may have to "improvise" a bit when using one of these scripts if you need to ask questions that aren't in it.

Determining Changes

Many equipment or connectivity issues occur because something changes. This could be a device that's been replaced, a component that failed and had to be repaired, or even a configuration change on the network or host. In any event, if something on the network or the host changes, it could cause a host or device to not communicate on the network properly or cause sporadic failures.

You'll want to start this step in the troubleshooting process by talking to people who represent the areas in question. For instance, you may want to talk to the network administrator or engineer to see if something on the network has changed, such as a device configuration, a new installation, or even a recent failure of a network device. You also may need to find out from the folks responsible for maintaining the hosts on the network if anything about a user's device recently changed, such as the addition of new software, a configuration change, a repair, or something of that nature. There also may be other specialized areas that you may need to look at, such as email servers, firewalls, file servers, and other network infrastructure devices. In addition to questioning the different points of contact for those areas, it would probably be worth your while to look into any trouble management or ticketing system that's used by technicians. This might tell you if someone has reported any similar problems recently, or if any problems have occurred that might have caused the issues you are running into now.

After determining if anything has changed on the network or host, this information, along with documenting any symptoms, questioning the user, and gathering any other information you can about the issues and the devices involved, should help you in forming an idea about what the problem actually is. All of this information may point to a device failure, a network media problem, or even some other external problem that's outside of your span of control. That's when you progress to the next step in the troubleshooting process, which is establishing a theory about the probable cause of the problem.

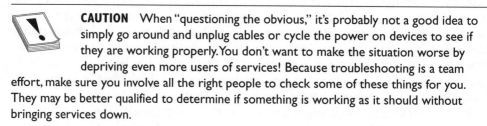

CAUTION Don't be tempted to short-change your time spent in identifying the problem! Gather as much information as you can, identify the symptoms, question the users, and find out what changed on the network. If you don't get enough information out of this step, you'll wind up wasting time and eventually coming back to this part of the methodology!

Establish a Theory of Probable Cause

Now that you have all the information gathered, you're going to need to establish a working theory of what the problem is and what caused it. You may find that your initial theory isn't correct, but at least you'll be able to rule that particular problem or cause out as you go through the remainder of the troubleshooting methodology. If you gathered enough information, however, it's more likely that your first theory will be more often correct than not, or at least lead you farther toward what the real cause is. Now you can start determining if your theory is correct.

Question the Obvious

One of the primary things you're going to do when trying to establish a theory of probable cause is question the obvious. This means you're not going to just assume that something works the way it should or is connected the way it should be; you're actually going to go and check it to make sure that it is. It is one thing to say, "Well, the cable should be connected" and another thing to actually go and see if it really is connected. You may find that you waste a little bit of time in confirming that different devices are working the way they should, or that media is connected the way it should be, but in fact, you are ruling out potential causes of the problem. Eventually, you are likely to run across something that is actually *not* working the way it should be. And that is likely when you will have found your true problem.

CAUTION When "questioning the obvious," it's probably not a good idea to simply go around and unplug cables or cycle the power on devices to see if they are working properly. You don't want to make the situation worse by depriving even more users of services! Because troubleshooting is a team effort, make sure you involve all the right people to check some of these things for you. They may be better qualified to determine if something is working as it should without bringing services down.

Testing the Theory to Determine Cause

Once you've come up with the theory, you have to test the theory to see if it is the correct one. This may mean making a change to the host or network to see if it fixes the problem. There are a few ways you can do this. First, if the problem applies to a larger group of users or to an entire section of the network, try your solution on a smaller segment if possible, or for a small group of users. If this seems to fix the problem, allow the network to stabilize a bit, and then expand your solution to the rest of the affected users and network. If the problem isn't fixed, reverse any actions you took, and go back to the previous step. If the problem affects only one user or a smaller group of users, try the solution to see if it works on the local segment only. Remember to change only one thing at a time if you can help it.

Confirming the Theory

If your solution fixes the problem, leave it in place and monitor the situation for a while to ensure that everything is working as it should. Have users, or even technicians, test different actions to ensure that the host, its applications, and its connectivity over the network function as they should. You may have to implement only a temporary solution to test the theory and confirm it in lieu of a more permanent fix. This is especially true if you require new equipment, or extensive recabling, for example.

Disproving the Theory

If your solution does not correct the problem, or offers only a sporadic fix, this may mean that your initial theory was incorrect. It can also mean that there are additional problems as well. In any case, you should reverse your actions and restore the network or host back to the way it was. You should also go back a few steps in the process to ensure that you have all of the information and symptoms you need. You may have to gather additional information, or ask more questions of the users involved.

Resolving the Problem

While the goal is to correct the problem, of course, resolving the problem unfortunately doesn't always mean the problem is fixed. In some cases it may mean that you can't solve the problem at your level due to knowledge or experience, decision-making authority, or area of control. For any of these reasons, you may have to escalate the problem to a higher level of troubleshooting. You also may have to get a more experienced technician or engineer involved. In some rare cases, you may even have to get a third-party troubleshooting technician involved for specific pieces of equipment, especially if they are under warranty or very complex. In any case, the next step is to establish a plan of action.

Establish a Plan of Action

Establishing a plan of action could mean one of two things. If you test your theory of probable cause, and it appears that you have successfully identified the problem, you then have to establish a plan of action to implement the solution to the problem. The solution could be an easy fix, or it may be a temporary fix of some sort while you wait on equipment to be repaired or replaced, for example. Oftentimes, the plan of action may have to be approved by the supervisor or management.

If you have not successfully identified the problem, the plan of action may be to escalate to higher troubleshooting levels or to management so that decisions can be made on how to proceed. Escalation may involve further in-depth troubleshooting, or it may necessitate the purchase of new equipment or communication services. It is likely that management will have to decide what route it will take based upon the criticality of the services lost, or the severity of the problem.

Identifying Potential Effects

Regardless of whether you've identified the problem or not, in addition to a solution, you must identify what effects the ongoing problem, or even the solution, may have on the network and its users. For example, if the issue is with a piece of faulty equipment that must be replaced, potential effects may be that certain users can't access critical services. You may put a temporary piece of equipment in its place that is less robust and may provide diminished services. You'll have to decide what the effect will be on both the users and the services that you can provide. In some cases, you may have to make decisions about which services to temporarily suspend or eliminate so that critical services can be maintained. Keep in mind that it likely won't be only you that makes that decision; you likely have lots of help from other technicians, managers, and, of course, the users!

Implement

Once you have definitively identified the problem and the solution, it's time to implement it. It may be a five-minute, one technician job, or it may require several hours of downtime and several technicians. Either way, you should make sure that you have the right solution, the right tools and equipment, and the right people to implement it.

Implement the Solution

In implementing the solution, you may have to install a temporary fix until a more permanent one can be put in place. Or you may have to go ahead and implement a permanent fix that requires management approval due to cost or manpower. Whatever the circumstances, make sure that you manage the implementation both from an efficiency standpoint and from a customer service standpoint. Make sure you schedule downtime for any equipment assets involved, especially if it involves suspending services for customers for any length of time. In most large organizations, there are formal trouble ticketing and maintenance procedures that must be followed.

Escalate the Problem

As mentioned previously, if you haven't identified the problem, or need approval for the solution, escalation may be your next step. In most large organizations, there are procedures for escalation to the next-level technician or to supervisory levels that may have to approve of certain solutions. When escalating a problem, make sure you documented all the information you have concerning the symptoms and the problem itself, as well as any troubleshooting actions you've taken up to that point. Once you have escalated the issue, you should try to be available as much as needed to help continue to solve the problem with the more experienced technician, or to provide any technical information management may need in order to approve a solution.

CAUTION Don't be afraid to escalate a problem if it exceeds your knowledge, experience, or authority. This is much better than making bad decisions that could make the problem worse.

Verify

Let's think positively here. Let's assume that you have successfully identified the problem and the solution, and have already implemented it. The next step would be to verify that the solution resolved the problem. This means testing out the network connectivity and hosts to ensure that everything functions as it did before the problem surfaced. You should verify with the users, as well as other network personnel, that the solution not only solves the problem it was intended to, but that it did not create any additional problems for anyone else.

Verify Full System Functionality

Verifying the solution worked may mean monitoring the situation for a few minutes, hours, or even a couple of days. You should review the results of any logs, if possible, that may indicate that the solution worked or failed, or use any other metrics that may indicate the same. For example, if the problem involved a nightly data transfer that didn't complete properly, then monitor the same data transfer over a period of several nights to determine if the transfer completed the way it should have. Verifying that the solution you implemented is working is also a really good way to make sure that you correctly identified the problem in the first place.

Implement Preventative Measures

Once you've fixed the problem, obviously you want to make sure that it doesn't happen again. While you can almost never predict with 100 percent certainty the problem won't occur again, you can take preventive measures to reduce the likelihood somewhat. If the problem happened to have been a broken cable due to heavy foot traffic kicking it periodically, then you could probably prevent a future occurrence by either rerouting the cable, or rerouting the traffic path around the equipment or cables. If the problem was a bad board in a switch or router, you may not be able to prevent another board from going bad, but you can keep a spare or two on hand in case another one does. In any event, prevention can reduce the likelihood of a repeat occurrence of the same problem, or at least reduce the amount of time and resources spent troubleshooting it the next time it happens.

Documentation

As I mentioned earlier in this chapter, documentation is very important to the entire troubleshooting process. Documentation lets other technicians and management know what happened, from the beginning to the end of the problem, and what steps you took to resolve it. Documentation can help identify a future problem more quickly, as well as the solution to resolve the problem. The three things you should definitely document are findings, actions, and outcomes.

Document Findings

Documenting findings involves documenting the troubleshooting steps you took to discover the symptoms and determine the problem. You should document all the information you've gathered, its source, and its relevancy to the problem. It's also important to document what led you to determine the true problem. Summarize your interviews with users. List any symptoms that were experienced by the user, host, and the network devices. Capture information on any configuration settings and the connection state at the time the problem occurred. This will help others follow the same path if a similar problem occurs again.

Documenting Actions

You must also document any actions you took in troubleshooting and resolving the problem. This includes any tools or techniques you used, any questions you may have asked, configuration settings you may have examined, and any devices you checked. You should document any solutions you tested and what the results were. Also, document the actions you took in establishing the working solution. Note any escalations that had to happen or any additional help or resources you needed in determining the problem or solution, as well as implementing it.

Documenting Outcomes

Finally, you should document any outcomes associated with the problem and its solution. This includes solutions you tested that didn't exactly work, and why, as well as what solution ultimately worked and what led you to it. You should also document any preventative actions and measures you took to preclude the problem from occurring again. You may need to include any administrative information with your documentation, such as trouble ticket number, technician number, and so on. The documentation that you make could be in the form of an organizational trouble ticketing and resolution database that will be used in future troubleshooting efforts.

 EXAM TIP Make sure you document everything relating to the problem and its solution. This means the findings, actions, and outcomes.

 EXAM TIP Make sure you know what happens at each step in the troubleshooting methodology, and the order the steps are performed in for the exam.

Troubleshooting Network and Over-the-Air Connectivity

Now that you've established a working troubleshooting methodology, you can apply this methodology to a wide range of problems that you may encounter in your career as a mobile device and infrastructure technician. You'll find that a great deal of your problems are going to be associated with network connectivity, whether it's on the wired side of the network, or the wireless side. Each different type of media is known to have its own unique issues from time to time, but you'll find that they have many issues in common as well; they just may be manifested differently.

The next several sections discuss the different types of issues that can occur with regards to network connectivity. Many of these issues are directly associated with wireless, or over-the-air, connectivity, and may involve network infrastructure or the device itself. Some of these even relate to the different technologies that are implemented into the network.

Basic Connectivity Troubleshooting Tools

Several tools are available to the administrator to help troubleshoot network connectivity issues. Some of these are in the form of built-in utilities present on most operating systems. Others are third-party open source or proprietary software tools that can provide advanced network connectivity information. Still other tools come in the form of hardware devices that can be used to trace problems over both wired and wireless networks. I'll discuss a few of these tools over the next few sections.

Utilities

There are several utilities that are built into the typical operating systems that run on most devices. These utilities work on Windows, Mac OS X, Unix, Linux, and even some mobile operating systems, such as iOS and Android systems. Usually these utilities are built into the command line interfaces of most of the standard operating systems, or in the form of apps for mobile device OSes. For the most part, they are standardized, but you will see some variations occasionally between operating systems. Many of these utilities are useful for troubleshooting connectivity issues.

The first utility I will talk about is the *ping* utility. I mentioned this in Chapters 1 and 2 when discussing the OSI model and the TCP/IP stack. Ping is a utility that makes use of ICMP packets at the network layer of the OSI model. Recall that ICMP is a maintenance protocol and essentially is used to establish that the remote host is alive and connected to the network. In order to use the ping command for troubleshooting, you simply type the ping command at the command prompt, followed by the IP address of the host you're trying to reach. If you receive a response back in the form of a reply, the host is alive on the network. Any other reply could mean one of several things. It may mean that the host is down, that no route to the host can be found, or that the ICMP protocol is blocked by a network device between your host and the destination host. Any one of these issues requires further troubleshooting. Figure 10-2 shows a typical ping output when executed from a Linux host.

The next utility you need to be aware of is the traceroute command. Traceroute also uses ICMP, but continues on past several intermediary devices and records the time it takes to connect to those devices along the way. It can also show the names and IP addresses of these devices. Traceroute will travel along the network path through all these intermediary devices until it reaches the destination

```
 ^  v  ×  CompTIA Mobility+
File Edit View Terminal Help
root@bt:~# ping www.mheducation.com
PING www.mheducation.com (166.78.101.13) 56(84) bytes of data.
64 bytes from mhe.cgsrvr.com (166.78.101.13): icmp_seq=1 ttl=50 time=52.8 ms
64 bytes from mhe.cgsrvr.com (166.78.101.13): icmp_seq=2 ttl=52 time=59.0 ms
64 bytes from mhe.cgsrvr.com (166.78.101.13): icmp_seq=3 ttl=50 time=60.7 ms
64 bytes from 166.78.101.13: icmp_seq=4 ttl=52 time=57.9 ms
64 bytes from mhe.cgsrvr.com (166.78.101.13): icmp_seq=8 ttl=52 time=54.7 ms
^C
--- www.mheducation.com ping statistics ---
11 packets transmitted, 5 received, 54% packet loss, time 19104ms
rtt min/avg/max/mdev = 52.822/57.050/60.705/2.867 ms
root@bt:~# 
```

Figure 10-2   The output of the ping command in Linux

host. As with the ping command, if ICMP is blocked at any one of these hosts, the traceroute command may fail or return inconclusive results. Figure 10-3 shows traceroute run on a Linux system.

 TIP   The traceroute command is spelled differently on Windows-based hosts than it is on Unix, Linux, and Mac OS hosts. In Windows, the command is spelled *tracert*. In Unix-flavored hosts, it is spelled out as traceroute. It works the same way, regardless of which OS you use, however.

```
 ^  v  ×  CompTIA Mobility+
File Edit View Terminal Help
root@bt:~# traceroute www.google.com
traceroute to www.google.com (74.125.134.105), 30 hops max, 60 byte packets
 1  router.TBIRD (172.16.30.1)  4.806 ms  5.585 ms  6.309 ms
 2  * * *
 3  user-24-96-68-181.knology.net (24.96.68.181)  29.253 ms  28.184 ms  29.812 ms
 4  user-24-96-198-29.knology.net (24.96.198.29)  23.244 ms  24.063 ms  24.605 ms
 5  user-24-96-153-141.knology.net (24.96.153.141)  25.334 ms  25.947 ms  26.558 ms
 6  user-24-96-153-133.knology.net (24.96.153.133)  27.269 ms  18.171 ms  18.713 ms
 7  dynamic-76-73-147-205.knology.net (76.73.147.205)  23.576 ms  22.378 ms  28.242 ms
 8  cr02-mx-atln-tlx.knology.net (76.73.147.237)  29.539 ms  29.508 ms  35.233 ms
 9  user-24-96-35-130.knology.net (24.96.35.130)  34.997 ms  35.056 ms  34.824 ms
10  72.14.233.54 (72.14.233.54)  34.378 ms  35.013 ms  34.947 ms
11  66.249.94.22 (66.249.94.22)  33.120 ms 66.249.94.20 (66.249.94.20)  32.756 ms 66.249.94.24 (66.249.94
.24)  29.616 ms
12  209.85.254.247 (209.85.254.247)  21.744 ms  20.484 ms  25.874 ms
13  * * *
14  gg-in-f105.1e100.net (74.125.134.105)  26.259 ms  23.943 ms  23.810 ms
root@bt:~#
```

Figure 10-3   Traceroute output from a Linux system

Software Tools

While the `ping` and `traceroute` commands can show you basic connectivity between two hosts, some issues can better be resolved using more advanced software tools. Most of the software tools are either third-party open source or proprietary tools, and can give you a wide variety of information about the network, its configuration, and status. Beyond simple network connectivity, most of these tools are worth the investment in order to help you manage the network performance and devices. While there are several hundred software packages out there that can help you manage your network, you should pick the one that will best assist you based upon your network infrastructure. Most tools cover wired networks at the very least, while some of the more advanced tools also allow you to manage wireless networks.

Figure 10-4 shows an example of a popular network management tool suite from SolarWinds, which offers a wide variety of individual tools and utilities to help you manage and troubleshoot network performance and connectivity.

Another software tool that requires a bit more than basic knowledge is a network traffic sniffer. A sniffer is able to monitor all the traffic on a network, examining and recording each piece of data as it travels over the network. This allows a network technician to examine the traffic later to detect performance and security issues. Most sniffers available today can capture traffic on both wireless and wired networks, but it essentially depends upon what type of interface card is on the device running the sniffer. Sniffers can be dedicated hardware devices, or software that runs on top of an operating system. Some OSes, such as Linux, include sniffer packages in their distributions,

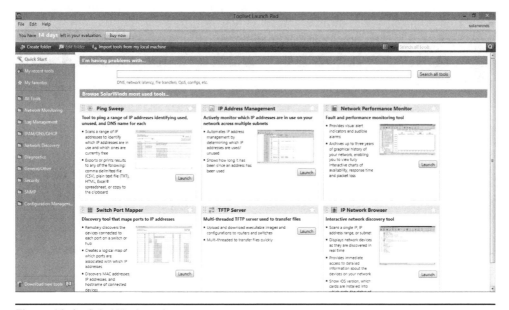

Figure 10-4 SolarWinds toolset

while others, such as Windows, have third-party software sniffers available that can be installed on them. One of the most popular software sniffers that is used across most common platforms is Wireshark. Wireshark is available for both Linux and Windows (some Linux distributions include it already installed), and is capable of sniffing both wired and wireless networks, depending upon the type of network card installed in the system, as well as the particular type of drivers available for the card. Figure 10-5 shows Wireshark on Linux sniffing a wireless LAN.

Hardware Tools

In addition to the basic utilities and software tools discussed in the previous sections, there are also a wide variety of hardware tools out there that can help you troubleshoot network performance and connectivity. Some of these tools are specific to the media involved; for example, there are hardware tools that check connectivity of Ethernet twisted pair cable, as well as those that check for breaks in fiber cable. These tools are usually called "cable testers" and "time-domain reflectometers" (TDRs), amongst others, and can be very simple in operation or quite complex. Some cable testers can not only tell you that there's a break in the cable, but even how far down the cable the break occurs. There are also hardware tools that can be used specifically for wireless LANs and frequently come in the form of basic radio signal detectors or advanced spectrum and frequency analyzers, such as the ones I discussed in Chapter 3.

Hardware tools are typically used when software tools and utilities tell you that there's a problem with connectivity or performance, but can't tell you exactly what the nature of the problem is. For example, a `ping` output may tell you that one host can't

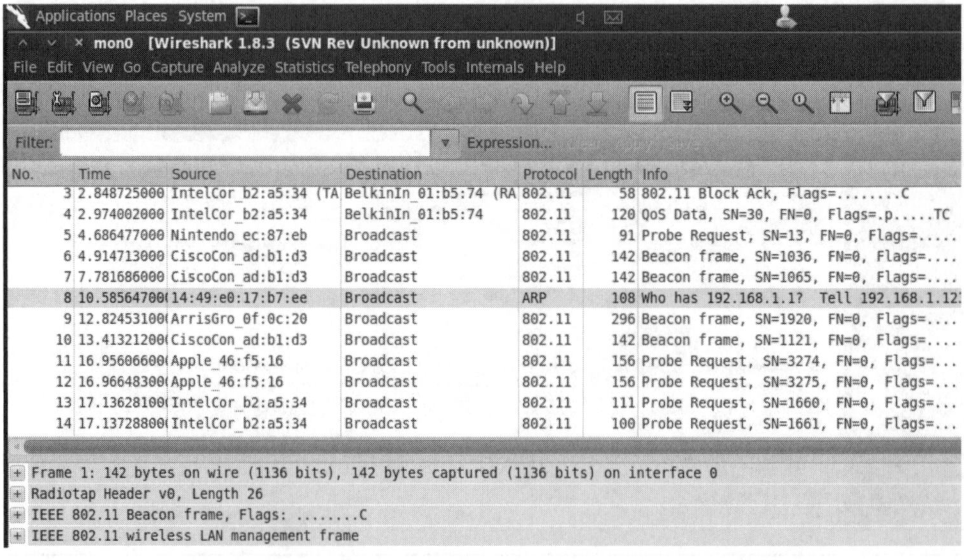

Figure 10-5 Wireshark sniffing a wireless LAN

contact another, but it doesn't tell you if the reason is a broken cable or not. A cable tester may be required to make that determination. Basic cable testers can be somewhat inexpensive, but other hardware tools, such as TDRs and spectrum analyzers, can run hundreds or even thousands of dollars, and can require some higher-level knowledge to operate them.

 EXAM TIP   There isn't any specific objective called out in the Mobility+ Exam Objectives for troubleshooting tools, but that doesn't mean you won't see anything about them on the exam. You likely won't be asked about specific tools, but `ping` and `traceroute` are still probably good to know for the exam.

Common Network Issues

While the focus of both this book and exam is really on mobile devices, you'll find that the line between wired and wireless networks, beyond the media type used, isn't always as clear-cut and different as you might think. While there are definitely some differences in the types of issues that wireless and wired networks can have, they also share common issues inherent to devices, applications, security, and the TCP/IP protocol stack in general. The next few sections discuss some common issues that may affect both wired and wireless networks alike, albeit sometimes in different ways. Then I'll focus more on unique over-the-air connectivity issues that are specific to devices that use wireless technologies, such as wireless LAN and cellular.

Keep in mind that network connectivity issues by and large mean physical types of problems, such as cabling, wireless signal issues, and host device or network equipment issues. However, you can also have issues with network-enabled technologies, such as email, VPN connectivity, and even certificate-based network authentication and encryption. There are also network service issues that can be disguised as connectivity issues, such as DNS and DHCP problems.

Latency

The first connectivity issue I'll discuss is latency. *Latency* is the delay, or lag, that can occur between message transmission and reception. Latency can occur over both wired and wireless networks. It can happen between specific points in the network, or across the entire network. It's generally cumulative, as multiple delays add up to increased latency. It can be caused by a wide variety of issues, including equipment or interface saturation, and even network routes the data takes from the sender to the receiver. For example, it can cause significant latency if a route takes a transmission through a major distribution point that is subject to a high volume of traffic. Other causes of latency could be slow or older network devices and interfaces, lower-grade cable, and weak wireless signal. The different points along the network path that latency may occur include customer equipment, the organizational distribution point, or even the upstream service provider.

Latency is typically shown as an increase in the time it takes for network traffic to travel between two points. Latency can be sporadic because network conditions can change frequently. Some of these sporadic problems happen randomly, but usually average out so there is no extended latency difference over time. They may turn out to be more of a performance issue than a loss of service issue, depending upon severity and frequency of occurrence. These types of sporadic problems can usually be cured by adding more efficient equipment, and optimizing the configuration of existing network devices. Serious latency problems, on the other hand, can show up as a definite significant delay in communications and a serious performance decrease to the point that it causes a loss of service. If you're troubleshooting latency between two hosts, a simple `ping` can help you. When trying to find latency over a longer network path, the `traceroute` command can show you all of the different connection points, or hops, the data traffic goes through between two points. Wireshark, or another network traffic sniffer, can show more detailed and advanced protocol options to help determine not only where the latency exists, but what might be causing it. The point in the path where there is significant latency will be shown by an increase in time. Figure 10-6 shows some latency in a `traceroute` output, as shown by the increased time between two hops.

 EXAM TIP Note that latency is an *increase* in the time it takes traffic to get from one point to another. So, the more time you see traffic take to travel between two hops, the greater the delay and the greater the latency.

No Network Connectivity

The issue of having no network connectivity at all is quite different from having slow performing or sporadic network connectivity. Slow or sporadic connectivity could be caused by the latency issues discussed previously, and may be more difficult to troubleshoot than having no network connectivity at all. In the case of a complete loss of network connectivity, you usually have to start from the host device and work your way out

```
^  ∨  ×  CompTIA Mobility+
File Edit View Terminal Help
root@bt:~# traceroute www.mheducation.com
traceroute to www.mheducation.com (166.78.101.13), 30 hops max, 60 byte packets
 1  router.TBIRD (172.16.30.1)  21.487 ms  50.035 ms  51.487 ms
 2  * * *
 3  user-24-96-68-181.knology.net (24.96.68.181)  51.500 ms  51.848 ms *
 4  user-24-96-198-29.knology.net (24.96.198.29)  51.849 ms  52.158 ms  59.284 ms
 5  user-24-96-153-141.knology.net (24.96.153.141)  59.358 ms  59.664 ms  74.077 ms
 6  user-24-96-153-133.knology.net (24.96.153.133)  74.241 ms  21.958 ms  21.913 ms
 7  dynamic-76-73-147-205.knology.net (76.73.147.205)  30.774 ms  30.501 ms  30.501 ms
 8  cr02-mx-atln-tlx.knology.net (76.73.147.237)  35.999 ms  36.001 ms  43.402 ms
 9  user-24-214-2-222.knology.net (24.214.2.222)  52.480 ms  51.022 ms  43.061 ms
10  xe-10-3-2.bar1.Tampa1.Level3.net (4.53.172.21)  43.473 ms  43.254 ms  41.326 ms
11  ae-0-11.bar2.Tampa1.Level3.net (4.69.137.110)  41.972 ms  42.630 ms  40.643 ms
12  ae-12-12.ebr1.Dallas1.Level3.net (4.69.137.118)  59.121 ms  59.765 ms  58.040 ms
13  ae-14-14.ebr2.Chicago2.Level3.net (4.69.151.117)  61.096 ms  52.853 ms  55.871 ms
14  ae-210-3610.edge1.Chicago2.Level3.net (4.69.158.229)  61.525 ms ae-209-3609.edge1.Chicago2.Level3.net (
4.69.158.225)  55.982 ms ae-212-3612.edge1.Chicago2.Level3.net (4.69.158.237)  60.927 ms
15  4.71.248.54 (4.71.248.54)  63.060 ms  60.980 ms  61.913 ms
16  coreb.ord1.rackspace.net (184.106.126.134)  53.892 ms  60.154 ms  61.605 ms
```

Figure 10-6 Latency between two hops

as far as you logically can away from the host to the point where you lose connectivity. For example, let's say a user's computer can't connect to the Internet. The problem may be a broken or loose cable (in the case of a wired network), a faulty wireless access point (in the case of wireless networks), a host configuration issue, or possibly a faulty network device. Once you check the cable, then you would make sure that the host can contact its default gateway, which is usually a router interface on its local subnet. If you `ping` the router interface and get no reply, then the issue is (from most to least likely) the network configuration on the computer, the network cable, or possibly the interface card on the host or even on the router itself. If you can successfully `ping` the default gateway, then you know that it's likely either a network device or service issue upstream away from the client, or possibly even a connection issue with the service provider.

In the case of a wireless or cellular data network, the `ping` and `traceroute` commands can still help you troubleshoot the problem. If it appears that the network is turned on, detected, and functioning on your wireless device, then use those commands to see if you can connect to the next higher device, which will still be your default gateway. This can be determined by viewing the TCP/IP configuration on the device. This is probably much easier done on a laptop or even a tablet device because they are more likely to have the typical TCP/IP utilities installed on them. For a smartphone, you may not have access to these utilities or all of the information you need, depending upon the device. In that case, you can try to contact the host device from an upstream network device.

Network Service Issues

There are several key network services that, if they do not function properly, can cause connectivity issues, and possibly lead you down the wrong path when troubleshooting. Two of the more common network services that can cause issues are DNS and DHCP. Each of these can affect connectivity in certain ways, so you may want to pay attention to them depending upon what the symptoms are.

DNS, as you recall from our chapters on network technologies, allows a computer to resolve Internet-friendly names, such as www.google.com, to IP addresses that most applications use. It uses both TCP and UDP ports for different actions and both are configured for port number 53. DNS issues crop up as the inability to contact an Internet or corporate host by its DNS name (or, as it's also called, its Fully Qualified Domain Name, or FQDN). This might be a web address, such as http://www.mheducation.com, for example, or something like mail.mycompany.com. The host could still be contacted and used by using its IP address, but some applications (and most users) may not be configured to use the IP address. Email clients, for example, may be configured with the server name instead of the IP address, and would rely on DNS to resolve it. In the event users complain that they can't get to Internet sites with their browsers, try running a `traceroute` on the site's FQDN. If it can't resolve the FQDN to the IP address, then you may have a DNS issue. Try running a `traceroute` against a known IP address (either on the local or a remote network) instead of a host name; if you receive a reply back, it could be a DNS problem.

DNS problems are usually client configuration issues. The client usually must have a correct DNS server IP address (and sometimes an alternate one) configured in its TCP/IP configuration, usually right along with the IP address, subnet mask, and default gateway information. If it is incorrect, or nonexistent, then it may have been configured incorrectly on the device or host by the administrator. If DHCP is used on the network, there may be an incorrect or missing entry in the DHCP server's database regarding the DNS server information that it sends to hosts. Figure 10-7 shows DNS configuration settings in Windows.

DNS server issues also occur, but are usually less frequent. The server could be offline, or suffering from a connectivity issue. If there is another DNS server on the network that functions as an alternate, and the client has the alternate's IP address information configured on it, it should failover to using the alternate server if it can't contact the primary one. If this doesn't happen, look at the device configuration to make sure the alternate's IP address is also listed.

TIP DNS configuration issues typically don't include incorrect port numbers because port numbers are not normally required to be entered on the client in order to communicate with a DNS server. Most operating systems and applications default to using port 53 by convention and design.

Figure 10-7
DNS configuration settings in Windows.

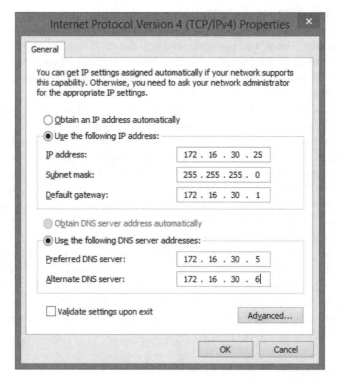

The other major network service that can affect client connectivity to the network is the DHCP service. Remember that DHCP is responsible for giving out IP addressing information to network clients automatically, assuming they have not been otherwise manually configured by an administrator. This information usually includes IP address, subnet mask, default gateway, and the primary and alternate DNS server addresses. DHCP service issues might manifest themselves in several ways. Usually the symptoms include clients that can't connect to the network or the Internet because they have no IP address, or a client may be using the APIPA range (169.254.x.x) I discussed in Chapter 2. This address range is non-routable, and a host using it may default to it if it can't get an IP address from a DHCP server. It would also only be able to contact other hosts that use the same range.

Because the DHCP server issues IP addresses on a lease basis, the IP address a client receives may be good for several hours or days before the client has to contact the server again to renew its lease on the address. Unfortunately, sometimes this means that DHCP problems may not be readily apparent with longer lease times. What typically happens is that a few clients at first may fail to connect to the network, and then gradually more and more will have issues. Examining only a few client configurations is usually enough to establish a DHCP server issue. The issue could be a lack of connectivity to the server, or an issue with how the server is configured. It can also be an issue with any routers that are placed between the DHCP server and the affected clients. DHCP messages from the clients are broadcast-based. Remember that routers don't ordinarily forward broadcast messages between networks. If a special exception is not programmed into the router to allow DHCP broadcast messages, the server will never receive the client's request to obtain or renew a lease.

EXAM TIP The two obvious symptoms of DHCP problems are when the client has no IP address assigned at all, or if the client uses the APIPA-assigned IP address (169.254.x.x). Either of these two conditions point to DHCP problems.

Email Issues

Email issues are common in the network. There are several different components in the email infrastructure that can cause problems sending and receiving email. First, on the client side, configuration can cause issues if the client has incorrect information, such as a wrong SMTP port number and server name or address. This might prevent the client device from sending email. Likewise, having an incorrect POP3 or IMAP port number and server address can prevent the client from receiving email. These basic configuration items cover the network issues that can occur in the client configuration, but there can also be security issues on the client that may cause email problems across the network. Authentication can be a primary security issue if the client is attempting to use incorrect user credentials, such as an incorrect username and password combination. In the event that the client uses certificates to authenticate to the email server, an expired, corrupt, suspended, or revoked certificate can prevent authentication. Another security issue can be encryption level set incorrectly on the client. The encryption level must be set to be identical to that of the server, or at least offer choices so that the client and server can negotiate a mutually compatible encryption method. These are all items that should be looked at on the client device in the event the user can't send or receive email securely. Figure 10-8 shows a configuration screen for sending email on an Android device. Note the SSL, port, and server address settings.

On the server side, most of the same issues that a client can potentially have could also affect the email server. This includes SMTP, IMAP, and POP3, and port numbers that could be configured incorrectly. It also may include different encryption methods that must be negotiated between the client and server. If the client and server are attempting to negotiate encryption and can't agree upon a mutually compatible method, communication will likely fail. Occasionally, there may also be issues with the users' mailbox configuration. If the email issue affects only one user, this is one area you may need to check. If the issue affects multiple users, then the network configuration or authentication configuration should be looked at as well.

Figure 10-8

Configuration settings for send-ing email on an Android device

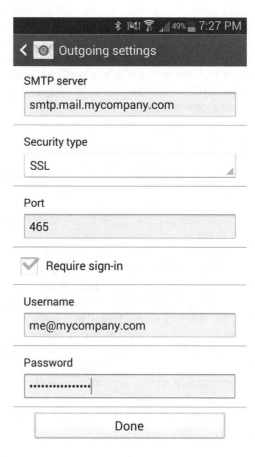

As far as server-side authentication goes, this issue might be unrelated to user credentials; rather, it could be the inability of the email server to contact an authentication server. If usernames and passwords are used, the email server must be able to contact a central authentication server in order to confirm the identity of the user. If certificate-based authentication is used, the issues could stem from an expired certificate on the server, or the inability of the server to verify the client's certificate with the authentication server.

Any one of these issues can be difficult to troubleshoot, but in the event of email issues, you should follow the standard troubleshooting methodology discussed earlier. Narrowing the problem to either the client or the server is an important part of troubleshooting email issues. Once you determine on which end the problem lies, following the troubleshooting steps and knowing what types of issues could be present on each should help you rapidly find the problem and solve it.

 EXAM TIP Remember the possible causes of email issues for the exam: incorrect port, incorrect or missing server address, and certificate issues. Also remember that it is rarely the server side; usually it is a client configuration issue.

VPN Issues

Early on in Chapters 1 and 2, I discussed virtual private networking and some of the technologies that make this possible. VPN problems can result from any one of several different potential issues. These issues could be on the client side or on the server/concentrator side. Issues on both sides could include encryption levels not set at compatible levels between the client and the VPN concentrator, authentication issues that may result from an inability to contact the authentication server, or even network configuration issues such as port and protocol settings.

There could also be issues with the network connection itself between the client and server. Because a VPN connection travels across an untrusted network, such as the Internet, there could be issues with the Internet connection as it travels across different devices between the client and server. This can include connectivity issues with switches and routers, or even security devices that may block the VPN ports and protocols needed to establish the connection.

In any case, you should troubleshoot the client side first, then the server side, and finally the connection between them, if possible. On the client, you should make sure that the VPN client itself is configured properly, to include ports, protocols, encryption levels, and authentication settings. In terms of mobile device VPN clients, several different options are available, depending upon the device itself, as well as the type of VPN client you are configuring. Some of these options are shown in Figure 10-9 for an Android smartphone, and include settings related to the type of VPN tunnel required (L2TP, PPTP, and so on), as well as the name server and other configuration details.

You should check the same items on the server side as well, including configuration items that allow the VPN concentrator to contact the authentication server. For connection errors that may be outside of the client and VPN infrastructure devices,

Figure 10-9 VPN configuration options for an Android smartphone

you may have to contact your service provider to determine if there are issues with the untrusted connection or network you are crossing.

EXAM TIP The three biggest issues with VPNs are connectivity, authentication, and encryption. Check for symptoms that may indicate one of these problems.

Certificate Issues

I've already mentioned a few of the issues you can run into with certificates. Many different technologies depend on client and server certificates, as well as user certificates, in order to function. Email, remote VPN connection, remote authentication, and other network functions may require certificate-based authentication methods in order to work. One of the issues you can encounter with certificate-based authentication is expired certificates. If either the client-side or the server-side certificates are expired, the security policy may not allow a connection and authentication to happen. The same issues could also happen if the certificate has been suspended or revoked. In these cases, it may not be a problem that's actually happening; it may be that the authentication server is doing its job in not allowing a suspended or revoked certificate to be used for authentication.

One other issue that can occur with certificates from time to time is that a certificate may be corrupt on the device. Sometimes simply reloading the certificate can solve the issue. On both client and server devices that use certificates, you should also check that the appropriate ports and protocols are used. This will usually be either SSL or TLS, and may involve TCP port 443 or one of the other application-specific ports that make use of certificate-based authentication. These are the issues with certificates that could affect network connectivity and authentication; in Chapter 11, when I cover troubleshooting specific security issues, I'll also cover some additional information about troubleshooting certificates.

Port Configuration

Network configuration is likely where you're going to see the large majority of your issues. One of the most common problems with network configuration is a client or server trying to use an incorrect TCP or UDP port. As with other configuration items, the port and protocol used for a particular service have to be configured identically on both the client and the server. An incorrect port configuration can cause a general lack of connectivity and authentication issues, and can make troubleshooting difficult. Most common protocols, such as HTTP, DNS, and DHCP, for example, default to their standard well-known ports, unless they are purposefully configured to use non-standard ones. If this is the case, you must make sure that both the client and server are configured for these non-standard ports. Users may also have to remember to include the non-standard port in the URL of the web site they are visiting if it's an HTTP request over a browser. It's a good idea to maintain a list of standard ports, protocols, and services that your network uses to assist in troubleshooting.

Network Saturation

Network saturation can be an issue that causes complete loss of connectivity at worse, or poor network performance at minimum. In either scenario, network saturation can be a difficult problem to track down because it's not often caused by a component that's broken; it may be that the network traffic exceeds the capacity or capability of the component. When the network is first designed, equipment may be able to handle the expected network load easily, but over time, as more devices, new protocols, and more traffic are introduced into the network, the load may exceed the capacity of the network. This could be in a major piece of equipment, or can even be in an interface on a switch or router, for example. This is one reason that performance baselining is so important on a network.

A performance baseline should be measured for the network after it is first put into operation, and periodically thereafter, especially after major network upgrades and additions. The performance baseline would tell you how well the network is handling the traffic load. There are various ways to do this, including the use of sniffers, network device logs, and even specialized devices that can measure traffic load. Most advanced network devices also have built-in methods of measuring traffic load and alerting an administrator if this traffic load exceeds a predetermined threshold.

In the event that network saturation is an issue, the logical long-term solution is to upgrade and replace older equipment that can't handle the load. For wired networks, you should look at replacing routers and switches, or upgrade the infrastructure by replacing copper cables with fiber, for example. For wireless communications, this may include conducting a site survey to account for changes in the wireless environment, and upgrading or adding access points (APs), as well as placement considerations for APs and antennas. You also should consider providing additional multiple paths and devices to help balance the traffic load. Segmenting networks further to help eliminate collision and broadcast domains is also a good step to take. In wireless networks, additional wireless LAN controllers and access points may help if the existing ones are trying to service too many clients. A short-term solution may be to impose bandwidth limitations on lower priority services and devices, or to reroute higher priority traffic. This is where the traffic shaping and QoS measures discussed in Chapter 8 may help.

Over-the-Air Connectivity Issues

Now that I've covered some of the common general network connectivity issues, let's switch gears and talk specifically about some basic over-the-air connection issues. These include problems with cellular signals, device activation, roaming, and application notification issues, such as those associated with the Apple Push Notification service (APNs), covered later. While I likely won't cover every unique issue a mobile device can have in terms of connectivity, these are the basic major ones that you will probably more often encounter, and the ones you'll need to be aware of for the exam.

In general, most of the problems associated with over-the-air connectivity involve the client device itself. A review of most manufacturers' troubleshooting web sites offers a wide variety of solutions to problems associated with cellular signal, roaming, activation, and so on. Many of these solutions are device dependent, of course, as well as operating system dependent. The range of troubleshooting solutions that the device vendor may offer includes cycling the power on the device, removing and reinserting the SIM, resetting the network or carrier configuration, and even drastic measures, such as wiping the device and reinstalling it from scratch. Although this last solution may be necessary every now and then, look at it as a last resort in solving a problem. Thankfully, the solutions I discuss here don't include that last one!

No Cellular Signal

An issue unique to cellular network connectivity is the loss of cellular signal. Very rarely will the issue be the cell tower itself or the carrier's equipment; usually this is a device issue. Range is the number one consideration when losing cellular signal. This usually happens as the device is traveling between cells and occasionally encounters what's known as a *dead spot* between cell coverage areas. The simple solution to this problem is for the user to bring the device back into a cell coverage area where the signal is stronger, or, if the user works in a remote area, a bit more involved solution may require a signal booster, such as a micro or picocell. Another common issue that affects cellular signal is obstacles. Tall buildings and other objects can block signals to the point that they are very weak or nonexistent. In this scenario as well, you could install a micro- or picocell at a strategic point that has better line-of-sight to a cell tower. Often, if a cell phone is inside a dense structure, signals will also be blocked. In some secure facilities, this is intentionally designed that way for security reasons. Usually, cellular phones are not allowed in those areas by policy.

Yet another cause of a weak or nonexistent cell signal may be a defective device or a device whose antenna is being interfered with somehow. This may occur because the user has put a case or some other accessory on the device that interferes with its transmission or reception. It also may be caused by a device that has a weak battery. These two potential issues are usually fairly easy to troubleshoot.

In order to rule out a defective device, you should have the user try reception from several different areas, including those in known good coverage areas. You also can have the user remove any extraneous accessories or cases to see if it affects cell reception. And finally, of course, have the user check to make sure the battery is fully charged. If none of these solutions work, you may have to issue a known good device to the user and bring the old one in for repair or replacement from the vendor.

 EXAM TIP Issues with weak signals almost always involve distance from the cell coverage area, obstacles that block the signal, or a device issue.

Roaming Issues

Several problems stem from roaming. Some of these problems may be particular to cellular data networks versus wireless LAN type of networks. In either case, potential sources of problems can be distance or coverage area from an access point or cell, hand-off between access points or cell areas, or the necessity to transition between cells of different carriers. Problems that can occur because of roaming issues include access tolls or charges, especially when roaming internationally, as well as signal loss.

Roaming between carrier coverage areas may cause a temporary loss of signal if the coverage areas are somewhat apart, and the device may struggle to lock in on another carrier's signal. The device may not automatically acquire a new carrier's network; you may have to manually configure it in the device's network settings. In some cases, the device may not be set to roam beyond its home area so the user may lose their signal completely when leaving their home carrier's network. Make sure the user is also aware of roaming charges that can apply, especially when the device is used in international travel. Some of the technical issues you may encounter when roaming internationally may require you to reset the device's time to match the time zone you're in if it doesn't automatically change the device time through the network. The time zone the phone is set for can affect roaming adversely if it doesn't match the local time zone. You may also have to change the signaling method for the phone from international GSM to international CDMA (or vice versa, as the case may be) if you are using a multiband phone. In some cases, the SIM can be incompatible with the local carrier in the country you're in. If the device is incompatible with the destination country's cellular network, or if the roaming charges are too high, it may be a good idea simply to turn off the device for the duration of travel. You could issue the user a temporary smartphone for the trip that uses the destination nation's home carrier to save on charges and ensure compatibility. Figure 10-10 shows how to turn off roaming for an iPhone.

Cellular Activation

Most consumer-level devices are usually configured and activated at the place of purchase by the carrier or vendor; however, devices that are part of an enterprise infrastructure may come preconfigured, or have to be configured by the appropriate enterprise mobile device management personnel. For the most part, activating a device on a cellular network should pretty much go off without a hitch. However, things do happen, so occasionally there may be issues with cellular activation. While the problem could conceivably be at the provider's network level, usually the problem will be at the device level in some way.

Figure 10-10
Turning off voice
and data roaming
on an iPhone

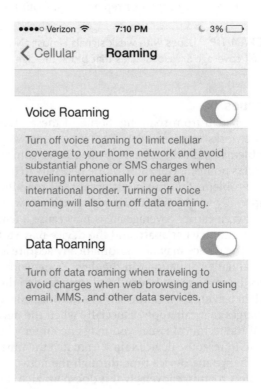

Assuming the device has been designed and configured to function on the particular carrier's network, most cellular devices are configured to automatically attempt to connect to the provider network and configure themselves. The user may need only to dial a certain number on the cellular phone, and the rest would be done automatically between the device and the network. Because very little user intervention is required to activate a cellular device, one of the issues that could occur involves a SIM that has been damaged or inserted incorrectly, or is even invalid. An invalid SIM will need to be brought back to the provider to be reprogrammed or replaced, as would a damaged one. Another issue that may affect cellular activation, although not really related to connectivity, is the account itself. In very rare cases, a new device may not activate because of issues with the user's account with the carrier. Corporate accounts should rarely ever have this issue, but it's worth checking if you have major activation problems.

Apple Push Notification Service Issues

The last topic I discuss in the chapter is the Apple Push Notification service. In other chapters I discuss the APNs from an enterprise development perspective, so here I will limit the discussion to connectivity issues only. Apple push notifications are configured to use cellular connections by default. If the device is using Wi-Fi and can't get a cellular signal, it may try to get push notifications from the wireless network. In this event, if there is a firewall or proxy server on the WLAN that is filtering traffic, the correct ports and protocols may have to be opened. Like any network service, the Apple Push Notification service suffers from some of the same problems that other services do, including port and protocol configurations on the device or the server, as well as communications interruption between the device and the network. In troubleshooting APNs issues, pay attention to ports. The APNs uses TCP port 2195, and the Feedback service uses TCP 2196. These ports should be opened up on the firewall or other security device to ensure that your devices can send and receive traffic through them to the APNs servers.

Although there is very little APNs troubleshooting you can do on the device level, the basic configuration step that you can actually perform is to set different notification options for individual apps. One of the interesting issues that can occur with the APNs is that you don't receive notifications for an app. This may not actually be a connectivity issue; it may actually be a configuration issue. When an app is first installed, and you are asked whether you would like to receive notifications, answering "No" to the prompt will turn off the notification service for that app. Unfortunately, under certain circumstances, you may not get the opportunity to turn notifications back on for the app. Sometimes, the only way to fix this problem is to uninstall the app, wait 24 hours, and then reinstall it. Figure 10-11 shows the limited client configuration settings for APNs in the Notification Center.

The APNs also has some very particular configuration requirements that you have to pay attention to with regards to the IP address range of the APNs servers. For example, one of the troubleshooting steps recommended by the Apple developer site is the inclusion of the entire class A network assigned to Apple, which is 17.x.x.x, in

Figure 10-11
The limited configuration settings available for APNs in the Notification Center

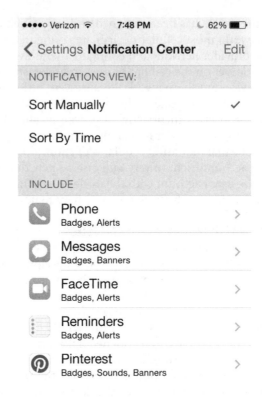

the firewall configuration. This is to avoid any issues, remote as they may be, with the device's ability to contact the service if it changes from one server to another due to load balancing (see Apple's iOS developer site at https://developer.apple.com/library/ios/technotes/tn2265/_index.html#//apple_ref/doc/uid/DTS40010376-CH1-TNTAG41 for details). There are, of course other troubleshooting issues involved with APNs as a service, but those are discussed in greater detail when you learn about troubleshooting application problems in Chapter 12.

Chapter Review

This chapter covered the importance of troubleshooting. I began by discussing an overall troubleshooting methodology composed of several steps that you can use to solve problems occurring within your network infrastructure. The methodology I discussed was a seven-step model that begins with identifying the problem. In order to identify the problem, you first have to gather information, and this can come from a variety of sources. You can get information from users, configuration files, network documentation, and real-time monitoring applications. The types of information you want to get include symptoms of the problem, what was going on the network when the problem occurred, and what the user was doing when he first saw the problem symptoms occur. You also want to determine what has changed on the network or the host that could have caused the problem to occur. Be careful not to confuse the symptoms with the problem, and take the time to follow the methodology in order to discover what the real problem is.

The next step in the methodology is to establish a theory of probable cause. You would use the information you've gathered during the first step regarding symptoms of the problem, but also questioning the obvious. This means to actually check the things you assume are working correctly to make sure they really are. This could include cables, network devices, and applications on hosts, as well as their respective configurations. The next step involves testing the theory to see if it actually holds as the problem. This involves either confirming or disproving your theory. Ways to do this include temporarily changing configurations, changing to new cables, or even swapping out network devices. Be careful that you only make one change to the network at a time so you know whether it definitively affects the problem or not, and change it back to the original state after you've tested it. If your theory is confirmed, continue to monitor the situation and determine the permanent solution. If it is disproved, restore the network back the way it was and go back into the previous steps in the process.

The next step is to resolve the problem by establishing a plan of action. This may mean fixing the issue or escalating it up to higher technical or supervisory levels if it can't be fixed at your level. In either case, a key thing you should accomplish here is identifying any potential effects on the infrastructure, impacts to the mission, or user services. You must determine what the effects on the infrastructure will be not only from the problem, but from the solution itself. Some solutions may temporarily interrupt services as they are implemented. Other effects include having to redesign or reengineer the network in order to put the solution in place. The next major step in the methodology is to implement the solution. This is done by installing either a temporary fix if needed, or a more permanent one if approved. If the problem has not been solved, or the solution cannot be implemented for some reason, you may need to escalate the problem to other technicians, or management. Once a solution has been implemented, you need to verify that it has solved the problem and not caused any other issues on the network. You should verify the functionality of all network services and devices, and this may take some time. You should also take this time to implement any preventative measures that will help keep the problem from occurring again.

The final step discussed in the troubleshooting methodology was documentation. Documentation is extremely important because it is the means to record the findings, actions, and outcomes from the troubleshooting process. This information can be useful if the problem occurs again in the future. Documentation can be implemented through a centralized trouble ticketing or maintenance record system.

The second part of this chapter dealt with network and over-the-air connectivity troubleshooting. First I discussed basic connectivity troubleshooting tools, which include built-in utilities, third-party software, and hardware tools. Some of these include the `ping` and `traceroute` commands, as well as network monitoring software and traffic sniffers. Hardware tools include cable testers, TDRs, and spectrum or frequency analyzers. Then I discussed general networking connectivity issues, including latency, loss of network connectivity, and network service issues. These problems can be traced to a wide variety of causes, including network device issues, client configuration issues, and service failure. The chapter also covered some problems that can occur with network-enabled applications and connectivity services, such as email and VPN services. I discussed basic problems that may be caused by certificate-based authentication, as well as port and protocol configuration issues and network saturation.

I then focused on specific over-the-air connectivity issues that affect cellular and wireless users. Some of the more common issues primarily affect the end-user mobile devices themselves rather than infrastructure equipment. The issues discussed include loss of cellular signal, issues involving roaming between carrier networks, cellular device activation, and finally, issues with the Apple Push Notification Service, which is typically a network service problem rather than a client configuration issue.

Questions

1. The seven steps of the troubleshooting methodology are listed here, out of order. From the choices provided, select the one that best describes the proper order:

    1. Establish a theory of probable cause.

    2. Identify the problem.

    3. Document findings, actions, and outcomes.

    4. Verify full system functionality and implement preventative measures.

    5. Establish a plan of action to resolve the problem and identify potential effects.

    6. Implement the solution or escalate as necessary.

    7. Test the theory to determine cause.

    A. 1, 5, 7, 2, 6, 4, 3

    B. 2, 1, 7, 5, 6, 4, 3

    C. 2, 3, 1, 7, 5, 4, 6

    D. 5, 1, 7, 2, 3, 4, 6

2. Which of the following steps are involved in identifying the problem? (Choose two.)

    A. Identify the symptoms.

    B. Question the obvious.

    C. Determine what has changed.

    D. Take steps to resolve the problem.

3. You are trying to resolve a connectivity issue on the network. The user calls and states that she can't connect to the Internet. Your first action should be to:

   A. Gather information.

   B. Question the obvious.

   C. Establish a theory of probable cause.

   D. Reboot the DNS server.

4. Which of the following is considered a problem, not a symptom?

   A. A client computer can't `ping` the default gateway.

   B. A user can't connect to a web site by its DNS name.

   C. A router has a malfunctioning interface card.

   D. A cellular phone can't lock onto a clear signal.

5. To test a theory of probable cause for a network connectivity problem, which of the following should you do?

   A. Replace the network cable and reboot the switch.

   B. `Ping` the default gateway.

   C. Change the DNS server address and the IP address of the host.

   D. Change out the client network card and replace the network cable.

6. All of the following should be documented while troubleshooting a problem EXCEPT:

   A. Actions

   B. Findings

   C. Questions

   D. Outcomes

7. Which of the following steps may help reduce the likelihood of the problem recurring?

   A. Recable key network segments.

   B. Identify the effects on the network.

   C. Replace key network devices.

   D. Implement preventative measures.

8. You have solved a problem with nightly backups that were not being accomplished as expected. How would you verify that the problem is solved and the solution has worked?

   A. Run a sniffer program on the network to ensure all of the data was transferred to the backup device successfully.

   B. Run a checksum on the data both before and after the backup.

    C. Randomly perform a full restore on a server from one of the backups.

    D. Monitor the backup process for several nights to ensure the backups are being accomplished successfully.

9. Which of the following situations may require that you escalate a problem? (Choose two.)

    A. The host requires reinstallation of the operating system.

    B. The solution requires expenditure of a large amount of money and man-hours.

    C. A new cable has to be installed for a router.

    D. The problem is beyond your experience and knowledge.

10. Which of the following is a good source for determining if the problem has occurred previously?

    A. Trouble ticketing system

    B. Questioning the users

    C. Questioning the network administrator

    D. Reviewing the logs

11. Which of the following troubleshooting tools can show the live devices along a path to a destination host, as well as the time it takes to reach them?

    A. `Ping`

    B. `Tracert`

    C. TDR

    D. Sniffer

12. All of the following are potential causes of network latency EXCEPT:

    A. Slow-performing network devices

    B. Network saturation

    C. High-speed data connections that are underutilized

    D. Unusually high volumes of traffic

13. Which of the following could cause complete loss of network connectivity for a host? (Choose two.)

    A. DNS server issue

    B. Defective wireless access point

    C. No configured default gateway

    D. Broken cable

14. Which of the following network services could cause connectivity issues gradually if it fails?

    A. SMTP

    B. DNS

    C. DHCP

    D. FTP

15. You have a mobile client that is having email connectivity issues. They can receive email, but cannot send it. The device shows that the correct SMTP port is configured, but no other SMTP information is present. Which of the following is also needed to send email?

    A. The correct SMTP server name or IP address

    B. A valid PKI certificate used for authentication to the SMTP server

    C. The valid POP3 or IMAP port number

    D. A valid DNS server address

16. Which of the following would cause VPN connectivity issues? (Choose two.)

    A. Incorrect protocol and port number

    B. Incompatible encryption settings

    C. Incorrect DNS server address

    D. Proxy server issues

17. A mobile client is having authentication issues when attempting to connect to secure email services. Which of the following could cause the problem?

    A. Incorrect port number

    B. Incorrect SMTP server address

    C. Corrupt or expired certificate

    D. Corrupt mailbox

18. All of the following are issues that can occur with roaming EXCEPT:

    A. Excessive charges

    B. Signal loss

    C. Cellular activation

    D. Weak signal

19. You are helping a user at a geographically remote site to resolve persistent signal issues with their cellular mobile device. After determining that the device itself is configured and working properly, what might be a potential solution for the remote user's consistently weak signal?

    A. Installing a microcell or picocell

    B. Moving the remote site closer to a stronger cell coverage area

    C. Asking the carrier to boost their transmit power

    D. Installing a more powerful wireless LAN controller at the site

20. You are troubleshooting a network issue that is affecting the APNs. Which of the following may cause APNs connectivity issues?

    A. The DNS server can't resolve the www.apple.com URL.

    B. The notification options for some apps on the client device are turned off.

    C. The firewall is not allowing traffic to the 18.x.x.x network.

    D. The firewall is not allowing traffic on TCP ports 2195 and 2196.

Answers

1. **B.** The correct order is 2, 1, 7, 5, 6, 4, and 3.

2. **A, C.** Identifying the symptoms and determining if anything has changed are two parts of identifying the problem.

3. **A.** Gathering information should be your first step.

4. **C.** A router that has a malfunctioning interface card is a definite problem that can be solved. The other choices are symptoms of undetermined problems.

5. **B.** `Ping` the default gateway. All other choices require changing two things at once, which isn't a wise course of action.

6. **C.** Questions to the user do not have to be specifically documented; their responses should be summarized and documented under Findings.

7. **D.** Implementing preventative measures, whatever they may be, will help reduce the possibility of a recurrence of the problem.

8. **D.** Monitor the backup process for several nights to ensure the backups are being accomplished successfully. This may mean reviewing logs and the backup files to ensure they are complete.

9. **B, D.** The solution requires expenditure of a large amount of money and man-hours, or the problem is beyond your experience and knowledge. The other two solutions many require coordination, but typically do not require escalation.

10. **A.** The organization's trouble ticketing system should be a good source of documentation to review if the problem has previously occurred.

11. **B.** `Tracert` (from a Windows system) or `traceroute` (from a Linux system) can be used to determine the live devices along a route to a destination host and the time it takes to reach them.

12. **C.** High-speed data connections that are underutilized usually don't cause latency; however, if they are better utilized, they can help prevent it.

13. **B, D.** A defective wireless access point could cause a complete loss of connectivity for wireless clients, and a broken cable could do the same for a wired client. The other two choices could severely limit connectivity to one degree or another, but will not cause a complete loss.

14. **C.** DHCP can cause gradual issues if the service fails, due to the time it may take a client to attempt to renew its IP address lease. The other choices would cause immediate issues, and not necessarily with only network connectivity.

15. **A.** A correct SMTP server name or IP address is also required in the SMTP configuration settings.

16. **A, B.** An incorrect protocol and port number, as well as incompatible encryption settings between the client and VPN server, would cause connectivity issues.

17. **C.** A corrupt or expired certificate could cause authentication issues with the network services that use certificates, such as secure email.

18. **C.** Cellular activation issues are not normally caused by roaming.

19. **A.** Installing a microcell or picocell is the best solution in this case. The other alternatives are either impractical or ineffective at solving the problem.

20. **D.** The firewall is not allowing traffic on TCP ports 2195 and 2196, which are required to be open for the APNs to work properly.

Monitoring and Troubleshooting Mobile Security

In this chapter, you will

- Learn about monitoring and reporting techniques used on mobile devices to ensure security
- Examine how to troubleshoot mobile security issues

In terms of security, throughout this book I have discussed different security issues that you may encounter in your day-to-day work as a mobile device professional. You've looked at security risks, as well as security technologies that can be employed to reduce or mitigate those risks. This chapter covers the different methods and techniques used to monitor security on your devices, as well as report their security status. This is useful in determining if there are any security problems within your infrastructure so that you'll be able to solve them more quickly.

I will also discuss troubleshooting security issues when they occur. In Chapter 10, I discussed developing and following a troubleshooting methodology, which you can also use in solving security issues and problems. I'll focus more on different security-specific issues and their possible causes and solutions, however, than the methodology itself in this chapter, but you will find that in your overall troubleshooting efforts, you can apply the methodology to determine the nature of and solution to security problems as well; this discussion focuses on only one part of the methodology.

The CompTIA Mobility+ exam objectives covered in this chapter are 4.3, "Explain monitoring and reporting techniques to address security requirements," and 5.5, "Given a scenario, troubleshoot common security problems." Completing these two objectives will help you apply your knowledge of mobile device security, giving you the knowledge you need to monitor security issues and troubleshoot any security problems you might encounter in your own mobile device infrastructure. I will open the chapter by discussing various monitoring and reporting strategies used to ensure security of the mobile device infrastructure.

Monitoring and Reporting

From a security perspective, monitoring in the enterprise is almost a necessity. While the idea of monitoring employees' use of the corporate network and computing resources makes many people uncomfortable, it is necessary because it helps to protect the security of the organization's data infrastructure. On the other hand, monitoring should be restricted to only those activities and data necessary to ensure the security of the organization. The Bring Your Own Device (BYOD) explosion in corporate infrastructures has forced security personnel to examine personal data privacy in a way that they have not previously had to when dealing with personal data on corporate-owned or controlled assets. This part of the chapter takes a look at device and network use monitoring, as well as compliance reporting, and I discuss how you can strike a balance between personal data privacy and your critical role as protector of the corporate data.

Device Compliance and Audit Information Reporting

Monitoring is used to determine compliance with security policy and controls, as well as detect problems with security. In traditional networks, employers often monitor actions such as Internet use, web surfing, data access, and so on in order to ensure that employees are obeying acceptable rules of use with regards to corporate assets. Monitoring also serves to ensure that data and systems are being accessed only by those authorized to do so, for authorized purposes. Monitoring and reporting can help security professionals in the organization detect data loss or theft, hacking attempts, or misuse.

Monitoring Policy

As with everything else in the security world, monitoring and reporting programs should be defined in policy. Corporate policy should define the purpose of monitoring, what will be monitored, and to some extent, why it will be monitored. Policy should spell out the conditions under which mobile device use will be monitored and reported, such as violations of acceptable use policy, suspicious data use, undesirable network connections, and so on. In the case of BYOD in the infrastructure, policies should also spell out under what conditions, and what type of data, the device is *not* subject to monitoring, if any. The policy may say that in order to use personal devices on the corporate infrastructure, all data under all conditions will be monitored, collected, and stored by the organization. Whatever degree of monitoring is enforced on mobile devices, whether they are corporate-owned or personally owned, it must be specified in the policy. Users must be required to have received, read, and accepted the policy prior to allowing them to use mobile devices connected to the corporate network in order for the policy to have any legally enforceable stance.

Security policies on monitoring should include statements to the effect that all the users' activity on the enterprise network will be monitored, including web surfing, data access, and app usage; data from the mobile device, possibly including personal data, will be collected and archived by the organization; and that circumventing monitoring processes or software is a violation of policy and is prohibited. Obviously, corporate leadership will have to make decisions based upon how these statements affect personally owned devices in a BYOD environment. For example, the policy may stipulate that personal data is subject to being collected from an employee-owned device, or even that a personal device will not be monitored when it is not connected to the corporate infrastructure. Keep in mind that the security policies of the organization will be based upon its tolerance for risk, and whether it tends to be more restrictive or more permissive with regards to the mobile device use by the employee.

EXAM TIP Remember that everything in information security starts with policy, and monitoring is no exception. Understand the role policy plays in security monitoring and reporting.

Monitoring Mobile Devices

In the mobile device world, monitoring is not so different from that in the wired network world. There's still monitoring software, in the form of server software and client software, such as agents installed on the device, and there are still specific actions and events that are audited and reported. Users are still required to sign acceptable use agreements detailing what they can and can't do with corporate computing assets. In most cases, users are still restricted from installing unapproved or unacceptable software (in this case, mobile apps) on their devices, just as they would be on their desktop computers. So, there are a lot of similarities between monitoring mobile devices and traditional wired desktops. There also several important differences.

One difference, obviously, is the mobile aspect of the device and its user. Unless actually connected to the corporate network, it may be difficult to monitor the device and the actions of the user in real-time. An *agent* on the device (an installed piece of monitoring software) may collect certain usage data regarding connections, app installations, and so on, and report that data back to the server at a later time when the device is connected to the network. Another difference is the degree of personal data allowed on the device. Mobile devices are ubiquitous in today's world, and it's very difficult for most people to separate their personal lives from their work lives, especially when mobile devices have extended an employer's reach to the worker even after they have left the company facility. Many users don't want to take the time and trouble to switch devices when they are conducting work business versus personal business. Most people find that having to carry multiple devices around with them is very inconvenient and prefer to use one device for everything. Given the fact that many employers do allow some limited personal use of organizational computing devices, there is bound to be personal data that winds up on a company smartphone, laptop, or tablet. These two issues, mobility and personal data, make for interesting challenges in security monitoring and reporting.

What Should Be Monitored?

Privacy discussions aside for the moment, there are several things that should be monitored on a mobile device. These things should be monitored to ensure compliance with acceptable use policies, network connection policies, and software policies. Apps should be monitored for license management and control, piracy prevention, and malware prevention. Apps should be periodically inventoried to ensure that the device has only the company approved software on it. Restrictions on what types of apps and from where they can be downloaded should be enforced and monitored to ensure compliance. The mobile device should also be monitored for unapproved connections to unsecure networks.

Content on mobile devices should also be monitored, again, to ensure compliance with acceptable use policy as well as to stem any illegal activity. For example, music files such as MP3s may be restricted on the device if company policy prohibits the user from having that content on it. If the user copies any MP3s to the device, this action could be logged and reported to a central mobile device management (MDM) monitoring server. Content that an organization may want to control and monitor includes MP3s or other music files, video files, photographs, text messages, emails, and data files used by mobile apps. Of course, all this depends upon the policy itself and whether that policy is very restrictive toward mobile content, or permissive and allows the user some degree of freedom in the use of the device. One possible exception to monitoring could be the use of a device for *privileged communications*. These are usually personal or legally protected communications between an individual and his or her lawyer, clergy, or health professional, for example. Both the law and corporate policy may protect these communications (for example, calls, emails, and text messages) from being monitored and stored by the organization.

One item that you should consider when monitoring an organization's IT resources to include mobile devices, is the use of warning banners and other documents that let the user know they are being actively monitored. This has several uses, including deterrence from performing unauthorized activities, but the primary one is from a legal perspective, so the organization can assert its due diligence and care by notifying the user of monitoring activities and what their legal expectation of privacy is (if any). This can prevent legal issues down the road if a user states that they did not know of or consent to monitoring by the organization. An example of a rather intensive warning banner, used in U.S. Government systems, is shown in Figure 11-1.

 CAUTION Your organization should have a warning banner that reminds people at various times (upon log in, when connecting to a corporate web site, or simply when connecting to the corporate network, for example) that monitoring is used by the organization to ensure security, and that users may have no reasonable expectation of privacy, unless otherwise specified in policy.

Device Monitoring Applications and SIEM

Device monitoring, from the technical perspective, involves several different elements. There are also different ways to implement device monitoring in the enterprise. The first way is through installation of a monitoring agent on the device itself that reports back to the corporate mobile device management infrastructure. The agent is usually a small piece of software that runs in the background on the mobile device. It shouldn't

Figure 11-1

The standard
U.S. Government
warning banner

YOU ARE ACCESSING A U.S. GOVERNMENT (USG) INFORMATION SYSTEM (IS)
THAT IS PROVIDED FOR USG-AUTHORIZED USE ONLY.

By using this IS (which includes any device attached to this IS), you consent to the
following conditions:

- The USG routinely intercepts and monitors communications on this IS for purposes
  including, but not limited to, penetration testing, COMSEC monitoring, network
  operations and defense, personnel misconduct (PM), law enforcement (LE), and
  counterintelligence (CI) investigations.
- At any time, the USG may inspect and seize data stored on this IS.
- Communications using, or data stored on, this IS are not private, are subject to
  routine monitoring, interception, and search, and may be disclosed or used for any
  USG-authorized purpose.
- This IS includes security measures (e.g., authentication and access controls) to
  protect USG interests--not for your personal benefit or privacy.
- Notwithstanding the above, using this IS does not constitute consent to PM, LE or
  CI investigative searching or monitoring of the content of privileged
  communications, or work product, related to personal representation or services by
  attorneys, psychotherapists, or clergy, and their assistants. Such communications
  and work product are private and confidential. See User Agreement for details.

[I Accept]

normally interfere with the operation of the device, and it should not use up too much
memory or CPU processing time. It should, however, be able to detect any activity or
event that is not compliant with corporate policy. When it detects an event, it should be
able to respond in a few possible ways, according to policy. It may simply log the event, it
may alert the user or the administrator, or it may even be programmed in some way to
prevent the event. It also may be programmed to take an action based upon the event.
For example, if the organization requires that all wireless network communications be
encrypted, the agent could automatically negotiate encryption with any wireless network the device connects to. It could also be used to prevent connection to any wireless
network that does not use encryption.

A second way to implement device monitoring might be through the use of specific
software installed on the device that notifies the user of activities that go against security
policy. These may be activities that the user is unaware of, or unintentionally engages
in, such as accidentally visiting a restricted web site, or activities that the user may
intentionally be attempting, such as connection to an unapproved wireless network.
In any case, the agent would notify the user and take action based upon the corporate
policy installed on the device.

Another way to implement device monitoring would be through the use of network-
based applications that would monitor the device, as well as obtain device and log
information, while the device is actually connected to the corporate network. This
could be an agentless form of monitoring and may not require software on the device.

Network monitoring software may monitor network traffic to and from the device, or, using administrative credentials, access the device logs and configuration remotely.

 EXAM TIP Understand the differences between agent-based and agentless monitoring.

Regardless of your monitoring strategy, there are some things you should consider when monitoring mobile devices. First of all, you should probably select an MDM product that allows you to monitor and collect information from a wide range of devices, especially since you likely have different devices from different manufacturers and the network. You should also be able to assign different policy elements to devices for monitoring, depending upon what it is you want to monitor, such as apps, Internet usage, device compliance, and so on. The monitoring software that you use should be able to provide alerts and notifications to the mobile device support staff and/or the user in case a serious or critical security event occurs on the device. Another thing you should consider is using some type of centralized monitoring, logging, and reporting tool, whether it is built into the MDM product or comes as a third-party product. This tool would be used when collecting information from all devices and categorizing it in different ways for easy access and analysis. You should be able to search by device, of course, but also by timeframe, and you should be able to look at multiple device events at once so you can see trends or detect security issues that are occurring across the enterprise. One such useful tool for being able to do all of this is a Security Information Event Management (SIEM) infrastructure.

SIEM is an industry paradigm that has developed and matured over the past decade or so that allows an organization comprehensive control over security events. It allows real-time alerts and notifications, of course, but also in-depth analysis of data from a multitude of sources over time that may affect the security of the infrastructure. Think of SIEM as one-stop shopping for complete enterprise-level security event management. SIEM involves the use of aggregating data over time from a wide variety of sources, including system event logs, firewall and intrusion detection system logs, and network device, application, and other security logs. SIEM looks for relationships between all the different data points and events in these sources in an effort to correlate any connections between seemingly unrelated events. For example, an innocuous random bandwidth spike from one network device that may happen periodically over weeks or even months could be correlated to a security event on another device to detect an activity such as a hacking attempt. Figure 11-2 illustrates how all of these diverse data sources are rolled up into SIEM.

 EXAM TIP While you likely will not be asked any questions on any particular SIEM package, understand the purpose and use of SIEM in the enterprise.

Figure 11-2
Relationships
between data
sources and SIEM

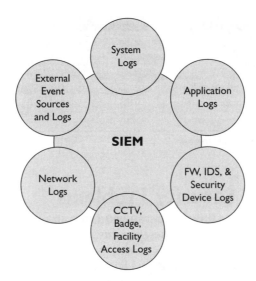

Monitoring Mobile Device Security Logs

Collecting and reviewing logs for mobile devices is similar to the practice of auditing logs in traditional devices and networks. While collecting the logs is a matter of ease, the analysis of them could be considered a lost art because most of it is probably done using automated tools these days, with little human intervention. Whether you are using an automated tool or reviewing logs visually, there are things that should be considered. First, the log should have a timestamp that indicates the exact time and date that the event or activity occurred. The timestamp should be based upon a centralized network-based time source so that all activity logs in the infrastructure reflect the same time, with little to no deviation from each other. This makes event correlation across the enterprise from multiple sources much more efficient, and supports forensics analysis of security events. The log should also include the source device, as well as information about the user that was logged into the device at the time the event occurred. If the log involves a network connection of some sort, then the source and destination host names and/or IP addresses should be listed. Network connections should also include the source and destination ports and protocols used in connection. If the log entry involves any kind of data object access, such as file or folder access, then the entry should also include what action was taken with the data object, by whom, and whether it was allowed or denied. All of this information helps an administrator reconstruct a timeline and sequence of events, and when combined with other log or event sources, can help build a complete picture of an event, such as a data breach, for example. Figure 11-3 shows a snippet from a log file that contains some of the critical elements I discussed.

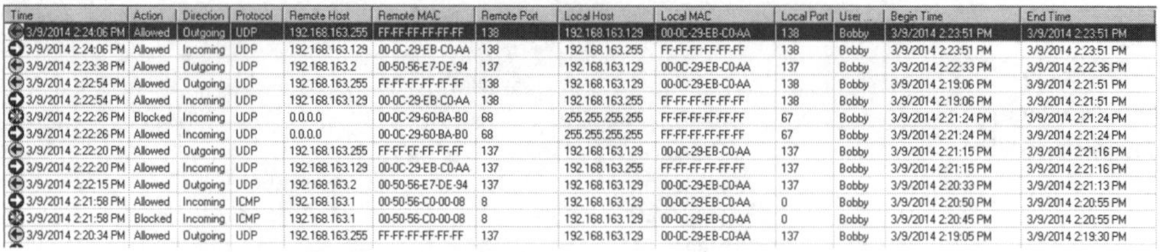

Time	Action	Direction	Protocol	Remote Host	Remote MAC	Remote Port	Local Host	Local MAC	Local Port	User	Begin Time	End Time
3/9/2014 2:24:06 PM	Allowed	Outgoing	UDP	192.168.163.255	FF-FF-FF-FF-FF-FF	138	192.168.163.129	00-0C-29-EB-C0-AA	138	Bobby	3/9/2014 2:23:51 PM	3/9/2014 2:23:51 PM
3/9/2014 2:24:06 PM	Allowed	Incoming	UDP	192.168.163.129	00-0C-29-EB-C0-AA	138	192.168.163.255	FF-FF-FF-FF-FF-FF	138	Bobby	3/9/2014 2:23:51 PM	3/9/2014 2:23:51 PM
3/9/2014 2:23:38 PM	Allowed	Outgoing	UDP	192.168.163.2	00-50-56-E7-DE-94	137	192.168.163.129	00-0C-29-EB-C0-AA	137	Bobby	3/9/2014 2:22:33 PM	3/9/2014 2:22:36 PM
3/9/2014 2:22:54 PM	Allowed	Outgoing	UDP	192.168.163.255	FF-FF-FF-FF-FF-FF	138	192.168.163.129	00-0C-29-EB-C0-AA	138	Bobby	3/9/2014 2:19:06 PM	3/9/2014 2:21:51 PM
3/9/2014 2:22:54 PM	Allowed	Incoming	UDP	192.168.163.129	00-0C-29-EB-C0-AA	138	192.168.163.255	FF-FF-FF-FF-FF-FF	138	Bobby	3/9/2014 2:19:06 PM	3/9/2014 2:21:51 PM
3/9/2014 2:22:26 PM	Blocked	Incoming	UDP	0.0.0.0	00-0C-29-60-8A-B0	68	255.255.255.255	FF-FF-FF-FF-FF-FF	67	Bobby	3/9/2014 2:21:24 PM	3/9/2014 2:21:24 PM
3/9/2014 2:22:26 PM	Allowed	Incoming	UDP	0.0.0.0	00-0C-29-60-8A-B0	68	255.255.255.255	FF-FF-FF-FF-FF-FF	67	Bobby	3/9/2014 2:21:24 PM	3/9/2014 2:21:24 PM
3/9/2014 2:22:20 PM	Allowed	Outgoing	UDP	192.168.163.255	FF-FF-FF-FF-FF-FF	137	192.168.163.129	00-0C-29-EB-C0-AA	137	Bobby	3/9/2014 2:21:15 PM	3/9/2014 2:21:16 PM
3/9/2014 2:22:20 PM	Allowed	Incoming	UDP	192.168.163.129	00-0C-29-EB-C0-AA	137	192.168.163.255	FF-FF-FF-FF-FF-FF	137	Bobby	3/9/2014 2:21:15 PM	3/9/2014 2:21:16 PM
3/9/2014 2:22:15 PM	Allowed	Outgoing	UDP	192.168.163.2	00-50-56-E7-DE-94	137	192.168.163.129	00-0C-29-EB-C0-AA	137	Bobby	3/9/2014 2:20:33 PM	3/9/2014 2:21:13 PM
3/9/2014 2:21:58 PM	Allowed	Incoming	ICMP	192.168.163.1	00-50-56-C0-00-08	8	192.168.163.129	00-0C-29-EB-C0-AA	0	Bobby	3/9/2014 2:20:50 PM	3/9/2014 2:20:55 PM
3/9/2014 2:21:58 PM	Blocked	Incoming	ICMP	192.168.163.1	00-50-56-C0-00-08	8	192.168.163.129	00-0C-29-EB-C0-AA	0	Bobby	3/9/2014 2:20:45 PM	3/9/2014 2:20:55 PM
3/9/2014 2:20:34 PM	Allowed	Outgoing	UDP	192.168.163.255	FF-FF-FF-FF-FF-FF	137	192.168.163.129	00-0C-29-EB-C0-AA	137	Bobby	3/9/2014 2:19:05 PM	3/9/2014 2:19:30 PM

Figure 11-3 A log file snippet

Common Security Problems

A wide variety of security issues can occur in your mobile device infrastructure, as well as your enterprise network. These security issues likely won't affect just mobile devices, but other parts of your network as well, and could even affect the entire infrastructure. Fortunately, most of these issues are fairly common, and when a security issue becomes apparent, usually you will see that it is one of the issues I discussed here or something closely related. Security issues can occur in various parts of the network infrastructure, including the host device itself, the transmission media, and the network infrastructure devices. Security issues can also come from the data that is received, processed, stored, and transmitted, as well as the apps that use the data. And of course, the user can be a source of security issues as well.

In the next several sections, I will discuss several of these security issues from a perspective of both technologies and practices. In addition to the exam objectives, I will also discuss several other sources of security issues that you may encounter as a mobile device professional. You'll often find that some issues are caused by several factors and not just a single one. I'll discuss these as well.

Certificate Problems

Some of the main security problems you'll encounter with secure communications, including authentication and encryption, are issues with digital certificates. Remember that digital certificates are used for a variety of purposes, including authentication, sending and receiving secure email, encryption, and digitally signing emails, documents, and even software. Most of the problems you'll encounter with certificates involve lifecycle issues, such as certificate expiration or revocation, and you may occasionally encounter issues with a corrupt certificate itself. The next few sections cover different certificate issues that can pop up on your mobile device infrastructure.

Expired Certificate

An expired certificate is actually an easy issue to both detect and solve. It simply requires that you be aware of when your certificates will expire, and know that you have to renew them prior to their expiration date in order to continue using them. You may depend upon the individual user to help you do this, although for the most part that's not the most reliable means of tracking certificate expiration. Most users, especially the non-technical ones, really don't make a point of monitoring when their certificates expire. To make it more difficult, most users on the receiving end of the certificate, when it's presented to them to verify the identity of a user or server, tend to ignore any warnings about certificate expiration and accept the certificate anyway.

The issues with certificate expiration include the inability to authenticate or digitally sign or encrypt documents once a certificate has expired. The simple solution is to maintain a database, usually through an automated mechanism built into the network or mobile device infrastructures, which will monitor the expiration of any organizationally issued certificates, and alert both the user and the administrator when the expiration date is within a certain amount of time. For certificates that are issued by your organization, the certificate management infrastructure will typically have facilities that allow this to be an easy task. For certificates that are issued by third parties, this may be more of a manual task in that the administrator must store the certificate and possibly record expiration data into the mobile device management infrastructure database. In either case, when the certificate reaches a certain timeframe threshold, both the user and administrator should be notified of impending certificate expiration. In the event this doesn't happen, or a certificate is not renewed within that timeframe, the certificate will become invalid and the user or administrator will typically make it a higher priority to reissue the certificate. Figure 11-4 shows an example of an expired certificate.

Figure 11-4
An expired
certificate

Certificate

General | Details | Certification Path

Certificate Information

This certificate has expired or is not yet valid.

Issued to: Class 3 Public Primary Certification Authority

Issued by: Class 3 Public Primary Certification Authority

Valid from 1/28/1996 **to** 1/7/2004

Issuer Statement

OK

Revoked or Suspended Certificates

Other lifecycle issues include revoked or suspended certificates. Remember from the discussions in Chapter 9 that a suspended certificate is usually the result of a temporary condition where the administrator or organization suspends a certificate and makes it unavailable for use by the user during an extended leave, investigation, or other circumstance. Whatever the condition is, the certificate can be reinstated when the organization chooses to. A revocation, on the other hand, means that the certificate is no longer any good and can't be used for any purpose. Typically, a revoked certificate cannot be reissued; a completely new certificate must be issued in accordance with organizational procedures. Certificate revocations are typically caused by conditions such as the employee leaving the organization permanently (through normal attrition, or by getting fired) or because a certificate has been compromised. Figure 11-5 is an example of a certificate that has been revoked. In this case, it is the infamous example of a Microsoft software signing certificate that was obtained fraudulently from VeriSign and later revoked after the fraud was discovered.

Revoked or suspended certificates will typically manifest themselves through warnings or error messages when the user attempts to authenticate, sign, or encrypt

Figure 11-5

A revoked certificate

with them. Error messages may also appear to the user that receives a certificate for authentication purposes. A revoked certificate will show up on a Certificate Revocation List (CRL), published by the organization. Both the organization and any other entity that typically accepts the organization's certificates must be able to access the CRL in order to determine if there are any applicable revoked certificates on it. The CRL is usually published by the organization to an internal server or to a public server, or on an as-requested basis by the client during normal updates or transactions from the client's operating system or security applications. CRLs are published using the Online Certificate Security Protocol (OCSP). Figure 11-6 is a sample of a CRL.

 EXAM TIP Know the possible conditions associated with a certificate that may cause issues: an expired certificate, a suspended certificate, and a revoked one.

Corrupt Keys

Occasionally, public and private keys can become corrupt because they are really nothing more than electronic files. Corruption could occur in transmission if network connections aren't reliable or if error conditions are present. The files could also become corrupt on storage media, or even when stored in an application, because of disk corruption,

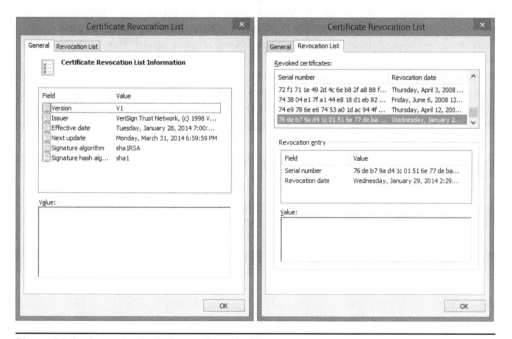

Figure 11-6 A sample of a CRL entry from VeriSign

hardware errors, and so on. Public keys, for the most part, can simply be resent to the other party, if necessary, or downloaded again from the server. Private keys, on the other hand, can be problematic to restore, unless they have been securely saved and stored on another media, such as a USB storage device, logical encrypted container, and so on. To restore a private key from secure storage usually requires the user to input a passphrase or pin in order to decrypt a private key that has been securely encrypted for protection. Once restored, the private key can be used as before.

Symptoms of corrupt keys usually include the inability to send or receive encrypted or digitally signed messages or files. Various error messages could also crop up in the form of file read or write errors when using encryption keys. Rarely, in case keys have not been securely stored, will a set of keys have to be reissued to the user by the Certificate Authority, or in the case of self-generated keys, have to be re-created.

 TIP Corrupt keys and certificates can be troublesome to track down. Usually, you will go through several other possible causes first before determining that the problem is a corrupt key or certificate.

Compromised Keys and Certificates

Compromised keys and certificates are a serious security issue. If a key or certificate has been compromised, that means someone else has access to the private key and may be able to not only encrypt files or messages, but also decrypt data that they are not authorized to access. In the case of digital signatures, a compromised key could mean that someone could impersonate an authorized user by digitally signing traffic or files with the compromised key. Once keys have been compromised, they should not be trusted. Any data, such as files or messages, that have been encrypted or signed after the key compromise should be treated as suspect. If the compromised key has been used to sign a digital certificate that may have been issued from a certification authority, then any certificates issued with that key should also be treated as compromised or suspect. This could affect the entire chain of trust, depending upon which key up the chain has been compromised. If the root certificate or private key is ever compromised, this could conceivably render the entire certificate chain as invalid. The entire chain of certificates may have to be revoked and re-issued.

In the event of a key compromise, the keys should be revoked as soon as it is determined that there has been a security issue. The revocation action should be published to the CRL as soon as possible, and pushed to appropriate clients, applications, or retrieval sites to prevent the compromised certificates from being accepted by anyone. If there is any data that has been encrypted with the keys, it should be decrypted as soon as possible. Once the keys have been revoked, they should be destroyed to prevent further use. New keys and certificates should be issued in order to re-encrypt any sensitive data that requires it.

 EXAM TIP If a certificate is suspected of being compromised, both it and any certificates signed by it in the hierarchy should be immediately revoked.

Authentication Issues

Some of the most common security issues in enterprise networks can be attributed to authentication issues. Authentication issues could stem from faulty security mechanisms, password issues, PKI issues, and even valid security problems. Usually authentication issues come in the form of users that can't log into the network or system, or even security events caused by unauthorized persons that are somehow able to authenticate to the network. The next few sections will examine some of the things that can cause authentication issues, and how those issues affect the ability of users to access enterprise resources. I'll also discuss some of the common reasons why authentication problems happen, what the ramifications are from these issues, and some basic troubleshooting tips you can use to solve them.

Authentication Failure

Authentication failure can manifest itself in several ways. First, the most obvious way is that users are unable to authenticate to the network or log into a device or system. This may be caused by incorrect credentials or even poor network connections. These are likely the first things an administrator would check before anything else. Figure 11-7 shows a typical error message when using incorrect credentials to authenticate.

If you can rule out these two conditions, then you might look at configuration issues with security and authentication mechanisms. This might mean troubleshooting problems with the authentication server itself, or making sure that the correct encryption

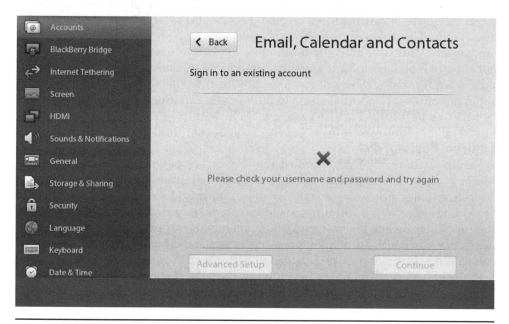

Figure 11-7   Password authentication error message

mechanisms and levels are configured on the system or device. If the device and the server are using two totally different encryption algorithms, for example, this would definitely cause authentication problems. Remember that on both ends of a communications session, the client and server must be able to negotiate a mutually acceptable encryption algorithm and strength. They each may be configured with several possible encryption or authentication methods that they can use; the important point is that they must be able to negotiate the strongest one they have in common. If only one user or device is having authentication issues, then start with the device and work your way up the network infrastructure to intermediate level devices, connection points, security mechanisms, servers, and so on. If multiple users and devices are having authentication issues, then you may want to start with the network-based authentication services. In a Windows Active Directory environment, this would mean looking at the domain controllers and making sure that they are replicating properly and are configured correctly for secure authentication. In other centralized authentication schemes, you must look at the central authentication server and its configuration.

Sometimes authorization issues can mask themselves as authentication problems. For example, suppose a user can authenticate to the network but repeatedly receives a prompt to input a username and password when accessing a particular shared resource. In this case, authentication is not realty the problem; it may just be that the user is not authorized to access that particular resource. This can usually be resolved by examining the object's access control list and granting the user the appropriate permissions to the resource.

Another authentication failure issue is one that may not be as obvious until a security problem happens. This could occur when an unauthorized user is able to access the system or network when he should not be able to. This could mean a security breach caused by stolen or compromised credentials, or even failure of authentication mechanisms. This issue could be much more difficult to track down in that you may not know of it until it's too late and the unauthorized user has already accessed data he should not. Typically, conscientious review of log files and a robust security auditing and monitoring system will alert you to these issues. Unfortunately, sometimes system damage or data loss is what alerts you to them instead.

Expired Passwords

Typically, an individual gets notified that his or her password is about to expire within a certain amount of days . This notification may arrive in a number of different ways, depending on how you set it up. You can set up a 14-day notification, for example, presenting a user with a pop-up message, or you can set up automated system emails starting 14 days before the password expiration date. If the user changes her password within that timeframe, the notification messages stop and the user's password expiration date is reset. If the user does not change her password within the timeframe, she might not be allowed to log on and authenticate to the system or network, unless she changes her password first. Figure 11-8 shows a typical password expiration "nag" message!

Figure 11-8

A typical password expiration warning message

ⓘ Consider changing your password   ↖ ✕
Your password expires today.
To change your password, press CTRL+ALT+DELETE and
then click "Change a password...".

Of course, it's a good idea for the user to change a password before the time expires. It's also a good idea to make the change from the local network itself rather than remotely because some remote connections and applications may or may not effectively manage the password change process very well. The user could change his password remotely, and assume that it is a successful change, and then still have authentication issues the next time he attempts to log in. In this instance, it may be best to just go ahead and have the user change it again when he is logged in to the local network, or have the administrator reset it to a temporary password, allowing the user to change it himself when he logs in again.

Passwords should be expired after a certain amount of time based upon policy settings, simply to ensure that the passwords don't exist long enough for an attacker to obtain the password hashes and attempt an off-line password attack. Given a complex password, an off-line attack may take months or even years; however, frequent password changes help to prevent the success of off-line attacks as an additional measure. Expiring passwords is one way to ensure frequent password changes.

Non-Expiring Passwords

Passwords that do not expire on the network present a security issue in that, given enough time, theoretically a hacker could conceivably crack a password and then use it to gain unauthorized access to the network. Password expiration options are usually set in the local policy for a standalone system, or in a network database for central authentication mechanisms, such as Active Directory or other LDAP-based systems. Even accounts that have been disabled because of infrequent use should have their passwords set to expire and occasionally reset in case they have been compromised in some way. Figure 11-9 shows the Local Security Policy account Password Policy controlling account password expiration. Notice that the expiration (maximum password age) is set to 42 days, which is the default for the policy and should be reset based upon the organization's security policies.

One possible exception to password expiration policies may be in the use of service accounts that have been created on a system to specifically run certain services. Administrators often set up these services with non-expiring passwords because an expired password would cause the service to stop and whatever functions or programs that rely on the service would no longer work. While the use of non-expiring passwords is still not an ideal situation, there may be mitigations that the organization can put in place in order to reduce the risk incurred by the use of these passwords for longer

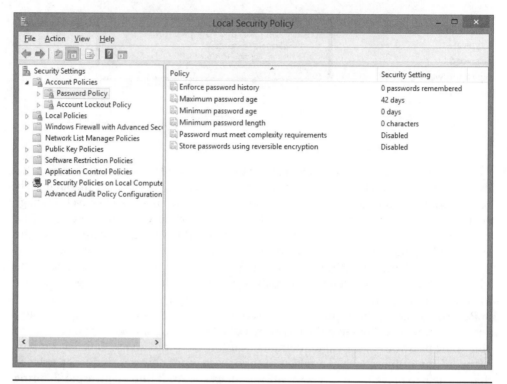

Figure 11-9 Default account Password Policy

periods. For example, the passwords could be created to be very complex and lengthy, making them difficult to remember, and ensuring that no administrator actually has the password or knows it. If necessary, the password could be stored in a locked physical container, such as a safe, or even in an encrypted logical container in a secure portion of the network. Accessing the password in order to perform maintenance on the service account should require a practice called two-person integrity, whereby policy requires two people to sign a log when the password is accessed or removed from safekeeping, as well as when account maintenance is performed on the service accounts. Another mitigation could be the use of the *M of N control* concept. M of N control requires at least a minimum number of possible people authorized to access a password or encryption key belonging to another account. For example, if five people are authorized to know or access the password, policy may require that at least three of the five (3 of 5) be physically present to access it from a safe or from a third-party escrow service. Yet another mitigation may be the use of a multipart password, where each designated administrator has only one portion of the password. By itself, it's not the complete password and can't be used to access the account; however, when combined with the other parts of the password that other trusted administrators possess, the full password can be input and used to administer the service account. Again, while non-expiring passwords are not ideal, and should be kept to a minimum, other mitigations may help to lower the risk involved if the organization determines that they will use service account passwords for extended lengths of time.

 EXAM TIP When troubleshooting service failures, it may be a good idea to check and see if the service's password is set to expire or not.

Network Device Problems

In addition to the issues that can be caused by applications and individual hosts on the network, network devices themselves can cause issues relating to secure connectivity. Most of the issues that network devices can cause can be traced to incorrect configurations, as well as incompatible security settings. In order to communicate securely between them, network devices must have common security settings enabling them to authenticate with each other, as well as exchange encrypted or signed data appropriately. The next few sections present some examples of network device security issues you may encounter in your professional travels.

Firewall Misconfiguration

Configuration issues are probably some of the most common found in network device problems. The ubiquitous network device, the firewall, is frequently blamed for almost anything that goes wrong on the network in terms of security or connectivity by most users, and unfortunately, sometimes those users are correct. A misconfigured firewall can cause issues that are sometimes difficult to track down. These issues usually manifest themselves at first as connectivity issues, and they may cause the administrator to look at the various network devices between the user's host and the external network. Often, the firewall is the last device looked at in a logical troubleshooting methodology. Users may complain that they can't access a certain web site or can't access the Internet at all.

After the normal troubleshooting steps and establishing that connectivity is working as it should, the next step may be to examine the firewall to determine if it is blocking, or even allowing, traffic that it shouldn't be. When examining the configuration on firewall, the administrator should first determine if there have been any rule changes recently that may have contributed to any connection problems to resources. Administrators may need to also look at the rule set for any rule that may be configured incorrectly. For example, an incorrect IP address or domain name in a rule may prevent users from connecting to an otherwise authorized resource on the Internet. Or, a mistyped port number could make the difference between not accessing telnet (port 23) or not accessing email (port 25), for example. Administrators should carefully read each rule, particularly new ones, to make sure they have been entered correctly and achieve the desired effect. Administrators should also pay attention to any rules that completely block a particular type of traffic or domain, to make sure they haven't inadvertently configured the rule to block more than they should. And, of course, any default deny or default allow rules should be checked to make sure they are configured properly and are located in the right place in the rule set. Remember that explicit default deny rules

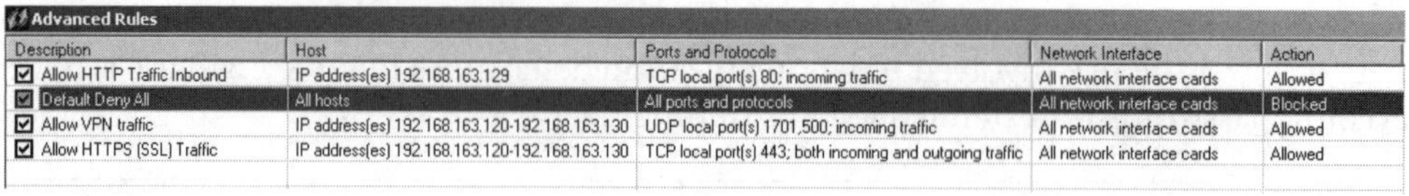

Advanced Rules				
Description	Host	Ports and Protocols	Network Interface	Action
☑ Allow HTTP Traffic Inbound	IP address(es) 192.168.163.129	TCP local port(s) 80; incoming traffic	All network interface cards	Allowed
☑ Default Deny All	All hosts	All ports and protocols	All network interface cards	Blocked
☑ Allow VPN traffic	IP address(es) 192.168.163.120-192.168.163.130	UDP local port(s) 1701,500; incoming traffic	All network interface cards	Allowed
☑ Allow HTTPS (SSL) Traffic	IP address(es) 192.168.163.120-192.168.163.130	TCP local port(s) 443; both incoming and outgoing traffic	All network interface cards	Allowed

Figure 11-10 A default deny rule in the wrong place in the rule set

should be located at the *end* of the rule set, as a catch-all for any traffic that is not otherwise permitted. Figure 11-10 shows an example of a misconfigured firewall rule.

Some firewalls and other security appliances, such as proxy servers, may also have configuration settings that allow or deny traffic based upon user accounts or groups. Rule sets may apply only to specific users and not to others, so a user may have to be placed in a group that is allowed to access certain traffic. Still other configuration settings may include certain applications, file types, and other content. Check these configuration settings as well when troubleshooting firewall configuration problems.

VPN Issues

VPN issues are typically fairly straightforward to troubleshoot, and usually involve disconnects between the type of authentication or encryption mechanisms used on the client and VPN server or concentrator. This is where configuring both the tunneling and encryption protocols correctly is important; you must make sure that a VPN server that uses L2TP is also using IPsec configured correctly, for example. You should also make sure that IPsec is configured identically on both ends of the VPN connection, to include authentication mechanisms, encryption algorithms, and encryption strength. Additionally, certificate-based authentication can be problematic because of any issues with certificate trust, key expiration or revocation, or even corrupt keys, especially when using SSL-based VPN solutions. Network issues, such as the correct IP addressing scheme and subnet mask, even though they are not necessarily security problems, should also be looked at in the event the VPN connection fails.

VLAN Issues

Remember from the discussions on VLANs in Chapters 1 and 2 that network switches working at the OSI Model's Layer 3 help create VLANs in order to reduce collision and broadcast domains, without having to use a hardware router. Beyond the improvement in network traffic that a switch can be used for, VLANs create security by segregating hosts from each other, and segmenting traffic, affording it better protection. In creating VLANs, you must pay attention to several configuration items, including logical IP addressing, security, and port configuration. One of the most common security issues with VLANs is placing hosts in an incorrect VLAN. This may prevent them from communicating with other hosts, as well as authorized resources. Administrators should make a point of recording what VLANs different hosts belong to, as well as what security restrictions the different VLANs may have. Most VLAN issues will typically be connection or even routing issues instead of security issues, however, and may have to be approached from a network troubleshooting perspective.

False Positives and Negatives

Despite the accuracy of monitoring and reporting systems, occasionally they make mistakes. They may inaccurately report that a potential security incident has occurred, or worse, not detect one and report that everything is okay. Either case requires some fine-tuning of the security mechanisms in place as well as the detection and reporting mechanism used in the infrastructure. This may require updating the baseline of what is normal or not normal use for access to resources. For example, earlier in the book I discussed how anomaly-based intrusion detection systems work, where the administrator has to run the system for an extended period of time so it will "learn" and develop a normal baseline of network activity. After it develops this baseline, if it picks up unusual network traffic at an odd hour, for example, it may report that traffic as a security event. If it turns out that the traffic is legitimate, possibly due to maintenance or another accepted event, then the administrator must determine whether or not to fine-tune the IDS. If the event is a one-of-a-kind or is not likely to occur again, the administrator may choose to not alter the IDS's baseline. However, if this new event will be recurring, then the administrator may want to include that event in the IDS baseline so that it will not alert on it again. This is just one example of how security mechanisms may report false positives or negatives. In the next couple of sections, I will define exactly what a false positive and false negative is, and how this may affect your network infrastructure.

False Positives

The concept of a *false positive* is fairly easy to explain. If the system detects something as a security event, and it turns out that it is actually not a security issue, then this is commonly called a false positive. A false positive can be triggered in all manner of security systems: the example of intrusion detection systems given previously, or even in systems such as biometric authentication systems, physical alarm systems, and other complex systems. False positives can be dangerous in that a high number of them can cause administrators to be complacent, simply because they get used to seeing them and may ignore actual security events in the process. A false positive can be remedied by retuning the baseline of the security system, to include additional security events or parameters that are considered acceptable. In some systems this also may mean decreasing the sensitivity of the system; in other words, configuring the system so that it is not as sensitive to events as it might be. The difficulty is in knowing the right amount of sensitivity the system requires to accurately detect false positives and false negatives.

False Negatives

Where a false positive is actually a non-event, a *false negative* could be a worse situation simply because it is a security event that should be detected and acted upon, but isn't for some reason. For example, let's say that someone uses a fake security badge to enter a facility. The human guard or receptionist may not be able to tell the difference between an actual badge and one that has been cleverly faked, so an unauthorized person may conceivably get through a security access point. This would be considered a false negative. An automated system using a more precise means of electronic badge access would likely be more difficult to fool and would probably catch an individual trying to use a fake badge. False negatives are serious in that they could allow security breaches and unauthorized access to systems and data. Again, tuning the system in some way will usually reduce the number of false negatives, but this has to be balanced so that the system is not so sensitive that it will report everything it encounters, resulting in increased false positives.

So essentially, the more sensitive the system is, the fewer false negatives it will produce, but more false positives could be reported. And conversely, a system that is less sensitive will not report as many false positives, but will, unfortunately, allow more security breaches without detection or action. Balancing between the sensitivity levels to lower both the false negative and false positive rates is a key challenge with most security systems. Some systems, such as biometric authentication systems (for example, fingerprint readers) are characterized by a "sweet spot," if you will, where the number of false reads is reduced to the best level possible; this "sweet spot" is called the *crossover error rate*.

Crossover Errors

The crossover error rate is the ideal point where the system is tuned such that false positives and false negatives are both reduced to the maximum point possible. Beyond this ideal point, either false positives or false negatives would increase significantly because they typically have an inverse numerical relationship with each other. The ideal point is usually where they are in balance and equal each other. Security systems that allow fine-tuning enable you to reduce both of these types of errors using a crossover error rate measurement. An illustration of this concept appears in Figure 11-11.

Figure 11-11
Crossover error
rate

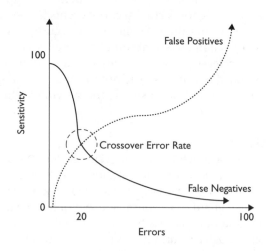

Other Security Issues

No single book could cover the myriad of security issues that can plague a network infrastructure or even a mobile device infrastructure. This chapter attempts to cover the security issues that you will likely see on the exam, in accordance with the exam objectives, as well as some of the more popular issues that you are likely to see in real life. Some of these issues involve the client device and some involve the network infrastructure itself. Some issues could definitely cover both ends of the connection. In the remaining sections of this chapter, I discuss some other miscellaneous issues that you should know how to troubleshoot.

Misconfigured Content Filtering

Try as you might, accurately configuring your network and mobile devices to restrict or allow certain types of content into your infrastructure can be a challenge. In previous chapters, I discussed how to use content filtering to prevent unauthorized content, such as malicious files, prohibited content, and so on from entering the network. I also talked about how to use content filtering to restrict data leaving the network, using this concept in data loss prevention. We all know that content filtering can protect the network, but sometimes it can also prevent legitimate content from passing to the network that users require in order to do their jobs. For example, proxy servers that restrict content based upon filenames or extensions may inadvertently limit legitimate content if the file has been named incorrectly or if the device does not have the appropriate file extension whitelisted (allowed) in its rule set. Content filters that prohibit traffic based upon keywords, such as those in a URL or typed into a search engine, may inadvertently allow content into the network when the keywords are misspelled in such a way that the filters don't catch them. Indeed, a common trick of both hackers and malicious users is to intentionally misspell URLs, keywords, or search terms in an effort to circumvent content filtering systems. This may result in illicit content entering the network, so rule sets may have to be written to take this into account.

Context also plays a part in content filtering. An organization, such as a private company, may restrict users from searching on or downloading content related to certain things, such as terrorist activities, for example. However, another organization, such as a U.S. military unit, may have to access those same sites and download that content to be able to more effectively identify terrorist activities. While the content filtering rule sets for both these organizations would obviously be different, many organizations purchase premade content filters for their enterprise network devices. These premade content filters may have words and phrases relating to terrorism restricted, so the one organization would have to leave those filters in place, while the U.S. military organization, using those same premade rule sets, would have to alter them to allow that content into their network. The points here are that filtering should be based upon organizational context, and that there is no one-size-fits-all solution for content filtering, even when it comes to these predefined rule-set subscriptions that come with an enterprise filtering device.

More advanced content filtering devices, in addition to those methods described, may also do content inspection, whereby the contents of a file are examined for prohibited traffic. Usually these types of devices are more complex and require more careful configuration, but are typically more accurate and able to both detect prohibited content more efficiently, as well as make smarter decisions regarding allowed content entering and leaving the network. However, they may also introduce latency into the network, as deep content inspection may require more computing and network resources, as well as additional time.

Incorrectly configuring content filters is usually one of those security issues that you will find out about very quickly, regardless of whether you have configured them to be too restrictive or too permissive. Users are your best source of information regarding misconfigured content filters, obviously, because if you have configured them to be overly restrictive and they can't send or receive valid content, such as compressed ("zipped") or encrypted files, for example, you usually hear about it quickly. The same is true if you have configured the filters too loosely, because users will eventually download content they shouldn't. Content filtering, like all security mechanisms, has to be fine-tuned to the point where it provides the level of security the organization requires, yet allows the organization's users to send and receive content required to do their jobs.

 EXAM TIP   If users can't download or email a certain kind of file used in their jobs, content filtering may be misconfigured and should be checked.

Encryption Issues

I discussed encryption several times throughout this book, so you know the components that make up encryption, such as algorithms and keys. These two basic components of encryption are also actually what cause most of the problems you see with encryption. When two devices or applications communicate with each other and use encryption, they both have to be configured to use identical encryption algorithms and mechanisms. For example, in order for two hosts to communicate using IPsec transport mode, the security associations have to have been configured identically on both hosts so that they can negotiate a mutually acceptable level of security. This means algorithms have to be configured the same, such as using AES-128 versus AES-256, for example. If the two hosts cannot negotiate a mutually usable algorithm, then they may not communicate at all or, possibly worse, default to a non-encrypted state for further communication, depending upon how they are configured. In Figure 11-12, for example, you see two hosts trying to communicate over SSH, but that can't establish a secure connection.

```
^  v  x  CompTIA Mobility+
File  Edit  View  Terminal  Help
root@bt:~# ssh bobby@192.168.163.128
Read from socket failed: Connection reset by peer
root@bt:~#
```

Figure 11-12   Two hosts that are having SSH connection issues

You may have to examine the configuration files on both hosts to determine why they can't communicate. Upon examining their SSH configuration files, you see from Figure 11-13 that one host does not support the algorithm strength (in this case AES) that the other is trying to use to communicate with. The solution to this problem is to add the algorithm or method to the configuration file of the other host.

As another example, suppose a client browser is configured to use only SSL of a certain version, but the server is only configured to use the latest version of TLS. Unless both hosts are configured with a variety of options allowing them to negotiate the highest mutually acceptable level of encryption, they may not be able communicate with each other, and an encrypted session may not be established.

In addition to encryption mechanisms and algorithms, key issues can present a problem. The most obvious problem is an incorrect shared key, passphrase, password, or PIN, making symmetric encryption impossible. Neither side would be able to decrypt what the other side encrypted, simply because they would not have the correct key because of an incomplete key exchange, a corrupt or damaged key exchanged between two hosts, or a key change on one side that was not implemented on the other. In any event, this issue will usually manifest itself as an inability to encrypt and decrypt. The usual solution is to delete the existing keys and load new ones, or simply share new keys or passphrases.

Encryption issues aren't limited to just symmetric encryption methods. Public key cryptography can have issues as well. Some of these issues can be more difficult to track down, however, because of the use of multiple keys by both parties. For example, if an individual's private key is corrupt, she may not be able to decrypt encrypted data that another party has sent to her using her own public key. The individual may also not be able to digitally sign a message or file with corrupt keys. The solution to this problem is usually to try re-exchanging public keys between the two parties, and possibly even for each individual to restore their private key from a secure backup. If this doesn't work, then the last resort may be to re-issue the public–private key pair of one of the parties.

In addition to keys and encryption configuration, applications that use encryption methods can cause issues if they are not configured properly, or if they can't support the method of encryption used. This may require a simple patch or upgrade, or

Figure 11-13

Neither of these hosts support the same strength of AES.

```
#    Port 22
#    Protocol 2,1
#    Cipher 3des
Ciphers aes128-ctr,aes192-ctr
#    MACs hmac-md5,hmac-sha1,umac-64@openssh.com,hmac-ripemd160
```

/etc/ssh/ssh_config file from 192.168.163.143

```
#    Port 22
#    Protocol 2,1
#    Cipher 3des
Ciphers aes256-ctr,arcfour256,arcfour128,aes128-cbc,3des-cbc
#    MACs hmac-md5,hmac-sha1,umac-64@openssh.com,hmac-ripemd160
```

/etc/ssh/ssh_config file from 192.168.163.128

reconfiguring the application itself. This should definitely be looked at before reissuing key pairs or reloading symmetric keys.

 EXAM TIP When looking for encryption issues, you should examine the configuration on both sides of the connection first to make sure both ends are using compatible encryption algorithms and methods. If those seem correct, look for incorrect or corrupt keys.

Anti-Malware Problems

Anti-malware issues can crop up for a variety of reasons. The most common issue is that the anti-malware definitions don't get updated on a periodic basis, reducing the effectiveness of the anti-malware solution in detecting the latest threats. While the enterprise-level anti-malware solution may be updated on a continuing automated basis, some mobile devices in the infrastructure may not necessarily receive those updates, depending upon how often they connect to the corporate infrastructure. In some cases, the users themselves may be responsible for the updates on their own devices, and unfortunately, users can't always be counted on to make sure the updates happen. This could result in devices not only being infected by malware, but spreading it to other devices that also haven't had their signature definitions updated.

Another common issue is when the anti-malware solution has actually been deactivated or turned off for a device, either accidentally by the user or even sometimes intentionally in order to access prohibited content that the anti-malware solution may otherwise detect. Some particularly nasty pieces of malware are actually able to prevent anti-malware programs from updating, or are capable of turning them off entirely. This is where security monitoring and reporting really come into play, as these scenarios could easily be detected and fixed through centralized or frequent monitoring of the device.

For these issues, centralized control of anti-malware throughout the enterprise is the ultimate solution, taking control out of the hands of the users whenever possible. This allows administrators to make sure that updates are performed on a frequent basis on all corporate devices. It also allows the mobile device administrators to ensure that the anti-malware solution is functioning properly on the device itself. Another solution is the use of network access control (NAC) devices, which would prevent a mobile device from connecting to the network unless it has the most current anti-malware signatures loaded on it, as well as any other security policy requirements, such as patch levels, for instance. In this scenario, the NAC device would impose security policy changes and even anti-malware updates on the device as soon as it attempts to connect to the network.

Beyond the problems of signature updates or anti-malware programs not functioning on devices, symptoms of anti-malware issues would be those associated with a malware or virus infection, including slow performing devices, devices that may have to be rebooted on a continual basis, data loss or corruption, network connection interference, or even data loss. These symptoms, in the course of normal troubleshooting, could indicate a malfunctioning mobile or network device, but these conditions are usually quickly eliminated. In the event the issue can't be narrowed down to a malfunctioning device, a malware or virus scan should be initiated to determine if an infection could be the problem.

Chapter Review

This chapter completes our discussion of mobile device infrastructure security by tackling monitoring and troubleshooting security. First, I discussed the importance of security monitoring and reporting. Some of the key items to consider with mobile devices are compliance with corporate policy and auditing security configuration information from these devices. Monitoring and reporting are concerned with items such as web and device usage, app installations and use, connections to unauthorized networks, and other potential security issues such as malware, misuse, and signs of external attack. Security monitoring and reporting must be done whether the devices are corporately owned or used in a BYOD environment. As with all security programs, policy set and enforced by the organization drives the degree of monitoring and how it will be implemented with a BYOD infrastructure. I discussed the types of items that should be monitored, including device use as well as content stored on the device itself. I also covered technical aspects of device monitoring, which could include individual agents installed on the devices that report back to the corporate infrastructure, as well as Security Information Event Management (SIEM) tools used to provide a unified approach to device monitoring across the enterprise. SIEM aggregates data across the enterprise including system event logs, network device logs, and security logs to provide an entire end-to-end picture of the security posture of the enterprise infrastructure. I also discussed the usefulness of audit logs, and exactly what should be audited and how.

We then turned our attention to troubleshooting security issues, including common security problems such as certificate issues, authentication issues, and network device problems. Many of the common security issues that you see include corrupt, expired, or revoked certificates and keys, which may prevent users from encrypting or digitally signing data. In terms of authentication issues, authentication failures that prevent authorized users from accessing network resources are probably the most common types of problems associated with authentication. Authentication issues could be caused by things as simple as forgotten passwords or much more complex problems such as incompatible encryption settings between hosts, or authentication mechanisms that are incorrectly configured. Network device issues include configuration issues with security devices, such as firewalls and VPN concentrators. These issues also may include some of the same encryption issues seen elsewhere between users and hosts. I also discussed VLAN issues, including incorrect VLAN settings, routing problems, and so on.

This chapter then covered the differences between false positives and false negatives, and how they adversely affect the security of the enterprise. You know that false positives are acceptable events that are incorrectly interpreted as security issues, and can result in complacency as well as a waste of resources in trying to track them down and resolve them. False negatives, on the other hand, can be true security issues, simply because they are the potentially harmful issues that are not identified as a security event. This could result in unauthorized access or a serious security breach in the infrastructure. Both false positives and false negatives can be reduced by simply fine-tuning the network and security infrastructure in the organization. The desired tuning level is referred to as the crossover error rate; it is the point where both false positives and negatives are reduced as much as possible without increasing each other.

Another security issue discussed in this chapter is misconfigured content filtering, which can allow undesireable content into the network at one extreme, or prevent authorized users from completing their work due to overzealous filtering on the part of security administrators and network devices on the other end of the spectrum. I also discussed a security issue that actually affects some of the other security problems mentioned in the book: encryption issues. Encryption issues include incompatible algorithm or encryption strength levels set on different hosts that may prevent the hosts from communicating with each other. Hosts and applications must be able to negotiate a commonly accepted level of encryption between them before secure communications will happen. Finally, another security issue I touched on both in this chapter and others is the problem of malware and how it affects the security of the enterprise as well as individual hosts and mobile devices.

Questions

1. What is the primary element in mobile device monitoring and reporting?

    A. Device agent

    B. Policy

    C. SIEM tools

    D. Acceptable use agreement

2. Which of the following are challenges involved with monitoring and reporting? (Choose two.)

    A. Legal compliance

    B. Personal use of the device

    C. Corporate policy

    D. Mobile nature of the device

3. All of the following should be subject to security monitoring and reporting, EXCEPT:

    A. Policy compliance

    B. App usage

    C. Network connections

    D. Privileged communications

4. Which of the following is considered a requirement of a device agent?

    A. It should always run on top of all other applications.

    B. It should be able to halt processing on the device when it detects an action against corporate policy.

    C. It should be able to report data back to the corporate infrastructure through logs and alerts.

    D. It should require the dedicated use of system resources in order to perform its monitoring and reporting functions.

5. What type of monitoring does not require any specific software installed on the device?

   A. Agentless monitoring

   B. SIEM monitoring

   C. HIDS monitoring

   D. Logless monitoring

6. You have a large mobile device population in the company and need to implement a way to monitor and report security compliance on all of the different devices in use. You recommend to management the use of what kind of monitoring and reporting infrastructure?

   A. Decentralized

   B. Centralized

   C. Agentless

   D. User-based

7. All of the following are elements that support mobile device centralized monitoring and reporting EXCEPT:

   A. Integration into MDM infrastructure

   B. Ability to assign different policy elements to different devices

   C. User-configurable policy elements and alerts

   D. Support for a wide range of devices

8. You are integrating a Security Information Event Management (SIEM) solution into your infrastructure. Which of the following is a key characteristic of SIEM that you want to ensure is present?

   A. Ability to aggregate data from multiple event sources

   B. Ability to manage only one authoritative source of data at a time

   C. Increased false positives

   D. Increased false negatives

9. Which of the following elements should always be included in log files and other audit trails? (Choose two.)

   A. Timestamp

   B. Event description

   C. Email content

   D. File content

10. What part of a certificate lifecycle is involved when a certificate stops working because its expected life has ended?

    A. Certificate retirement

    B. Certificate revocation

    C. Certificate suspension

    D. Certificate expiration

11. A certificate that can no longer be used permanently because an employee has been fired or left the company is said to be _____?

    A. Revoked

    B. Expired

    C. Suspended

    D. Retired

12. Which of the following are potential reasons a certificate could be suspended? (Choose two.)

    A. Investigation

    B. Long-term leave of absence

    C. Termination

    D. Possible certificate corruption

13. Which of the following protocols is used to publish a Certificate Revocation List (CRL)?

    A. IPsec

    B. OCSP

    C. RSA

    D. SSL

14. All of the following symptoms may indicate a corrupt key or digital certificate file, EXCEPT:

    A. Inability of the application to read the certificate file

    B. Inability to send or receive encrypted messages

    C. Failure to decrypt a message without the private key

    D. Inability to send or receive digitally signed messages

15. If a key or certificate is suspected of being compromised, it must be immediately _____.

    A. Suspended

    B. Revoked

    C. Investigated

    D. Reissued

16. Which of the following could cause authentication problems? (Choose two.)

    A. Incorrect user credentials

    B. Incompatible encryption settings

    C. Incorrect VPN server address

    D. Firewall server issues

17. A mobile client is having issues when attempting to connect to certain web sites using specific port numbers. Which of the following should be examined first as the most likely source of the problem?

    A. Firewall rule set configuration

    B. VPN server protocol issue

    C. Corrupt or expired certificate

    D. Encryption settings

18. An issue caused by a security mechanism that falsely reports a piece of network traffic as acceptable is an example of a:

    A. False positive

    B. False neutral

    C. False negative

    D. Crossover error

19. You are helping a user troubleshoot issues with an encrypted connection to another host. The user can connect to other hosts on the network and successfully establish encrypted communications. Which two configuration items should you examine on both hosts to determine if they are compatible for encrypted communications? (Choose two.)

    A. User's digital certificate

    B. Encryption algorithm or method

    C. User's network connection

    D. Encryption strength

20. What type of system could be used to prevent a device with outdated anti-malware signatures from connecting to the network, unless they are updated?

    A. Network access controller

    B. VPN concentrator

    C. Firewall

    D. Content filtering device

Answers

1. **B.** Policy is the primary element in monitoring and reporting, and drives why and under what conditions it is accomplished.

2. **B, D.** Personal use of the device (especially when permitted by the organization) and the mobile nature of devices are both challenges when implementing monitoring and reporting.

3. **D.** Privileged communications may not be permitted for monitoring, depending upon legal guidance and compliance issues.

4. **C.** A device agent should be able to report data back to the corporate infrastructure through logs and alerts.

5. **A.** Agentless monitoring does not require an agent or specific monitoring software installed on the device. It may simply use the device's native logging facilities or network-based software to reach out to a device and gather information on it.

6. **B.** Centralized monitoring is the only choice that permits you to manage all mobile devices.

7. **C.** User-configurable policy elements and alerts run counter to the concept of centralized monitoring and reporting.

8. **A.** The ability to aggregate data from multiple event sources, and analyze that data, is a key characteristic of a good SIEM system.

9. **A, B.** At a minimum, a timestamp and the nature of the event itself should always be included in a log entry. Additional items should include host and destination addresses, user actions, and information about the objects accessed. It would be impractical to include the actual content of files or emails in log entries.

10. **D.** Certificate expiration occurs when a certificate has reached the end of its programmed use. Certificate retirement is not a term used with certificate lifecycles. The other two choices interrupt the normal life expectancy of a certificate.

11. **A.** A revoked certificate is one that is permanently disabled for use due to an event such as employee termination.

12. **A, B.** A certificate could be temporarily suspended from use by the organization during investigations or during long-term absences by its user.

13. **B.** The Online Certificate Security Protocol (OCSP) is used to publish CRLs.

14. **C.** Failure to decrypt a message without the private key is not a symptom of key corruption; it simply means you must own the private key to decrypt a message encrypted with the public key in the pair.

15. **B.** If a key or certificate is suspected of being compromised, it must be immediately revoked to prevent its use.

16. **A, B.** Incorrect user credentials, as well as incompatible encryption settings between hosts, could cause authentication issues. VPN or firewall issues may cause network connectivity issues that prevent authentication, but are usually not direct causes.

17. **A.** The firewall rule set configuration is the first likely problem, especially if the user can't access certain web sites using specific port numbers.

18. **C.** A false negative is when a system allows or reports something that is unacceptable as acceptable.

19. **B, D.** In order to establish encrypted communications, the two hosts must be able to negotiate a mutually compatible encryption algorithm or method, as well as desired encryption strength. Either of these that are not configured properly on the two hosts will cause encrypted communications to fail.

20. **A.** A network access controller (NAC) would be responsible for preventing connections to the corporate infrastructure for devices that do not meet connection requirements, such as anti-malware signatures, security patch levels, and so on.

Troubleshooting Client Issues

In this chapter, you will

- Learn how to troubleshoot various mobile device issues
- Learn how to troubleshoot common application and operating system issues

So far, this book has discussed troubleshooting from various perspectives, including developing a troubleshooting methodology, as well as troubleshooting the network infrastructure. You've also learned about troubleshooting a myriad of security issues that can plague a mobile device infrastructure. This last chapter will focus on moving from the general to the specific, narrowing down the troubleshooting discussions to the mobile client level. You're going to take a look at troubleshooting common device problems, including power issues, synchronization, authentication setup, and even issues that occur when mobile devices crash. You'll also take a look at troubleshooting various application and operating system issues that can crop up. These include problems with application installation, update and configuration, app store issues, and issues that may be specific to the OS platform itself.

This chapter covers two more CompTIA Mobility+ exam objectives: 5.2, "Given a scenario, troubleshoot common device problems," and 5.3, "Given a scenario, troubleshoot common application problems." These two objectives will round out our discussions on troubleshooting, as you use the knowledge you've gained so far in the book and apply it to the client level.

Troubleshooting Device Problems

Let's now take a look at troubleshooting issues at the mobile client level. In some cases, troubleshooting the device is where you should probably start; however, in others, you may want to start with infrastructure itself, depending upon the nature the problem. In any case, the next few sections focus on issues that can occur with the client device and the applications that run on it. One thing to keep in mind is that, like the exam, I'm not going to go in depth on any single platform or OS; the issues discussed in this chapter can be applied equally to most types of mobile devices, regardless of OS, vendor, or manufacturer. Each platform has its own peculiarities and unique issues that can occur, but for the purposes of this chapter, this will be a more general discussion. While I obviously can't cover every single possible issue that could happen to a mobile device, these next few sections cover most of the things you will see on a day-to-day basis.

Power Issues

The first major issue with client devices is power. This may seem like a very simple issue to troubleshoot, and for the most part, the obvious issues can be simple. However, sometimes there are power issues that aren't as obvious and may not be as easy to troubleshoot and fix. For this topic, let's look at battery, power supply, and power outage issues.

Battery Life

Mobile devices obviously run on batteries. Batteries eventually run out of power. These two simple facts don't adequately address the issues that battery life can cause, and the issues that can cause problems with battery life itself. Mobile devices are rated differently in terms of how long the battery should power a device during normal use, how long the device can go between battery charges, and the levels of power that both the battery provides and requires in order to charge.

Most modern mobile devices use some variation of lithium-ion (Li-ion) batteries or nickel–metal hydride (NiMH), which allow for longer use and more power for intensive apps and hardware, and are considered more environmentally sound than the older nickel-cadmium (NiCd) batteries older devices used. Over the years, as device features have increased, displays have gotten bigger, mobile CPUs have gotten faster, and apps have become more graphics intensive, battery power has become a premium for mobile devices. Batteries have not necessarily kept up at the same pace with the other rapidly evolving hardware and features on mobile devices, however. While marketing ads do proclaim extended battery life as a feature for the latest and greatest model, newer, faster hardware and features in proportion seem to make up for that extended battery life in most cases.

Several factors can affect battery life. There are a few, however, that significantly reduce battery life during normal use. One of the main factors is network usage. Any time the device is connected to a network, such as a wireless network or even the cellular one, the device is constantly transmitting and receiving data, sometimes in the form of simple management messages that are sent back and forth to the network entry point device. Even when there is no real data being transmitted or received over the network, these management messages require power from the device, and they can, over time, significantly reduce the battery charge. Often, when traveling, a device such as a smartphone will constantly roam or try to pick up the strongest cellular or wireless signal it can find, and this constant search for stronger signals can significantly drain battery power. This behavior can be controlled with most smart devices however, through configuration changes that limit device roaming or searching for new wireless networks.

Another significant cause of battery charge reduction is the constant use of GPS or location services. When location services are turned on, the device is constantly receiving GPS data in order to fix its location. This could be for applications that require location data in the background, or for active applications, such as mapping software, that use the GPS receiver. In any event, the user will notice the significant reduction in battery charge, and over the life of the device this will reduce the life of the battery as well. The simple solution to this, of course, is to be judicious with networking location services in that when they are not required, they should be turned off. Of course most users don't necessarily remember to do this. Another tip is to review configuration settings for apps that use location services and disable the ones that you really don't need to have location data for. In the event that you may actually use an app and require location data, you can always turn the setting back on and let the app get a location fix. Some examples of applications that require location data include those used to find restaurants and movie theaters, for example, as well as specialized apps, such as geocaching apps, and mapping and navigation apps.

Another factor that affects battery life is screen display and brightness control. A screen that is configured to constantly show the brightest level will significantly reduce the battery life. Users often mistakenly adjust the screen brightness to be constantly at the highest level, thinking that they will see it better in bright sunlight or in darkened indoor areas, when often this isn't the case. The display controls on a mobile device should typically be configured to automatically adjust the screen brightness based upon factors such as device sleep or suspension, app usage, or even smartphone call use. Figure 12-1 displays the battery usage for an Android smartphone. Notice how much of the battery is being drained from the screen configuration alone!

Figure 12-1

Battery usage for a smartphone

EXAM TIP   Be familiar with the factors that can reduce battery power and battery life.

Most users that have used both iPhones and other devices are well aware that iPhones, iPads, and other "iDevices" have batteries that are typically not user swappable and require a trip to a service facility. Unlike Apple devices, many other manufacturers do place a user removable battery within the device that can be changed when it goes bad, so the purchase of a spare battery may be a good investment. For those that don't, they also may require a trip to an authorized service facility. Figure 12-2 shows an example of a removable battery from a mobile device.

There are several apps that can be downloaded and installed on mobile devices that can help the user manage battery life. These include apps that can alert the user in case the battery power falls below a certain threshold, apps that can warn a user if

Figure 12-2
A removable
battery and its
device

Figure 12-3
A battery usage
app for Android

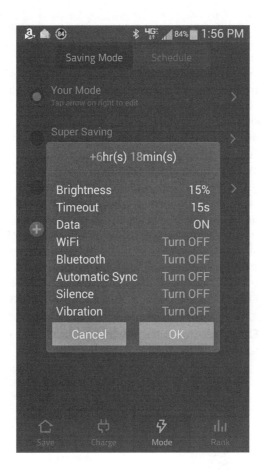

the battery has issues with charging or maintaining a charge, and even apps that can manage memory, CPU, and hardware usage, as well as app usage in the background to reduce battery usage and extend battery life. Figure 12-3 shows an example of one such app for an Android smartphone.

Power Supply Problems

The next piece of hardware in the chain of power for the device is the internal power supply for mobile devices. This is most frequently a piece of hardware that is not typically user serviceable, in that, for the most part, you can't simply crack open the case and swap the power supply out the way you would do a desktop unit. Power supply problems can range from intermittent issues causing the device to lose power and recycle, freeze, or lockup, to not powering up at all. Additionally, faulty power supplies can cause internal device batteries to not charge properly, resulting in partially charged batteries or devices that won't run except when connected to AC power. Because most power supplies built into mobile devices aren't user serviceable, the real solution is to take them to a professional repair facility operated by a technician authorized by the vendor. In this case, the vendor could possibly change power supplies out, but this is sometimes not very cost-effective. In a large number of cases, it may be more cost-effective to simply buy a new device, usually an upgrade, and restore the device content from a recent backup.

External power supplies, such as the type that laptops, tablets, and smartphones use to charge their internal Li-ion or NiMH batteries, also go bad, but the big difference here is that they are usually easily replaceable, albeit sometimes at a little bit higher cost, depending upon the particular device. An external power supply/charger can run upwards of $100+ for a genuine manufacturer-branded unit for laptops, and can also be quite expensive for devices such as iPhones and other Apple devices, because they use unique connectors. Some mobile devices have done away with proprietary connectors and can be recharged using standard USB cables. Generic power supplies can cost considerably less, but you must make sure that the electrical specifications, such as voltage and wattage, are correct for your device, as well as the unique power connectors some devices have. Figure 12-4 shows three examples of external power supply cables that are used to charge mobile device batteries.

 EXAM TIP   You will not be required to know voltage and wattage specifications for different batteries on the exam.

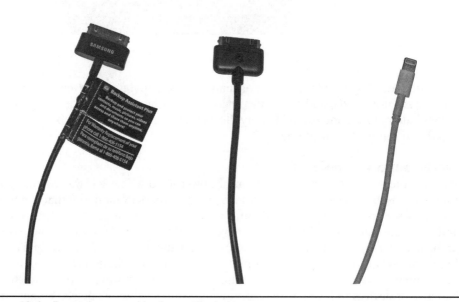

Figure 12-4   External power cables for mobile devices

Power Outages

Power outages can have interesting effects on mobile devices. Some devices may go into a sleep or standby mode when they lose power, but this is typically only for short periods of time. Their running state may be saved to an internal storage device or disk so that when they are powered back up, they remain in the same state they were in before they powered down. This may include open applications, unsaved data, and other user-state information. However, after a period of time, the device may have no power at all to it, as in the case of a completely dead battery, and may lose the saved state information. This will almost always result in corrupt user data, especially if any user data files were open when the device lost power. It also may result in corrupt applications that may have been open and in use at that time as well.

As with batteries, most mobile devices have apps built-in (or available from an app store) that can monitor power usage and alert you if there are power issues. There are also diagnostic applications available that service technicians can obtain to troubleshoot power issues. If the organization has an in-house service department that has the knowledge and equipment to troubleshoot mobile devices used by the organization, they should have access to these apps, as well as spare batteries and external power supplies in stock to assist in troubleshooting and repair.

 CAUTION You should never attempt to disassemble batteries, power supplies, or any component that has electricity flowing through it because of the risk of electric shock. When troubleshooting power issues, it's best to leave component disassembly up to the professionals at an authorized repair facility.

Device Issues

Beyond power issues, there are several problems and issues that could affect a device, and this includes synchronization issues, authentication issues, and device hardware issues. I discuss several of these in the next few sections. Understand that this book obviously can't cover every single issue that could affect a mobile device, but it will examine the most common ones. Before we jump into specific issues, it's a good idea to examine configuration changes in general as they can affect all of the other issues discussed here.

Configuration Change Issues

Two aspects of configuration change issues are discussed in this section. Because one of them is more specific to the enterprise level I won't spend as much time on it. At the enterprise level, I'm talking about major changes that really affect all or most of the devices in the enterprise. This could be an enterprise-level upgrade in enterprise standard OSes, or even in devices used. At the enterprise level, configuration changes are significant and require careful planning and execution. Changes should be managed through a formal configuration change process that must be documented and approved prior to being implemented. The reason to bring this up at all in this section is that it can affect all of the users' mobile devices significantly, and sometimes differently, depending upon the disparate devices, operating system versions, and so on in the enterprise. When implementing these major changes, you should make sure that users are aware of them and how they will affect their devices. That way, when the changes are implemented across the enterprise, you can minimize the onslaught of user phone calls to the help desk, and have a manageable way of helping them with their devices when features and apps change with mass configuration changes.

At the device level, configuration changes can cause issues in several different ways. First of all, upgrading the operating system or even a single app can cause the device's configuration to change. Authentication and encryption settings, network settings, location services, and many other items can all be changed based upon an app or an OS change. In some cases this may have unintended consequences, such as the user not being able to connect to different enterprise services, such as email. In other cases, the user may not be able to update her apps or OS patches or even antivirus signature updates. These different configuration changes can have both usability and security impacts on the device, so they have to be approached with care. For this reason, all the different issues I discuss in the following sections regarding device, operating system, and app changes, as well as troubleshooting those issues, are important to consider.

Configuration changes are almost always the first place you should look when a device or its apps begin to have issues. As you learned early on in the troubleshooting methodology, one of the first things you should do when a user or device starts having issues is to determine what has changed on the device since it last worked properly. This could be as innocuous as an app upgrade or new installation, or even an OS patch. In any case, configuration changes are often what lead to device or app issues. It's a good idea to document the baseline configuration for your mobile devices and update it periodically so you are aware of how app or OS changes have changed the configuration of the device itself. In an enterprise environment, this may be easier because of the capabilities of most modern enterprise management software, including both MDM (Mobile Device Management) and MAM software. In infrastructures that make heavy use of BYOD (Bring Your Own Device), or lack centralized MDM, this would be somewhat more difficult and may have to be performed on a per device basis, if it can even be accomplished at all due to the scope and size of the enterprise and the level of device usage within the organization.

 EXAM TIP Always look for configuration changes on a device when trouble-shooting issues with it. As with our troubleshooting methodology from Chapter 10, you are looking for what changed on the device since the last time it worked properly.

Synchronization Issues

Synchronization refers to several processes and actions the mobile device can perform in order to update itself with user-related information, such as contacts, email, media files, and even app updates. Various mobile devices are synced differently, depending upon the device vendor and software required. For example, (almost) everyone is familiar with ubiquitous iTunes software, which is not only used for playing media and purchasing it from the iTunes store, but is also used to sync iPods, iPhones, iPads, and the iTouch. Although there are other applications that can similarly sync Apple devices, iTunes is the preferred method, obviously, because it offers the most complete range of features, including the all-important software and patch update feature. Figure 12-5 shows an example of iTunes updating an iPhone's software version.

Most other mobile devices, including BlackBerry, Android, and Windows devices, also have a similar app store, like Apple has for iTunes, which can sync certain configuration settings, apps, software upgrades, and so on. In some cases, these other mobile devices— Android devices, for example—don't necessarily require any particular application to synchronize with, and may use individual apps to synchronize their parts of the device. For example, an email app may be perfectly capable of synchronizing its data to include email and contacts. Like the later versions of iOS, most of these other devices are capable of syncing over the air using Wi-Fi or cellular technologies. In an enterprise environment, syncing devices in the infrastructure may involve using the vendor's commercial app store for some device aspects, while other aspects, such as enterprise-specific apps, security policies, and configuration settings, may be synced with the MDM infrastructure.

Synchronization issues include incomplete sync of data due to connectivity issues, device issues, or even remote infrastructure issues. Sometimes synchronization issues

Figure 12-5   iTunes updating an iPhone's software

can cause an incomplete downloading of email or even duplicate email, for example, as the device may retry the synchronization process over and over if it can't get a good connection, continually downloading the same email messages. A device may attempt to sync to download an OS patch or update and may fail. The most likely culprit is connectivity issues with Wi-Fi or cellular connections, and the problem can usually be resolved by moving the device to an area with a stronger signal. Of course, this doesn't prevent upstream connectivity issues, which may also have to be examined.

In some cases, there may be other issues that prevent synchronization. These can be a wide range of problems, including authentication issues, OS version issues, or incorrect configuration settings. If a device won't sync even after getting it to a stronger, more stable connection, these are some of the things that should be examined. Another problem may be the remote end of the connection. This may be the enterprise email server, or even the entry point into the enterprise network, such as a NAC device or a VPN concentrator. Failure to properly authenticate or meet the requirements of the entry device may prevent a device from synchronizing.

One other issue you may want to examine when you have synchronization issues is that in some cases synchronization can occur from multiple sources. A device can synchronize from an enterprise app store, for example, as well as the vendor app store, personal email services, such as Gmail and Yahoo, and even from third-party providers of "whatever-as-a-service" and cloud storage. So in troubleshooting synchronization issues, you may have to take into account that different providers may have different configuration settings, to include encryption and network settings, and in turn these configuration settings could conflict. In the enterprise environment, it's probably incumbent upon the mobile device management team to put together both a management and technical strategy that will ensure minimal conflict between different synchronization sources.

 EXAM TIP Understand the symptoms and issues related to device synchronization, including network connectivity and authentication issues.

Authentication and Password Issues

At the device level, authentication issues are certainly a problem and can be traced to several different possibilities. Previous chapters discussed what can go wrong with certificate-based authentication, as well as with secure authentication, that may involve different authentication or encryption settings between an enterprise device and a user's mobile device. In troubleshooting authentication issues, obviously you want to ensure that the proper certificate is loaded on the device, make sure that it is still valid (not expired, suspended, or revoked), and ensure that its trust chain is loaded on the device as well. You also want to make sure that the certificate is not corrupt, as this can occasionally (but rarely) happen, and this may require downloading and reinstalling a new certificate. You want to check that any apps that use authentication and communicate with the enterprise are configured appropriately with the correct authentication and encryption settings, as well as network settings, such as ports and protocols. These are items that you will have to troubleshoot, of course, on both ends of the connection, but on mobile devices it's probably even more important simply because enterprise configuration options are less likely to change on a regular basis than device options might. This is usually because the "user-assisted" configuration changes occur, new applications are loaded, and so on.

Passwords can also cause authentication issues simply because it's still likely they will be present on the device, in addition to any other authentication methods used in the enterprise. Not every single app or every single service is capable of using certificate-based authentication so there are still likely to be some services that require a username and password combination or a PIN. In fact, unlocking a device typically requires a password or a PIN. And if you are using a BYOD infrastructure, then the user likely has his own passwords and PINs he uses on the device for personal data and apps. In any case, passwords can cause issues because they can be forgotten or entered incorrectly. They can also become corrupt if the user has stored them permanently in the app so that he doesn't have to type them in every single time he accesses that app. Configuration settings could cause the app to overwrite password settings and the user may get password prompts repeatedly until he enters the correct password. This can happen with email passwords, for example, as well as wireless network connections if the device has "forgotten" the passphrase or it has become corrupt. In some cases, a bad network connection can cause the password to be corrupt during transmission and the service or app may reject it because of that. The user may be prompted again to enter the password.

Troubleshooting password issues includes making sure, obviously, that the user has the correct password or PIN, and resetting stored passwords in apps when required. A user that repeatedly can't access a network service because of password issues certainly should check to make sure she is entering the correct password, but after that has been ruled out, the network connections should also be checked. And of course, any configuration settings on the app and the network service should be checked to make sure that the password type or password encryption settings match. For example, if the network service requires a password encrypted a certain way or that excludes or requires certain characters, the app should be configured similarly.

 EXAM TIP Always check to make sure the user is using the correct password. Sometimes the most obvious solutions are the correct ones.

As a last-ditch solution, if no other troubleshooting step works, the password should be reset. In a password reset scenario, typically the administrator would set up a temporary password and force the user to change it upon her first successful login. If the app or the services are not configured to do this, then the user should be advised to manually change it after she successfully accesses the service or app. If the password is reset, the app may need to be reconfigured appropriately, especially if it uses a stored password or requires different settings based upon the password change. It should also be synced with the relevant service to ensure that the passwords are synced up and changed successfully.

Profile Authentication and Authorization Issues

Profiles can cause issues as well, simply because profiles contain all the necessary configuration security settings to access services from the enterprise. Profiles can contain configuration settings such as email server settings, network connection settings, encryption settings, and even authentication settings. Users also may have multiple profiles for different network or device-based services and apps, although the preference is to have a single profile that tries to accommodate all of the enterprise services and applications that the user may connect to. It's possible that profile settings can become corrupt or change over time, especially if new services or apps have been installed or have made changes to the profile. Symptoms of corrupt profiles or other profile issues could include the inability to authenticate to the enterprise completely, or inability to authenticate only to certain services. Symptoms could also include problems with the connections or using the services.

Solutions for profile issues may involve examining all the configuration settings in the profile and making sure they are reset back to what they should be, based on the enterprise requirements. If any apps or services have made configuration changes to the profile that are required for their use, the profile may have to be altered to allow for those configuration changes. This may also involve changing configuration settings on the enterprise app or service. This should be a last resort, however, because most enterprise services are configured for multiple users, rather than only for a single user; making changes to the services themselves may cause problems for other users who don't have profile issues. Another last-resort solution is to delete the user's existing profile and start from scratch with a new one, possibly from an enterprise template, or from very basic configuration settings, reinstalling and reconfiguring new apps or services until you discover what caused the problem. As with most other services I discussed that involve authentication, checking to make sure the profile's encryption and authentication settings match those of the enterprise or target service is an important troubleshooting step as well. You should also check network connection settings in profiles to make sure they are configured as they should be.

 TIP A corrupt profile is likely to cause multiple connection and authentication issues for a user, so it should be high on the list of items to troubleshoot when you see multiple problems.

Device Crashes and Hardware Issues

You learned about hardware issues in previous chapters, but mostly from an enterprise or network device perspective. Hardware issues obviously can affect mobile devices as well, so they are worth discussing here. The most severe type of hardware failure is a device crash. The crash could be temporary, where the device simply locks up or freezes, or it could also reboot itself occasionally or on a continual basis. The worst type of device crash occurs when the device crashes and never powers on or boots back up again without significant errors.

Most often, the tendency is to immediately think that there's something wrong with the hardware in the device, and this could very well be the case sometimes. In other cases, the problem could be a corrupt or misconfigured operating system, or even a faulty app that has been allowed to access the hardware in ways it shouldn't. I discuss those other cases a bit later in the chapter. For now, however, let's focus on hardware issues. Hardware issues could include battery and power issues that were previously mentioned, but could also include a device that has had a corrupt firmware issue (possibly due to a faulty firmware upgrade) or simply a piece of hardware that is no longer functioning properly. Firmware issues can be resolved by researching the device vendor's documentation, to include blogs, web sites, support forums, and so forth, for any known issues, and then possibly upgrading to a later version or, occasionally, even downgrading to a previous version of firmware.

Troubleshooting device hardware issues can be a little problematic in that the end user may not have the skills or knowledge to do this, and the enterprise technicians may not have the diagnostic equipment or software to do so. Often, a device will be sealed and doesn't offer a user the ability to easily open or service the device. Sometimes the device vendor may be the only entity that can properly diagnose and repair a hardware issue. This might mean returning the device to an authorized repair facility or completely replacing the device. There are often rudimentary diagnostic apps that can be downloaded and used on the device itself. There are also more advanced applications that can be installed on a desktop system and used to access the device when it is connected by a cable to diagnose hardware issues. In some cases, these apps may be able to generalize a hardware issue, but not provide enough specific information to fully diagnose it or repair it.

General hardware issues that can affect mobile devices include processor and memory issues, faulty or damaged circuit boards or connections, broken parts due to excessive force when the device has been dropped, or screen damage. Screen damage is one of those hardware issues that is quite easily diagnosed but not always easy to fix. Frequently, the device must be replaced or at least returned to an authorized repair facility to replace the screen. Figure 12-6 shows an older iPhone 3G with a damaged screen.

In the event the device has been dropped or otherwise been physically damaged, it may be beyond repair and have to be replaced. Again, usually only an authorized repair facility could make this determination. Rarely, a device may simply be defective and have to be repaired because it was not assembled correctly, or even designed and engineered properly, and may have issues with faulty components or design because of this. In that event, the device will likely have to be replaced, and there may in fact be a vendor notice disseminated describing the issues and the possible resolutions.

OS Issues

Operating system issues can cause several device problems, including device malfunction, application malfunction, and the inability to properly connect to and communicate with enterprise services. There can also be security issues caused by the operating

Figure 12-6

An older iPhone 3G with a damaged screen

system, including configuration or patching issues, and authentication and encryption problems. Like standard desktop systems and servers, mobile device operating systems should be periodically checked for configuration settings as well as patches and updates. Over time, installation of different apps or changes to the operating system may cause issues with the OS configuration. Sometimes these changes can lead to incorrect configuration settings or settings that are incompatible with other apps, services, or security requirements. Also, just like the operating systems on traditional devices, OS files and configuration settings can become corrupted over time through unintentional user action, faulty OS files and patches, or simply due to installation of incompatible apps. Fortunately, unlike desktop operating systems, mobile device OSs are usually far more restrictive in terms of user, app, and hardware interaction with the OS, so these issues happen much more infrequently. When they do occur, however, they can cause data corruption, app issues, and sometimes even device crashes.

Mobile device operating system issues may present symptoms such as the inability to connect to a network, application problems such as an app that freezes up or won't start, or even device freezes and crashes. When troubleshooting OS issues, you should proceed from the general to the specific, in much the same way that you would troubleshoot other device or network issues. You should determine if the symptom you are seeing affects only one app or service, or more than one. If it affects more than one app or service, then it could be an OS issue. Device issues such as freezes or crashes could be caused by hardware, apps, or the operating system itself. The OS could be mismanaging hardware, such as the CPU, memory usage, or in some cases, even battery usage, and creating issues for the device. Most of the symptoms, however, are common to app, device, or OS issues so sometimes they can be difficult to track down. When troubleshooting any of these, eliminating the others is probably the best course of action. For example, when eliminating device or app issues, you want to know if symptoms affect only one or multiple apps, which apps might be running when the device freezes or crashes, what the memory or CPU usage is when problems occur, and so on. Examining these and other symptoms may narrow the problem down to the operating system, if it can't be pinned on a particular app or hardware problem.

Resolving operating system issues may be troublesome as well. This is another one of those cases where the user may not have the knowledge or ability to troubleshoot this problem (and, in an enterprise infrastructure, you probably wouldn't want them to anyway). If you have an enterprise-level dedicated group of technicians who can troubleshoot and fix OS problems, that may be the better solution. If the enterprise doesn't have this capability in-house, then of course a third-party approved vendor repair facility maybe the answer.

Applying a patch or reinstalling one or several apps may correct the issue in some cases; one common solution is to reinstall the device from a known good backup. I've discussed the value of good backups previously in the book a few times, and this is one instance where you can see the value of ensuring that you have good device backups. In the case of an iPhone or iPad or other Apple device, restoring the device from a known good backup typically restores the operating system, apps, and even user data, as of the date of the last known good backup. In the case of other devices, such as Android smartphones or Windows tablets, you have the ability to back up data such as contacts, email, media files, and so forth a bit more independently. You may find that wiping Android and Windows devices and reinstalling them back to factory settings is a much more effective way in those cases, and you can then restore your data from a good backup and reinstall your applications from the app store. There are also good third-party backup solutions for those types of devices that can back up and restore all of the data, including the OS, apps, and user data.

 EXAM TIP   While simply reinstalling the OS is one solution to correct OS problems on the device, make sure you have a good backup before restoring it, as data loss can occur.

Troubleshooting Common Application Problems

Now that we've covered device, hardware, and operating system troubleshooting with a mobile device, it's time to examine another source of problems the user could have with the mobile device: app issues. These can be more difficult to track down in some cases, simply because you have to troubleshoot and determine whether an issue is caused by the device configuration or hardware, problems with the operating system that may apply to multiple apps, or issues that plague an individual app itself. Using the troubleshooting methodology described in Chapter 10 will help you narrow down the issues to one of those device components. The next few sections discuss a wide range of problems that can occur with application installation and configuration, as well as app problems that can be caused by issues with reinstalling, upgrading, or even replacing an app. I also discuss issues that can affect apps coming from the enterprise or vendor app store.

Application Installation and Configuration

A great many of the application issues that you will have on a mobile device come from configuration settings that are introduced or changed when an app is first installed, replaced, reinstalled, or even removed from the device. Some of these new configuration settings can be caused by the operating system or configuration settings on the device itself, and some could be the result of being changed or overwritten by other apps. I examine each of these cases in the sections that follow, and I discuss a few key apps and services that can cause many problems, such as location services and email. Keep in mind that a lot of these scenarios assume that the administrator or user is attempting to install, upgrade, or remove apps from the device itself versus remotely using MDM or MAM enterprise capabilities. Although some of these same issues can occur, usually using enterprise-level app management utilities will result in a much more controlled, efficient, and cleaner installation or change action.

Issues with Installing Apps

Problems with apps can start occurring even when they are first installed. This can be caused by numerous issues up front, including incompatibility with your device or operating system, lack of a needed patch before installing the app, or even lack of proper permissions or authorization to get the app installed and configured correctly. When first installing an app, the user must ensure they have a good, solid network connection to the app store through either Wi-Fi or cellular services. Some apps, in fact, will only download and install when the device is connected to Wi-Fi because of its size or network speed. In some cases an app will download first in its entirety and then install, but sometimes it appears as if the app is downloading and installing simultaneously. The key is, of course, to have a good network connection that can support the bandwidth needed to download and install the app.

Once the app is downloaded, the user may be guided through a mini-setup wizard or be prompted to input certain information such as authentication information (be it certificate or username and password), encryption settings, or service and server settings. Additionally, the user will sometimes be asked for permission for the app to use different services or hardware on the mobile device. In any of these steps during setup, there can be issues if the user does not have this information or is not authorized to install the app or give permission for the needed configuration settings. If the device is completely managed by MDM or MAM, it may be prudent to install apps only when the administrators are available to help with the settings or are able to preconfigure them so that app installation goes easily. For the most part, regardless of OS or device vendor, app installation is a fairly straightforward process.

Symptoms of a bad app installation usually pop up when the user first tries to use the app. The user may get an error or warning message, or the app simply doesn't work. The first step is to try to go into the app configuration settings and make sure that they are correct and match the network or enterprise service settings the app may require in order to function properly. You should also make sure that the app has the required permissions to the hardware device it needs to work. Also check, of course, encryption and authentication settings if required. If none of these things seem to solve the problem, then the most logical step is to reinstall the app itself. If this still doesn't solve the problem, you may want to look at the app vendor's documentation from the vendor's support web site and see if there are any known issues with the app or patches that have to be installed prior to use. You may have to report issues yourself to the vendor in order to figure what the next step is as you search for a solution. In the case of in-house apps, troubleshooting will probably be a much more efficient process as you will have direct contact with app developers.

Upgrade Issues

Upgrading an app from a previous version to a newer one can have its own issues. Sometimes the new version of an app must be configured differently or requires different permissions or settings in order to work. Just as with installing an app for the first time, the user may have to know these different configuration settings and have the ability to grant permissions to the app for the use of different device hardware or services. Sometimes issues can come up when upgrading an app because, in addition to different configuration settings, user data from the legacy app may not compatible with the new version. For example, the new app may format data differently, or use new file formats. It also may use a different encryption algorithm or scheme that the old data must be adapted to use. With modern app development, this usually doesn't happen, but it could occasionally.

In any case, symptoms of upgrade issues may mirror the issues that you would have when installing a new app for the first time, in that it may be incompatible with the device or simply not function. You may get standard errors or warning messages, or in some cases you may not be able to access data saved by the old application. As with new apps, looking at configuration settings and permissions is probably your first step. As these configuration settings are corrected or changed, you should check to see if the app is working properly. You may also want to review the vendor's documentation regarding changes to the new version that may prevent it from working on your device or with your existing data. Sometimes the answer is to simply uninstall the old version of the app and reinstall the new version, keeping in mind that you may or may not be able to preserve existing data unless it is independently backed up first.

 EXAM TIP When troubleshooting issues with installing or upgrading apps, the app vendor's support site can provide valuable information. Don't neglect to check it out for solutions to app problems.

Problems with Replacing and Reinstalling Apps

I mention this as a separate troubleshooting section from installing and upgrading an app, simply because, often, users may actually replace an app with a newer version that may not be the next higher version, but several versions newer, or even replace an app with a totally different one, expecting to use the same data files with the new app. There are also issues that can happen when reinstalling an app, as configuration settings may be inadvertently retained between the installations and may cause issues. In either case, most of the issues that you'll see with replacing or reinstalling an app are the same types of issues you may see with the initial install or with upgrading one. Symptoms are similar to the ones just mentioned, such as a nonfunctioning app or error messages; troubleshooting steps would also be similar. This might include checking configuration settings, of course, as well as making sure that data is in the right format for the new version or new app. Obviously, before reinstalling or replacing an app, you want to make sure you have a good backup of your data before uninstalling the previous app. One other troubleshooting step you may want to take is to completely power off the device and power back on between uninstalling one app and installing the new version. This may clear out any data in persistent memory that may be used in configuring the new app. You also may want to check the app documentation to see if there are any persistent settings or configuration files that remain on the device in between the old and new apps that may cause issues. On some devices, these can be deleted; on others, it's more difficult to change or delete the settings. Again, consulting the documentation from the vendor, usually from the product support sections of the vendor's web site, is a good way to determine what issues may be present between app versions or when installing a completely different app that performs similar functions. For example, you may uninstall one app that manages a user's company travel expenses and install a similar one that purports to use the same data and format as the previous one, but then determine that there has to be data format or configuration changes in order to use the existing data.

Issues with Removing Apps

You normally don't have the same issues with removing apps as you would with installing or upgrading them. One of the main issues you may have is that remnants of data files or configuration changes can be left on the device by the app you are removing. For example, apps that make changes to authentication or encryption settings may leave those configuration settings in place even after they are removed. You should check the device configuration settings for these types of instances. Uninstalling an app by using the device's built-in uninstall capabilities, or by using the preferred app store utilities, is usually the best way to remove an app. Simply deleting its icon from the device screen doesn't uninstall the app itself, as some users might believe. The app would simply reappear again the next time the device is synchronized. In the case of an app used by the enterprise, the administrator would want to ensure that there are no remnant data files containing sensitive information that would no longer be needed on the device, and enterprise-specific apps are usually uninstalled as part of the MAM process, so these considerations should be taken into account from the infrastructure side.

In some cases, there may be problems actually uninstalling or removing the app because of user permissions. In this case, the user trying to remove the app (if this isn't something you're trying to do remotely through the MDM or MAM) may not have permission to do so, or may not have the ability to configure the device properly after removing the app. Additionally, another app may depend upon the app you are trying to remove or its data, and may cause issues when trying to uninstall the app. For example, think of an email app that uses data stored in a separate contacts app. Uninstalling the contacts app (and its related data) may cause issues for the email app because it now doesn't have a source of data for contacts to use in emails; there may be errors or warning messages during the uninstall process that reflect this. This may or may not allow the uninstall to successfully complete, depending upon the nature of both apps.

Missing Applications

Missing applications are an interesting issue, simply because of why they may be missing. You may get calls from users who complain that applications they installed are no longer present. You may have to explain to these users that the applications were unauthorized and were uninstalled remotely by your MDM/MAM infrastructure. Ensuring that users are educated on what they can and cannot install or use on the device is a good practice.

On the technical side, however, missing applications can occur for a variety of reasons. Synchronization issues can cause an app to disappear from the device simply because it wasn't synchronized properly and the MDM or app store infrastructure has not retained information on the app for the device. Another reason may be that the app was installed from an unauthorized source and was removed automatically for that reason. Sometimes, as devices are refreshed or restored from backup, an app may not be included in the refresh or backup if it was installed after the last known good backup. The user may have to reinstall the app, assuming it's authorized and comes from an authorized app store, or it may have to be included in the next configuration refresh that's pushed to the device.

App Store Problems

This section covers the different issues that can occur with app stores. In addition to the standard OS and vendor app stores, such as those that are maintained by Apple, BlackBerry, Google, Amazon, and Microsoft, I'm also talking about the enterprise-level app stores where individual enterprise-specific apps are developed and pushed out to mobile devices. I'll also mention the independent (and sometimes unauthorized) app stores from which users may occasionally download apps.

Chapter 6 covered the different app stores from the major mobile device vendors. They all have many things in common, as well as some significant differences. If you remember correctly, most of their differences involved the monolithic (or vertical) development and marketing models that Apple and BlackBerry have; both manufacturers completely control device, app, and OS development, unlike the more open, horizontal models used by Google and, to a lesser degree, Microsoft and Amazon. When you look at the methods of developing, implementing, and marketing apps, these differences are fairly important. From the device level, however, when you look at how they are installed, configured, and managed, the differences become fairly insignificant. What becomes more important at the device level are the common issues you could have with app stores regardless of their internal development schemes. That's really the focus of this section, but it helps to remember that the various vendors do have significant differences with control models over their apps and respective app stores.

Common app store issues that most users will see from the device level are problems connecting to, authenticating with, and downloading and installing apps and their updates, as well as operating system patches and updates. Obviously, network connection issues are a big player in app store issues. In order to successfully connect to and download apps, the device has to have a stable and strong network connection. For most apps, standard cellular technologies, such as 3G and 4G, are fairly sufficient. For other, larger apps, they may require a dedicated Wi-Fi connection that has better throughput. A weak, slow, or intermittent network connection may prevent a device from even connecting to an app store and may also prevent successful passing of authentication credentials for the user's app store account. Obviously, the solution to this problem is to ensure that the user has a good, stable connection. For internal enterprise-level app stores, administrators may advise (or require) the user to only download or update apps from the enterprise app store when they are connected to the enterprise infrastructure, for both network connectivity and security reasons.

The next issue with app stores is authentication. For users who have to download and install apps from the respective vendor app stores, these vendors require an individual account that either users set up for themselves or the corporate MDM administrators set up for them. In the case of BYOD organizations, there may even be separate accounts for the user's personal apps as well as enterprise-approved apps, although this may prove to be unwieldy. Obviously having incorrect credentials would cause authentication issues, but also having an unstable network connection may cause similar authentication issues because of network latency or data corruption. These two problems could result in the inability of the user to authenticate to the app store.

Once the user has established a good connection to the app store and has authenticated properly, other issues that may occur, depending upon which type of store they connect to, include a lack of authorization in terms of permissions to download and install certain apps. For example, even apps that are present in the iTunes store for download may not be authorized for use on corporate devices or on some BYOD devices that contain corporate data (depending upon the corporate policy) simply because they may represent a security risk. To further the example, the organization may have its preferred email app and not authorize a user to download, install, and use a different third-party app from the app store. This is where mobile device policies pushed down from the corporate MDM infrastructure would serve to control and limit apps and their use on the devices.

One other interesting issue with commercial app stores is the issue of who pays for and who owns the app, especially in the case of a BYOD infrastructure. The user may argue that because he owns the device, he can install whatever app he likes, and the organization has no say in it. Again, this is where corporate policy comes into play to define and control limitations on users. For this reason, it may prove to be smart (or not) to have separate accounts with commercial app stores for users who have devices that have both personal and corporate apps and data on them.

As far as corporate app stores go, keep in mind that an app store really is just a centralized place from where users can download apps. This could be a share on a server that the user accesses from a browser or a function of MDM/MAM that pushes apps to the device. Apps that come from corporate stores are usually those specially developed by the enterprise and are provided for use with corporate data. Other than the network connectivity and authentication issues discussed previously, the real issue with accessing these particular apps is to make sure that the servers that store them are available for the users to get to and access. Authentication may be provided by certificate-based services, using a single sign-on framework to access both the app store and other enterprise services such as email, file shares, and so forth. Any authentication issues the user encounters when accessing a corporate app store are also likely affecting these other services as well, so when troubleshooting authentication issues for enterprise app stores, you may be better served by looking at some of these other services to see if there are authentication problems with them as well. Remember the strategies discussed in Chapter 11 on how to troubleshoot authentication issues, particularly those associated with certificate-based authentication, and apply those techniques to troubleshooting app store authentication issues as well.

App stores that are not vendor specific or that are provided by the corporate enterprise network are considered independent or third-party stores and may or may not be authorized for the user to obtain apps from. Enterprise-level policies can be implemented and pushed down to the device, which prevents the user from accessing these independent app stores, but in some cases they may be authorized, based upon their legitimacy and usefulness to either the user or the organization. In such cases, the same issues would apply as those found in the other types of app stores; the enterprise would have to take into account network connectivity, authentication, account usage, and billing when developing policies allowing the use of these stores. The major issue with independent app stores is authorizing the user to download only certain types of apps that are considered safe and secure when dealing with corporate (and even in some cases personal) data. These should typically be allowed only on a case-by-case basis and restricted with a default-deny type of policy, allowing only certain exceptions, and denying all other apps. Unauthorized apps that are found on mobile devices during routine audits, or when they cause device function or security issues, are a source of trouble that should be both prevented and dealt with firmly when they happen. If these unauthorized apps show up on a device, the first troubleshooting step would be to look at the policy that is pushed down to the device (or lack thereof) and make sure that the independent app store is authorized, and that the correct apps are listed by exception in the policy. A little user education and training, obviously, is yet another "troubleshooting" step you could take as well.

Location Services Problems

Aside from some of the issues with devices, hardware, and the operating system itself, there are occasionally issues with location services on mobile devices. Most mobile devices today are made with some type of technology built-in in order to ascertain the device's location. Some devices use built-in GPS capabilities while others also use 3G and 4G cellular technologies as well as Wi-Fi in place of GPS (or to supplement it) to get location data for the device. Location data is used for some of the obvious things, such as map applications, GPS coordinates, and so forth, but it's also used for some things that are not so obvious to the user. Most apps use location data for a variety of reasons. In some cases, it's to assist the user with whatever functions the app performs, such as locating a restaurant or a bank, for example, near the user. In some cases, it may be used to provide services such as weather or local traffic conditions. And, for better or for worse, some apps use location data rather covertly for the purposes of marketing and reporting user data back to their respective vendors for various reasons. For the most part, this particular use is usually unknown by and undetected by the user. In any case, there many different uses for location data, so it's considered an important feature of most modern mobile devices including smartphones, tablets, laptops, and so on.

Location issues can cause some obvious problems, such as maps not working properly or an app not being able to determine the correct device location to better provide services to the user. This might result in a user getting lost, of course, but in actuality this may not be the worst thing that can happen to the user. Some apps may simply just not work properly if they can't ascertain the device's exact location. More serious location issues could prevent an organization from locating a device that has been lost or stolen, for example, or could prevent an organization from tracking a corporate device that is being misused, or that is somewhere it's not supposed to be if the organization restricts device use in some geographic areas or circumstances.

Symptoms of location issues include the obvious, such as a mapping not being able to identify the exact location of the device, or error or warning messages from the device, OS or various apps that rely on location data. A few symptoms may also include not-so-obvious issues from apps that may rely on location data, not for user assistance, but for how they operate internally. Most of the symptoms with location data will, however, be fairly obvious or at least easily identifiable.

Troubleshooting location problems begins with simple actions such as making sure that your GPS, cellular data, and Wi-Fi are turned on and functioning properly, as sometimes these services can be inadvertently turned off by the user or by an app, and have to be periodically reactivated. Typically, a warning message would indicate whether or not GPS or data services are turned off, so this would be an easy problem to identify and fix. Other issues may only affect specific apps because they may have been configured such that they are not allowed to access or use location services. This is usually a matter of going into the app configuration or location services settings and allowing the app to make use of those services. Figure 12-7 shows an example of allowing apps to use location services.

Figure 12-7

Allowing app
permissions for
location services

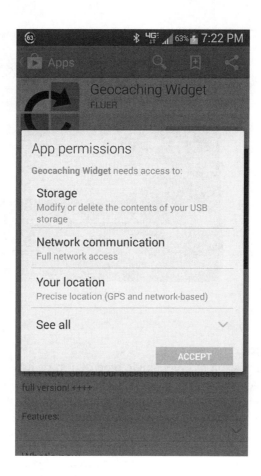

Of course there also other problems that may cause location issues that are related to the operating system and the hardware in the device. Along with the OS issues discussed previously, there could be OS configuration settings that are configured incorrectly for GPS, cellular, and Wi-Fi services that may prevent location services from functioning properly. These configuration items should be checked when location service symptoms are being seen with more than one app. Hardware problems may lie in the actual GPS or network hardware with the device, and should be treated as described in the previous sections on hardware. Although laptops and other mobile devices may have removable network or GPS modules in them, most of these components are not user serviceable and have to be replaced or repaired by authorized service technicians. Figure 12-8 shows an example of a netbook with a removable Wi-Fi module in the upper-right hand part of the device, below the battery.

Figure 12-8
A removable
Wi-Fi module in a
netbook

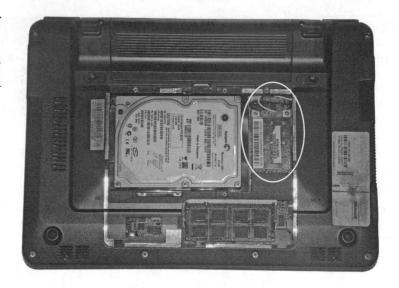

Email Issues

Again, we can't explore every single type of app or its issues that you could run into out in the corporate world, so I've chosen to focus on the particular ones you'll need to understand for both the exam objectives and for real-life problems you are likely to encounter with mobile devices. Email is another such app that we should take a look at and explore for both purposes. While the organization may have preferred email apps that are pushed to the mobile device and controlled through policy, several issues could crop up that you need to be aware of, as well as how to troubleshoot them. The first issue deals with the preferred apps themselves. The organization may want to impose a preferred email app on the user's device for several reasons. First, with the preferred email app, it would be much easier to configure, control, and maintain its configuration updates. A second reason is security, as different security settings required for the user could also be included in the app before it is pushed down to the user's device. Allowing the user to install and use a non-approved email app leads to the third reason. A third-party email app may not lend itself to being centrally managed, and this would enable the user to change required configuration and security settings needed to successfully communicate with the corporate email infrastructure on a secure basis. A third-party app may also permit other apps to access the corporate data contained in it, such as the actual emails, contacts, attachments, and so forth, resulting in data leakage outside the controllable confines of the organization.

Beyond the preference for a standardized, approved corporate email app, there are other issues with email the user could encounter. Chapter 10 covered troubleshooting network connectivity issues, and I discussed a few issues with email that could occur. Remember from the discussion on network troubleshooting that there are several configuration issues that must be looked at when connectivity is a problem. If the email issues are affecting multiple users, then you can quickly rule out individual device issues. If, on the other hand, an individual user is having email issues, you should start with the device itself and the relevant email app. The network configuration, such as the ports and protocols used to make the email connection, is the first thing you should look at. You should examine ports and protocols in the device's general configuration settings (where applicable) as well as those controlled by the app. You should also look at authentication and encryption settings, of course, as most corporate email implementations are likely to be secured by encryption and strong authentication (I would hope). Once you've ruled out both network and security configuration settings on the device and in the app, obviously you would want to go and look at the corporate infrastructure side of the connection and make sure that the email server is configured properly, along with the user's particular inbox and email settings that reside on the server.

 EXAM TIP   Recall from the discussions in earlier chapters the different ports and protocols that email services use, and make sure your devices and apps are configured appropriately.

Chapter Review

This final chapter of the book has gone from the general to the specific and moved away from troubleshooting the mobile network and security infrastructure to the client level, the device and apps that run on it. The chapter covered troubleshooting several device issues, and began by talking about both battery and power issues. We first looked at battery life, and the different factors that can reduce the battery charge on a mobile device, as well as reduce its overall useful lifespan. These factors include network reception, location services usage, foreground/background app usage, and screen display brightness and usage. I discussed ways to reduce all these factors in order to increase your mobile device battery life. I also discussed power supply usage, both from the perspective of the internal device power supply, as well as external power supplies used to power the device and charge its battery. I then discussed the effects that power outages have on mobile devices in general, such as data loss, as well as operating system and app corruption. I discussed different diagnostics utilities that may assist the user or a technician in troubleshooting both battery and power issues.

I discussed configuration change issues from both an enterprise perspective and a device perspective, and how these can affect the enterprise network as well as the users' device experience. We then turned our attention to several other device issues, including synchronization, authentication, and profile issues. I discussed the possible symptoms and causes of synchronization issues, and what to look for, as well as how to troubleshoot these issues. I continued the discussion on authentication issues through several chapters in the book, and during this chapter discussed, from a client-level perspective, different authentication issues, including certificate-based authentication and password issues. I also talked about some of the problems that could happen with passwords as well as password resets.

As devices are hardware, and hardware occasionally fails, I discussed some of the different hardware issues that can occur with mobile devices. Hardware, as well as software, issues can cause devices to freeze, lockup, reboot periodically, or even crash and not come back on again. I discussed a few different components of the mobile device that can cause issues and how most of them are typically not user repairable and must be looked at by an authorized service technician. I also examined the potential software issues on a mobile device, and this includes both the operating system, as well as the apps that are installed on the device. We looked at operating system issues, to include symptoms and troubleshooting, and how they may affect the device operation. We know that two of the primary issues with operating systems are patches and configuration problems. Mobile applications can also have those issues, and can suffer from other ones related to installing, upgrading, and removing them. I also examined the different issues that can result from configuration problems with apps.

I then examined app store problems, including those that could be associated with both commercial app stores, such as those operated by Apple and Google, as well as those associated with an enterprise-level internal app store. Network connectivity and authentication are two primary issues that can occur with app stores, as well as authorization to download, install, and use apps from both commercial and non-approved app stores. This chapter also examined some other issues that affect mobile devices and their users, including location services problems, as well as email problems. Location services issues can affect apps that make use of GPS, cellular, and Wi-Fi technologies to assist the user in determining their location or in finding points of interest near them. Some apps, however, may covertly use location data for marketing or vendor purposes. Issues with location services can include the deactivation of the GPS or network services, as well as permissions different apps have to use device location in their configuration settings.

Email issues include those that I have discussed previously related to the email infrastructure, to include both network connectivity as well as email server configuration. From the device perspective, configuration issues are usually the most common, including incorrect port and protocol configuration, as well as authentication and encryption settings. I also discussed the pros and cons of using a preferred email app versus a third-party app, and how that will affect both usability and security on the user device.

Questions

1. Which types of batteries do modern mobile devices primarily use? (Choose two.)

   A. Li-ion

   B. NiCd

   C. NiMH

   D. NiB

2. All of the following factors can significantly reduce battery life on a mobile device, EXCEPT:

   A. Display brightness

   B. Cellular roaming

   C. Wi-Fi management messages

   D. Standby or sleep mode

3. What is the first thing you should examine concerning most mobile device client issues?

   A. Vendor patches

   B. Configuration changes

   C. App updates

   D. Operating system patches

4. Which of the following are a symptom of synchronization issues?

   A. Failed OS updates

   B. Complete email downloads

   C. Notification of available app updates

   D. Strong Wi-Fi connection to the app store

5. You are troubleshooting a client authentication issue and suspect that it is due to a corrupt certificate on the device. Which of the following would be the best way to resolve this issue?

   A. Revoke and reissue the certificate.

   B. Reload the root certificate onto the device.

   C. Reinstall the user's certificate on the device.

   D. Renew the certificate.

6. You have reset a user's password after repeated password authentication problems. What is the next step you should take to ensure secure access by the user?

   A. Install a new root certificate.

   B. Synchronize the device.

   C. Require the user to change her password upon next login to the app or service.

   D. Change the encryption settings on the corporate infrastructure server.

7. You are resetting a user's profile and want to ensure that it contains all of the correct settings for accessing corporate services. Which of the following authentication settings would the profile contain that enable the user to securely connect to the enterprise network? (Choose two.)

   A. Plaintext passwords

   B. Encryption methods

   C. Client IP address settings

   D. Required authentication protocols

8. Which of the following could most likely cause a device crash?

   A. A deactivated Wi-Fi module

   B. A pending OS patch that is downloading but not yet installed

   C. An app that offers a feature upgrade

   D. Recently upgraded firmware

9. You have just updated a device with an OS patch, and now the user complains that it periodically freezes, forcing the user to have to power the device off and back on. What should be your first troubleshooting step?

   A. Upgrade the OS to the next higher version.

   B. Restore the device from a current backup.

   C. Reload the operating system from scratch.

   D. Nothing, the user is likely causing the problem by misusing the device.

10. Which of the following is most likely to cause an app to fail during installation?

    A. Intermittent network connection

    B. GPS inadvertently turned off

    C. Faulty firmware

    D. Corrupt certificate

11. All of the following may prevent a user from upgrading an app EXCEPT:

    A. Permissions

    B. Configuration settings

    C. Firmware

    D. Authentication

12. What is one thing you should ensure occurs before uninstalling and reinstalling apps?

    A. Device upgrade

    B. Device backup

    C. OS upgrade

    D. OS patching

13. All of the following could cause app store connection issues EXCEPT:

    A. Authentication settings

    B. Network settings

    C. Account settings

    D. Device storage encryption settings

14. What may be required to download an app that is very large in size or requires a faster connection?

    A. GPS connection

    B. 3G/4G connection

    C. Bluetooth connection

    D. Wi-Fi connection

15. Fill in the blank with the appropriate answer: Enterprise-level app stores are typically a function of _____.

    A. MDM

    B. Vendor app stores

    C. Independent app stores

    D. Email

16. You have multiple users that are having issues connecting to the enterprise-level app store and email services, but can connect to other corporate resources easily. Which of the following issues do you suspect is most likely the problem?

    A. Network connectivity issues

    B. Authentication issues

    C. OS upgrade issues

    D. Email app issues

17. Which of the following is an issue when users connect to and download apps from an independent or third-party app store?

    A. Synchronization

    B. Authentication

    C. Unauthorized use of the store or its apps

    D. Encryption

18. All of the following are symptoms of issues with location services, EXCEPT:

    A. Wrong device location

    B. Inability to synchronize to the enterprise app store

    C. Inability to get a location fix on the device

    D. No data for points of interest near the user

19. What other method can be used to get a location fix on the device if GPS services are not available?

    A. Cellular network services

    B. Bluetooth services

    C. Location services app

    D. Email from MDM server

20. Your CIO is evaluating the benefits of allowing users to download and use their favorite third-party apps for corporate email, and asks for your advice. Which of the following are good reasons to NOT allow the users to download third-party email apps, and to stay with a standardized enterprise solution? (Choose two.)

    A. Ease of use

    B. Higher-level security than enterprise apps

    C. Lack of centralized configuration and security

    D. Possible data leakage to other apps

Answers

1. **A, C.** Most modern mobile devices use some variation of lithium-ion batteries (Li-ion) or nickel-metal hydride (NiMH).

2. **D.** Standby or sleep mode helps to stretch battery life by reducing power drain on the battery.

3. **B.** Configuration changes are the first thing you should examine; all of the other answers can actually fall into this category because they all can alter the configuration of the device.

4. **A.** Failed OS updates are one symptom of synchronization issues.

5. **C.** Reinstall the user's certificate on the device if you suspect it has become corrupt.

6. **C.** After a password reset, you should require the user to change his or her password upon next login to the app or service.

7. **B, D.** A profile contains settings for secure authentication to the enterprise network, including required authentication protocols and encryption methods. Neither plaintext passwords nor IP address settings would allow a user to perform secure authentication to the network.

8. **D.** Faulty firmware can cause device crashes. None of the other options are likely to cause this problem.

9. **B.** You should restore the device from a current backup.

10. **A.** Downloading and installing an app requires a strong, stable network connection.

11. **C.** Device firmware is not typically an issue that would prevent a user from upgrading an app to a newer version.

12. **B.** Backing up the device is one thing you should ensure is accomplished before uninstalling or reinstalling an app.

13. **D.** Device storage encryption settings should not prevent you from connecting to an app store.

14. **D.** A Wi-Fi connection may have to be used for larger apps or those that require a faster, more stable connection.

15. **A.** Enterprise-level app stores are typically a function of Mobile Device Management (MDM).

16. **B.** Authentication issues are the most likely cause because the users can connect to other corporate services, which would rule out network connectivity issues. Operating system issues would cause other problems, and email app issues would prevent successful connection to the email server, not necessarily the enterprise app store.

17. **C.** The most common issue with independent app stores is that the user may be unauthorized to use the store or its apps.

18. **B.** The inability to synchronize to the enterprise app store is usually not a symptom of location services issues.

19. **A.** Cellular network services can supplement or replace GPS services in getting a relative fix on the device's location if GPS is not available or working. The other solutions either cannot be used to get a location fix or would rely on GPS or cellular services to do so.

20. **C, D.** Third-party apps do not always allow centralized configuration and security control, and can allow data leakage (email, contacts, and attachments, for example) to other apps on the device.

About the CD-ROM

The CD-ROM included with this book comes with Total Tester practice exam software with a pool of 200 questions, enough for two practice exams, and a PDF copy of the book.

Total Tester provides you with a simulation of the CompTIA Mobility+ exam. You can create practice exams from selected domains or chapters. You can further customize the number of questions and time allowed.

The software can be installed on any Windows XP/Vista/7/8 computer and must be installed to access the practice exam questions.

System Requirements

The software requires Windows XP or higher, Internet Explorer 5.5 or higher, or Firefox with Flash Player, and 30MB of hard disk space for full installation.

Installing and Running Total Tester

From the CD-ROM's main screen you may install Total Tester by clicking the Software Installers button, then selecting the Total Tester Certified CompTIA Mobility+ Exams button. This will begin the installation process and place an icon on your desktop and in your Start menu. To run Total Tester, navigate to Start | (All) Programs | Total Seminars or double-click the icon on your desktop.

To uninstall the Total Tester software, go to Start | Settings | Control Panel | Add/ Remove Programs (XP) or Programs and Features (Vista/7), and then select the CompTIA Mobility+ Total Tester program. Select Remove and Windows will completely uninstall the software.

About Total Tester

Total Tester provides you with a simulation of the CompTIA Mobility+ exam. There are 200 questions, enough for two complete practice exams. The exams can be taken in either Practice or Exam Simulation mode. Practice mode provides an assistance window with hints, references to the book, an explanation of the correct and incorrect answers, and the option to check your answer as you take the test. Both Practice and

Exam Simulation modes provide an overall grade and a grade broken down by certification objective. To take a test, launch the program, select a suite from the menu at the top, and then select an exam from the menu.

Free PDF Copy of the Book

The contents of this book are provided in PDF format on the CD-ROM. This file is viewable on your computer and many portable devices.

To view the electronic book on your computer, Adobe Reader is required and has been included on the CD-ROM.

 NOTE For more information on Adobe Reader and to check for the most recent version of the software, visit Adobe's web site at www.adobe.com and search for the free Adobe Reader or look for Adobe Reader on the product page.

To view the electronic book on a portable device, copy the PDF file to your computer from the CD-ROM, and then copy the file to your portable device using a USB or other connection. Adobe offers a mobile version of Adobe Reader, the Adobe Reader mobile app, which currently supports iOS and Android. The Adobe web site also has a list of recommended applications.

Technical Support

For questions regarding Total Tester software or operation of the CD-ROM, visit www.totalsem.com or e-mail support@totalsem.com.

For questions regarding the PDF copy of the book, e-mail techsolutions@mhedu .com or visit http://mhp.softwareassist.com.

For questions regarding book content, e-mail customer.service@mheducation.com. For customers outside the United States, e-mail international.cs@mheducation.com.

Acceptable Use Policy (AUP) Organizational policy that describes both acceptable and unacceptable actions when using organizational computing resources, as well as the consequences of unacceptable use.

access control list (ACL) Set of rules implemented on a network or security device to filter inbound and outbound network traffic.

access point (AP) A wireless device providing connectivity, authentication, and encryption for clients on infrastructure-based wireless networks.

access point name (APN) A configuration setting for mobile devices that points to a cellular network gateway.

Active Directory (AD) Microsoft's centralized authentication and directory services infrastructure, implementing the Lightweight Directory Access Protocol.

Apple Push Notification Service (APNs) Apple's network service responsible for sending alerts and notifications to Apple devices; it uses TCP ports 2195 and 2196.

Bring Your Own Device (BYOD) Mobile device environment in which employees use their personally owned devices to access, store, and process data belonging to the organization.

business-to-business (B2B) Term used to describe business-to-business transactions and applications involving data transfer between organizations.

Call Data Recording (CDR) The process of recording call metadata, to include subscriber information, as well as information about the recipient.

Carrier Sense Multiple Access with Collision Avoidance (CSMA/CA) A media access method in which a host will attempt to avoid data collisions by transmitting data over the media only when it senses that the transmission media is clear to do so; it is primarily used in wireless networks.

Carrier Sense Multiple Access with Collision Detection (CSMA/CD) A media access method in which the host detects a competing data transmission and sends a "jamming" signal to let other hosts know about the data collision, and then waits a random period of time before attempting retransmission.

Certificate Authority (CA) Entity responsible for issuing and managing digital certificates throughout the certificate lifecycle.

Certificate Revocation List (CRL)   Electronic file published by a certificate authority, which shows all certificates that have been revoked by that CA.

Certification Authority (CA)   Entity responsible for certifying that the security controls for a system are functioning properly and mitigating threats to that system as expected.

Circuit Switched Data (CSD)   A form of data transmission used with GSM networks that uses time-division multiple access (TDMA) methods and virtual circuits.

Code Division Multiple Access (CDMA)   A signaling technology that uses spread spectrum and coding in order to send multiple messages over a band of frequencies.

Common Configuration Enumeration (CCE)   A set of unique entries for preferred configuration settings and policies, and their corresponding guidance statements and controls. (Note that although still a part of the CompTIA Mobility+ exam objectives, CCEs have been archived as of 2013 by the National Institute for Standards and Technology.)

Coronal Mass Ejection (CME)   A large mass of solar wind and magnetism released into space by the sun. CMEs, like solar flares, can sometimes cause interference with RF signals.

Data Loss Prevention (DLP)   The combination of technologies, processes, and procedures used to prevent the release and loss of sensitive organizational data to unauthorized entities.

Demilitarized Zone (DMZ)   A network architecture situated between an untrusted network and a protected network, which acts as a protective buffer zone between the two networks.

Device Manager (DM)   An aspect of mobile device management that allows the administrator to manage and control all functions of a mobile device; it is also the name of a configuration tool in the Microsoft Windows operating systems that allows you to manage devices attached to the host.

Disaster Recovery (DR)   The process of reacting to an incident or disaster and recovering an organization, its personnel, and systems to a functioning state.

Domain Name Service (DNS)   The service and protocol associated with resolving Internet Protocol addresses to human-recognizable domain names; DNS uses TCP and UDP ports 53.

Dynamic Host Configuration Protocol (DHCP)   Service and protocol responsible for automatically providing IP addressing information to network hosts; it uses the UDP protocol on ports 67 and 68.

Elliptic Curve Cryptography (ECC)   A public key cryptography algorithm that uses the structures of elliptic curves over finite fields; ECC requires smaller amounts of computing power and memory, making it suitable for mobile devices.

End User License Agreement (EULA)   The software licensing agreement the organization and end user must agree to and abide by in order to use the software.

Evolution Data Optimized (EVDO)   A cellular communications technology that evolved from CDMA2000 technologies; it was used to bridge the gap between older CDMA and 3G standards.

Exchange ActiveSync (EAS)   A Microsoft protocol used to manage email provisioning and configuration on mobile devices, as well as certain limited aspects of the device; previously known as AirSync.

File Transfer Protocol (FTP)   An application-level protocol used to transfer files from one host to another; it is an unsecure protocol that does not encrypt its traffic, and uses TCP ports 20 and 21.

FTP over SSL (FTPS)   A version of FTP made secure by tunneling it over a Secure Sockets Layer (SSL) connection; it uses TCP port 990 and should not be confused with either SFTP (which is an SSH implementation of FTP), or Secure FTP, which normally involves tunneling ordinary FTP traffic over an SSH connection.

Galois/Counter Mode (GCM)   A secure mode of operation employed on symmetric key block ciphers; it is very efficient and fast, and requires 128-bit block sizes.

General Packet Radio Service (GPRS)   A transitional cellular technology used between 2G and 3G technologies, based upon the GSM family of cellular technologies.

Global System for Mobile Communications (GSM)   A second-generation (2G) cellular communications technology and standard developed and widely used in Europe; it eventually became the worldwide standard for cellular communications, and gave rise to subsequent generations of cellular standards, including 3G and current 4G technologies.

Google Cloud Messaging for Android (GCM)   Data push notification service used for Android operating systems.

High Availability (HA)   A term used to describe a system or network that must be kept at a significantly high and reliable level of availability for its users; typically measured as some form of a precise decimal percentage, such as 99.999 percent availability.

High Speed Packet Access (HSPA)   A transitional technology used to move cellular data networks from third-generation to fourth-generation technologies; it is a combination of two other protocols, the High-Speed Downlink Packet Access (HSDPA) and High Speed Uplink Packet Access (HSUPA) protocols, and allows higher data rates (up to 330 Mbps downlink and 34 Mbps uplink).

Hypertext Transfer Protocol (HTTP)   The markup language used to render and format data as a web page.

Industrial, Scientific, and Medical (ISM) Name given to the frequency band reserved for wireless devices operating in one of several radio frequency ranges, including the 2.4 GHz range used by IEEE 802.11 standard wireless devices.

Information Assurance Support Environment (IASE) Portion of the Department of Defense (DoD) Information Systems Agency (DISA) devoted to information assurance support for DoD systems.

Internet Engineering Task Force (IETF) An international organization that develops Internet standards, particularly those associated with the TCP/IP suite of protocols.

Internet Message Access Protocol (IMAP) A client-side messaging protocol used to receive email from a central server; IMAP uses TCP port 143.

Internet Protocol (IP) A protocol in the TCP/IP suite that is responsible for logical addressing and routing packets to different sub-networks; it resides at the network layer of the OSI model.

intrusion detection system (IDS) A system designed to detect network intrusions based upon traffic characteristics.

intrusion prevention system (IPS) A system that is not only responsible for detecting network attacks based upon certain traffic characteristics, but that also has the ability to prevent and stop them upon detection (see also *Network Intrusion Prevention System*, or NIPS).

Kerberos Constrained Delegation (KCD) A technique developed to restrict the level of delegation of privileges that could be used by services and their associated accounts, typically in Microsoft Windows Server products.

Lightweight Directory Access Protocol (LDAP) A protocol that is used in distributed directory services networks, such as Active Directory, to assist hosts in locating network resources; it replaced the older X.500 directory services protocol.

local area network (LAN) A group of hosts networked together in order to share data; they are usually all configured to be on the same logical subnetwork.

Long Term Evolution (LTE) Transitional (fourth-generation) technologies used for high-speed data access on cellular networks. It evolved from the GSM and UMTS family of technologies and is based on standards from the 3GPP and the ITU. Although LTE is based upon the same standards, it is not a true 4G technology. It is, however, being used to transition to true 4G technologies, such as LTE-Advanced.

Management Information Base (MIB) A hierarchical database that uses object identifiers to manage devices through the Simple Network Management Protocol (SNMP).

Message Digest 5 (MD5)   A hashing algorithm used to compute fixed-length message digests of variable-length pieces of texts.

Messaging Application Programming Interface (MAPI)   Application programming interface used by Microsoft products to enable use of email services in non-email-aware applications and devices.

Mobile Application Management (MAM)   Management structure and technologies used to centrally manage mobile device apps, rather than the entire mobile device.

Mobile Device Management (MDM)   The management structure and technologies used to centrally manage all aspects of mobile devices for an organization.

Mobile Enterprise Application Platform (MEAP)   A collection of mobile application development products and techniques used to manage mobile app development across the enterprise; it is typically used in a larger, multivendor mobile device environment.

Mobility as a Service (MaaS)   Third-party, cloud-based service that provides enterprise mobility management services to organizations. Note that while there are multiple industry definitions for the term MaaS, this one is used per the exam objectives.

Multiple Input Multiple Output (MIMO)   Technologies involving the simultaneous use of multiple transmit and receive antennas to optimize radio frequency communications, providing improved data throughput and range.

Multiple Mobile Channel Access (MMCA)   A term describing technologies that allow multiple devices to access several frequency bands or channels simultaneously; these technologies include Frequency Division Multiple Access (FDMA), Time Division Multiple Access (TDMA), and Code Division Multiple Access (CDMA), among others.

Near Field Communication (NFC)   A very short-range radio frequency technology that allows mobile devices to communicate with each other using small embedded chips, simply by touching them together or bringing them within a few inches of each other.

network access control (NAC)   Methods and technologies used to impose security and configuration settings on a mobile device before it is allowed into the corporate network; NAC is usually implemented through a hardware device or associated technologies.

Network Address Translation (NAT)   A method used by various network and security devices to translate an organization's public Internet Protocol addresses to its private Internet Protocol address space.

Network Intrusion Prevention System (NIPS)   A system that is responsible for not only detecting network attacks based upon certain traffic characteristics, but also has the ability to prevent and stop them upon detection (see also *intrusion prevention system*).

Network Operations Center (NOC)   A centralized control center, either on the organizational premise or outsourced to a third-party, which controls and manages all networking aspects of the enterprise infrastructure.

Online Certificate Security Protocol (OCSP)   A security protocol used by an organization to publish the revocation status of digital certificates in an electronic certificate revocation list.

Open Systems Interconnection (OSI) model   A model used to describe the theoretical process of how data is sent and received over a network, as well as the interactions of data on the network with devices and other types of data; it is used as a template from which to develop networking protocols.

operating system (OS)   Software used to access the interface between the user and the hardware itself, translating user interaction to larger operations; examples of mobile operating systems include Apple iOS, Android, and Windows Phone.

Original Equipment Manufacturer (OEM)   A company that produces equipment, either as the original manufacturer of that equipment, or as an authorized vendor contracted by the original manufacturer.

personal area network (PAN)   A very small-range network that uses personal devices and very low power, and short-range protocols, such as Bluetooth, NFC, or ZigBee.

Personal Information Manager (PIM)   A type of app that is used as a personal organizer, containing a wide variety of user information, such as contacts, financial information, personal notes, and so on.

Port Address Translation (PAT)   A form of address translation in which an organization's private IP addresses are translated into an organization's public IP addresses, but in which sending and receiving ports are also translated, in addition to the IP address itself.

Post Office Protocol (POP)   A client-side messaging protocol used to receive email from a central mail server; the current version is POP3, and it uses TCP port 110.

Power Line Ethernet (PLE)   A method of delivering data to a network device via standard electrical lines; this prevents the necessity of installing additional cabling because electrical cabling and outlets are usually already present near the network device.

Power over Ethernet (PoE)   A standard allowing electrical power to be passed over Ethernet cabling, reducing the amount of cabling required as it both powers devices and serves as a transmission medium for data.

Preferred Roaming List (PRL)   A list of preferred service providers and frequency bands that will be used by a cellular device when roaming between carriers.

Pretty Good Privacy (PGP)   A form of asymmetric encryption invented by Philip Zimmerman, which originally used a web-of-trust model (instead of a trusted hierarchy) for personal secure communications.

Quality of Service (QoS)   Methods and measurements used to both determine and ensure the level of performance of network traffic.

radio frequency (RF)   The number of changes, or cycles, in voltage of an electrical signal during a specific time period; it is usually measured in hertz (Hz), and a frequency of one hertz equals one cycle per second.

received signal strength indicator (RSSI)   A subjective measurement of the power level in a received radio signal by a device.

Recovery Point Objective (RPO)   The maximum tolerable time period in which data can be lost by the organization due to an incident or disaster.

Recovery Time Objective (RTO)   The maximum amount of time that can be allowed to pass between an incident and recovering the business to an operational state.

Remote Desktop Protocol (RDP)   A protocol used to access the graphical desktop on a remote host, typically a Windows computer; it uses TCP port 3389.

Secure Hashing Algorithm (SHA)   A set of cryptographic hashing algorithms developed and published by the National Institute of Standards and Technology; used for secure technologies such as password protection, message integrity, and digital signatures.

Secure IMAP (IMAPS)   A secure version of the IMAP protocol that uses SSL for encryption; it uses TCP port 993.

Secure SMTP (SSMTP)   An implementation of the Simple Mail Transfer Protocol (SMTP) that is sent over an SSL connection to provide authentication and encryption services; it uses TCP port 465.

Secure Socket Layer (SSL)   A secure application-layer protocol that relies on digital certificates and public/private keys to set up authentication and encryption services between two hosts; it is used to provide security services for various unsecure protocols, such as HTTP, SMTP, and FTP.

Security Information and Event Management (SIEM)   Refers to the technologies and products used to integrate security information management and security event management information into a centralized interface, providing real-time event correlation and analysis.

Self-Service Portal (SSP)   A method of provisioning and updating a mobile device, where the user is directed to a particular web-based application or URL from an SMS or email message; the self-service portal offers the user the opportunity to download configuration settings and apps, and it provides assistance in initially setting up a mobile device for the enterprise network.

Server Routing Protocol (SRP)   A network protocol used between a Black-Berry Enterprise Server (BES) and BlackBerry mobile devices; it is a proprietary protocol that primarily uses TCP port 3101.

service-level agreement (SLA)   A contractual agreement, signed by an organization and a third-party provider, that details the level of security, data availability, and other protections afforded the organization's data held by the third party.

Service Set Identifier (SSID)   A basic network name for a wireless network, including both Basic Service Set (BSS) networks, which use single access points, and Extended Service Set (ESS) networks, which use multiple access points.

Simple Mail Transfer Protocol (SMTP)   An unsecure messaging protocol used to send email messages to other hosts; it uses TCP port 25 by default.

Small Office Home Office (SOHO)   A term that refers to a small-to-medium network set up to support a limited number of users.

Software as a Service (SaaS)   A third-party cloud-based service that offers out-sourced use of software by an organization; this allows an organization to use licensed software at a lower cost than buying, installing, and maintaining software.

SSH File Transfer Protocol (SFTP)   One method for sending FTP traffic over a Secure Shell (SSH) session using native SSH commands and methods; note that this is not the same thing as *tunneling* regular FTP traffic over secure shell (referred to as FTP over SSH, which is called Secure FTP).

Subscriber Identity Module (SIM)   An integrated circuit chip, usually mounted on a small cardboard card, that stores information related to the account and identity of an authorized subscriber to a carrier network.

System Development Life Cycle (SDLC)   A framework describing the entire useful life of a system or software, which usually includes phases relating to requirements definition, design, development, acquisition, implementation, sustainability, and disposal.

Telecom Expense Management (TEM)   The processes and methods used to manage the expenses associated with maintaining telecommunications carrier services for an organization's mobile devices.

Time Division Multiple Access (TDMA)   A signaling method that uses defined time slices to allow access to a frequency band by multiple devices.

Transmission Control Protocol (TCP)   A transport-level protocol that establishes a defined connection with the sending and receiving hosts before a data segment transmission session begins; it also manages segment sequencing and retransmission of lost segments through the use of sequence numbers.

United States Cyber Command (USCYBERCOM)   A United States military command tasked with centralized command-and-control of U.S. Department of Defense cyberspace operations; it is a subordinate unit of the U.S. Strategic Command and is headquartered at Fort Meade, Maryland.

Universal Mobile Telecommunications System (UMTS)   A third-generation cellular signaling technology in the Global System for Mobile Communications (GSM) family of standards.

User Datagram Protocol (UDP)   A transport-layer protocol used to carry datagrams between two hosts; it does not rely on established connections, nor does it manage data sequencing or retransmission of lost data.

Virtual Private Network (VPN)   A technology used to securely connect to an organization's internal network by tunneling unsecure protocols and data over a secure connection through an unsecure external network, such as the Internet, to a secure device known as a VPN concentrator.

Voice over IP (VoIP)   A set of technologies used to send telephony and voice services over standard Internet Protocol networks.

Volume Purchase Program (VPP)   A licensing and purchasing program that allows an organization to negotiate software use for multiple users in the organization.

Wireless Fidelity (Wi-Fi)   The name given to the consumer and commercial wireless technologies using the Institute of Electrical and Electronics Engineers (IEEE) 802.11 wireless standards; the term "Wi-Fi" is a trademarked name belonging to the Wi-Fi Alliance.

Worldwide Interoperability for Microwave Access (WiMAX)   A wireless communication method used as a pseudo or transitional fourth-generation cellular technology; it provides data rates of 30 to 40 megabits per second (Mbps) and was originally intended as a "last-mile" technology for remote consumers and businesses unable to get access to cable and DSL technologies.

Numbers

1G, 124
2.5G, 124
2.75G, 124
2G, 124
3DES (Triple DES), 316
3G, 120–122, 124
3rd Generation Partnership Project (3GPP), 120
4G
    cellular technologies, 122–124
    GSM standards, 118–119
    IMEI numbers in GSM family of devices, 239–240
802 standards. *See* IEEE 802

A

absorption, in RF propagation, 90
acceptable use policy, in monitoring mobile
   devices, 390
access control
    access levels as risk mitigation strategy, 293
    application access, 333
    authentication, 330
    biometrics, 331
    data segregation and, 341–342
    device access, 333
    incident response policies, 299–300
    managing mobile devices, 249
    multifactor authentication, 331
    network access control. *See* network access
      control (NAC)
    overview of, 330
    permissions, 293–294
    PIN numbers, 332–333
    port based (802.1X), 156–157
    single-factor authentication, 330–331
    tokens, 332
    wireless networks and, 333
access points (APs)
    coverage planning, 159–160
    lightweight and autonomous, 50–51
    rogue, 276
    troubleshooting network saturation, 374
    Wi-Fi, 137
    wireless (WAPs), 139, 277
ACK flag, TCP segments and, 15

activation of mobile devices
    on cellular networks, 237
    OTA activation, 237–238
    overview of, 237
    troubleshooting, 376
Active Directory (AD)
    as directory service, 204
    LDAP and, 28
ad-hoc mode, operation modes for wireless networks,
   138–139
ad-hoc networks
    centralization vs. decentralization in network
      design, 57
    network topologies and, 56
adjacent channel interference, 161
administrative permissions, mobile devices
   and, 187–188
Advanced Encryption Standard (AES)
    as encryption algorithm, 317
    SRTP and, 325
    troubleshooting encryption issues, 408–409
    WPA2 and, 155, 324
AES-CCMP mode, in WPA2, 324
agent-based/agent-less monitoring, 392
AH (Authentication Header), IPsec, 321
Airsync, 28
algorithms
    encryption basics and, 314
    hashing, 318
    protecting data-at-rest, 315–318
    protecting data-in-transit, 318–326
    troubleshooting encryption, 408
amplitude modulation (AM) radio, 87–88
amplitude, of radio waves, 85–86
Android OS/Android devices
    antivirus app for, 290–291
    application store for, 196–197
    battery usage, 423
    cellular settings, 126
    comparing mobile platforms, 176–177
    containerization of data and, 342–343
    firewall app, 292
    history of mobile devices in enterprise, 174
    jailbreaking apps, 284
    synchronization issues, 427
    troubleshooting VPN configuration, 371–372

antennas
    in cellular communication, 112–113
    characteristics of, 92–96
    overview of, 92
    signal strength and, 160–161
    troubleshooting cellular networks, 374–375
    types of, 96–99
anti-malware
    solutions, 290–291
    troubleshooting, 410
antivirus solutions, 290–291
APIPA (Automatic Private IP Addressing)
    overview of, 66
    troubleshooting network service issues, 369
APNs. *See* Apple Push Notification service (APNs)
app stores
    software risks and, 281–282
    troubleshooting, 437–439
Apple. *See also* by individual Apple devices
    app store issues, 437
    application stores, 195–197
    comparing mobile platforms, 176
    history of mobile devices in enterprise, 174
    iOS. *See* iOS (Apple)
Apple Push Notification service (APNs)
    overview of, 29
    push notification technologies, 266
    troubleshooting, 377–378
    troubleshooting OTA connectivity, 374
application layer (layer 7), of OSI model, 10, 17
application stores, 195–197
apps/applications
    access control, 333
    centralizing management of, 246–249
    configuring and deploying mobile applications.
        See mobile applications, configuring and
        deploying
    content creation in mobile devices, 175
    data communication in app development,
        264–265
    decommissioning, 245
    default applications for mobile devices, 197–198
    for device monitoring, 390–392
    digital certificates and app development, 264
    in-house application requirements, 262
    installation issues, 434
    networked applications, 5–6
    platforms and, 263
    publishing, 262–263
    removal issues, 436–437
    sandboxing, 228–229, 295
    sideloading, 196
    synchronization issues, 426
    troubleshooting installation issues, 434–435

types of mobile applications, 260–262
    upgrade issues, 435
    vendor requirements, 263–264
architecture
    client-server architecture, 6
    MDM, 180
    network architecture. *See* network architectures
asset disposal, life-cycle operations, 255
asymmetric algorithms, 314
asymmetric key cryptography, 334
attenuation, in RF propagation, 91
audits, device compliance and, 388
authentication
    in access control, 330
    app store issues, 438
    application access, 333
    biometrics, 331
    certificate-based, 325
    data access, 342
    DoS attacks and, 276–277
    EAP and, 326
    expired passwords, 400–401
    failure of, 399–400
    hashing and, 318
    IPsec and, 321
    man-in-the-middle attacks and, 280
    multifactor, 331
    non-expiring passwords and, 401–403
    passphrase-based, 156
    PIN numbers in, 332–333
    single-factor, 330–331
    SSL and, 22–23, 319
    tokens, 332
    troubleshooting device issues, 428–430
    troubleshooting email issues, 369, 371
    troubleshooting generally, 399
    troubleshooting profile issues, 430
    WEP and, 154–155
    in wireless networking, 153–154
    WPA and, 155–156
Authentication Header (AH), IPsec, 321
authentication servers, in 802.1X, 157
authenticators, in 802.1X, 157
authorization
    app store issues, 438
    defined, 330
    incident response policies, 299–300
    troubleshooting authentication issues, 400
    troubleshooting profile issues, 430
Automatic Private IP Addressing (APIPA)
    overview of, 66
    troubleshooting network service issues, 369
azimuth, of radio antenna, 94–95

B

backbones, network, 69
backhauling traffic, in network management, 69
backups
    corporate data, 209–210
    device and server, 203–204
    frequency of, 204–207
    incident response policies, 299–300
    local backup, 210–211
    mobile devices, 209
    personal data, 210
    policies, 184
    testing, 211–212
    types of, 205
bandwidth (network capacity)
    managing, 70
    traffic shaping and routing and, 70–71
    troubleshooting network saturation, 374
bandwidth (radio)
    gain and, 95
    radio frequency band, 86
Base Station Controller (BSC), 114
base stations (BS)
    devices in cellular communication, 114
    roaming and switching and, 125
Basic Service Set (BSSID), 142
bastion host architecture, 59
batch provisioning, 242
battery life, troubleshooting, 420–423
beamwidth, of radio antennas, 93
bi-directional antennas, 98
BIA (business impact analysis), 200–201
biometrics
    in access control, 331
    device access and, 333
    single-factor authentication, 330–331
BitLocker Drive Encryption, 327, 329
BlackBerry
    app store issues, 437
    comparing mobile platforms, 176
    history of mobile devices in enterprise, 174
    synchronization issues, 427
block algorithms, 314–315
block-level encryption, 327
Bluetooth
    802.15 standard, 149–151
    data link layer and, 13
    overview of, 149
    personal area networks and, 151–153
bridges, 41
Bring Your Own Device (BYOD)
    app store issues, 438
    containerization of data, 228–229, 342
    establishing level of control over, 236

    IT and security policies for, 182–183
    monitoring, 388–389
    overview of, 181
    personal device security, 288
    privacy issues, 388
    troubleshooting, 429
broadcast domains, routers and, 44
broadcasting, compared with unicasting and
    multicasting, 14
BS (base stations)
    devices in cellular communication, 114
    roaming and switching and, 125
BSC (Base Station Controller), 114
BSSID (Basic Service Set), 142
bus networks, 52–53
business continuity
    backup frequency and, 204–205
    disaster recovery and, 199–200
business impact analysis (BIA), 200–201

C

CA (Certificate Authority), 338–340
cable testers, 364–365
cabling. *See* wiring/cabling
call hand-off, between cells, 111
capacity planning, in Wi-Fi site survey, 159
captive portal, security of mobile devices and,
    189–190
carrier waves, in radio transmission, 87
Category 5 (Cat 5) cabling. *See also* wiring/cabling, 71
causes of network issues
    creating theory of probable cause, 356
    testing theory of probable cause, 356–357
CCB (change control board), 254
CCMP (Counter-mode Cipher Block Chaining
    Message Authentication Code Protocol), 324–325
CDC (Cipher Block Chaining mode), DES, 316
CDC (Cipher Block Feedback mode), DES, 316
CDMA (Code Division Multiple Access)
    in history of cellular, 108
    signaling technology, 116
cells, 110–113
cellular networks
    CDMA, 116
    cells, 110–113
    CSD, 119
    device activation, 237
    fourth generation technologies, 122–124
    frequency ranges, 110
    generations (1G, 2G, 3G), 109
    GSM, 118–119
    history of, 108
    infrastructure, 113–115
    overview of, 107

cellular networks (*cont.*)
    Q&A, 128–132
    review, 126–128
    roaming and switching between network types, 125–126
    signaling technologies and standards, 115
    TDMA, 116–118
    third generation technologies, 120–122
    tower spoofing attacks, 281
    troubleshooting activation, 376
    troubleshooting app store issues, 438
    troubleshooting location services issues, 440
    troubleshooting OTA connectivity, 374–375
Certificate Authority (CA), 338–340
certificate expiration/renewal, life-cycle operations, 252–253
Certificate Revocation List (CRL), 397
change control board (CCB), 254
change management
    determined what changed in troubleshooting, 355
    life-cycle operations, 254
channels
    co-channel interference, 161
    wireless, 148–149
Cipher Block Chaining mode (CDC), DES, 316
Cipher Block Feedback mode (CDC), DES, 316
ciphertext, encryption and, 314
Circuit-Switched Data (CSD), 119
Class A addresses, IP address ranges, 64, 66
Class B addresses, IP address ranges, 64, 66
Class C addresses, IP address ranges, 64, 66
client-server architecture, 6
clients. *See also* mobile devices
    in networking, 5
    troubleshooting. *See* troubleshooting clients
    wireless clients, 135–136
cloning, device cloning as hardware risk, 286
cloud services
    distribution methods, 247–248
    Google Cloud Messaging, 198
    on-premise vs. cloud-based support for mobile devices, 192
clustering, for high availability, 207
co-channel interference, 161
Code Division Multiple Access (CDMA)
    in history of cellular, 108
    signaling technology, 116
cold sites, in disaster recovery, 202
collision domains, 42
collisions
    defined, 42
    on shared Ethernet connections, 52
collisions, in hashing, 318
compliance, audit information reports and, 388

computer processing units (CPUs), troubleshooting, 430–431, 433
configuration change, troubleshooting, 425–426
connectivity issues
    email and, 442–443
    hardware for testing, 364–365
    software for testing, 363–364
    troubleshooting generally, 366–367
    troubleshooting OTA connectivity, 360–361, 374–378
    utilities for testing, 361–362
contacts, synchronization issues, 426
containerization of data
    personal and corporate data, 228–229
    as risk mitigation strategy, 295–296
    as security technology, 342–343
content
    centralizing management of, 246–249
    creation vs. consumption in mobile devices, 175
    LDAP for content security, 249
    managing in off-boarding process, 245
    monitoring mobile devices, 390
    updating, 248
content filtering
    as risk mitigation strategy, 296–297
    troubleshooting, 407–408
corporate data
    containerization of, 228–229
    restoring, 212
Counter-mode Cipher Block Chaining Message Authentication Code Protocol (CCMP), 324–325
Counter mode (CTR), DES, 316
coverage planning, in Wi-Fi site survey, 159–160
CPUs (computer processing units), troubleshooting, 430–431, 433
crashes, troubleshooting device issues, 430–431, 432
credentials, in authentication, 330
CRL (Certificate Revocation List), 397
cross-certification trust, 340
crossover error rate, 406
cryptography. *See* encryption
cryptovariable, in encryption, 314
CSD (Circuit-Switched Data), 119
CTR (Counter mode), DES, 316

D

data
    app development and data communication, 264–265
    backing up, 209–210
    Data Loss Prevention (DLP), 297
    encryption of data-at-rest, 315–318
    encryption of data-in-transit, 318–326
    flow in OSI model, 10–11

incident response policies, 299–300
local backup, 210–211
protecting in mobile devices, 209
recovering, 211
redundancy for high availability, 207
restoring, 212
rules for data communication (protocols), 8
segregating. *See* data segregation
testing backups, 211–212
wiping from mobile devices, 289
Data Encryption Standard (DES), 314–316
data link layer (layer 2)
bridges operating at, 41
layers of OSI model, 10
of OSI model, 13
Data Loss Prevention (DLP), 297
data segregation
access controls, 341–342
containerization of, 228–229, 295–296, 342–343
overview of, 340
policies, 340–341
datagrams, UDP, 16
dB. *See* decibels (dB)
deactivating devices, 245
dead zones
in cellular networks, 111
troubleshooting OTA connectivity, 374
decentralized networks, vs. centralized, 6–8
decibels (dB)
measuring antenna power levels, 96
measuring radio signal power in, 94
measuring signal strength per distance, 160
decibels isotropic (dBi)
measuring antenna power levels, 96
overview of, 95
decommissioning applications, 245
decryption. *See also* encryption, 314
default applications, mobile devices, 197–198
default gateway, 48–49
defense-in-depth, 297–298
Demilitarized Zone (DMZ), 60–61
denial-of-service (DoS) attacks, 276–277
deployment practices
establishing level of control over devices, 236
overview of, 234
scaling devices and users, 234–235
deprovisioning devices. *See* off-boarding
DES (Data Encryption Standard), 314–316
destination ports
showing, 21–22
vs. source ports, 19
devices, for enhancing cellular coverage, 114–115
devices, mobile. *See* mobile devices
devices, network. *See* network devices

DHCP (Dynamic Host Configuration Protocol). *See* Dynamic Host Configuration Protocol (DHCP)
differential backups, 205–207
digital certificates
app development and, 264
compromised keys, 398
corrupt keys, 397–398
expiration/renewal, 252–253, 394–395
lifecycle of, 336–338
management infrastructure for, 338–340
overview of, 335–336
in PKI, 227–228
revoked or suspended, 396–397
troubleshooting, 372–373, 394
digital modulation, 88
digital signatures, 318
Digital Subscriber Lines (DSL), 134
dipole antennas, 96
direct sequence spread spectrum (DSSS)
802.11b and, 145
digital modulation and, 88
directory services (DS)
LDAP/AD, 28
mobile devices, 204
setting up, 227
disaster recovery
business continuity and, 199
locations, 202–203
principles, 201–202
disk drives, block-level encryption, 327
display (screen), demand on battery power, 421
distribution
centralizing management of content and
applications, 246–247
methods, 247–248
DLP (Data Loss Prevention), 297
DMZ (Demilitarized Zone), 60–61
DNS (Domain Name Service). *See* Domain Name
Service (DNS)
documentation
of MDM infrastructure, 233
of network issues, 359–360
of security incidents, 304–305
of Wi-Fi site survey, 163–164
Domain Name Service (DNS)
overview of, 23
troubleshooting network service issues, 367–368
UDP and, 16
DoS (denial-of-service) attacks, 276–277
DRP (disaster recovery principles), 201–202
DS (directory services). *See* directory services (DS)
DSL (Digital Subscriber Lines), 134
DSSS (direct sequence spread spectrum)
802.11g and, 145
digital modulation and, 88

Dynamic Host Configuration Protocol (DHCP)
  overview of, 23
  troubleshooting network service issues, 367–369
  UDP and, 16
dynamic range, ports, 20–22

E

EAP (Extensible Authentication Protocol). *See*
  Extensible Authentication Protocol (EAP)
EAS (Exchange ActiveSync), 266–267
ECB (Electronic Codebook mode), DES, 316
ECC (Elliptic Curve Cryptography), 317
EDGE (Enhanced Data Rates for GSM Evolution), 121
EDR (Enhanced Data Rate), Bluetooth, 150
EFS (Encrypting File System), 328–329
EIRP (Equivalent Isotropically Radiated Power), 96
electromagnetic (EM) spectrum
  frequency ranges for cellular communication, 110
  measuring radio frequency, 85
  radio waves in, 81–82
Electronic Codebook mode (ECB), DES, 316
elevation, of radio antennas, 94–95
Elliptic Curve Cryptography (ECC), 317
EM spectrum. *See* electromagnetic (EM) spectrum
email
  troubleshooting client issues, 442–443
  troubleshooting generally, 369–371
  troubleshooting synchronization issues, 426
employees, termination of, 244–245
Encapsulating Security Payload (ESP), IPsec, 321
encapsulation, of PDUs, 11
Encrypting File System (EFS), 328–329
encryption
  application access and, 333
  block-level, 327
  data access and, 341–342
  device access and, 333
  file-level, 328
  folder-level, 329
  full disk, 326–327
  implementing, 326
  IPsec and, 321
  man-in-the-middle attacks and, 280
  overview of, 313–315
  protecting data-at-rest, 315–318
  protecting data-in-transit, 318–326
  public key cryptography, 334–335
  removable media, 329
  risk of weak encryption keys, 279–280
  RSA algorithm, 325
  securing wireless networks, 333
  SSL and, 22–23, 319
  troubleshooting email issues, 370
  troubleshooting generally, 408–410
  WEP and, 154–155

Wi-Fi technologies, 154
  WPA and, 155–156
end-of-life, life-cycle operations, 255
end user licensing agreements (EULA), 228
Enhanced Data Rate (EDR), Bluetooth, 150
Enhanced Data Rates for GSM Evolution
  (EDGE), 121
enterprise
  802.1X in enterprise version of WPA/WPA2, 186
  enterprise mobility management, 174
  history of mobile devices in, 174–175
  infrastructure requirements for mobile
    devices, 186
ephemeral ports, 20–22
equipment, in network topologies, 58
Equivalent Isotropically Radiated Power (EIRP), 96
escalation, of network issues, 358
ESP (Encapsulating Security Payload), IPsec, 321
ESSID (Extended Service Set), 142
Ethernet
  collisions on shared connections, 52
  data link layer standards (802.3), 13
  port based access control (802.1X), 156–157
EULA (end user licensing agreements), 228
EVDO (Evolution-Data Optimized), 121
Evolution-Data Optimized (EVDO), 121
Exchange ActiveSync (EAS), 266–267
Extended Service Set (ESSID), 142
Extensible Authentication Protocol (EAP)
  in 802.1X, 157
  certificate-based authentication, 325
  securing data-in-transit, 326
  securing wireless networks, 333
eXtensible Markup Language (XML)
  configuration file, 241
  profiles in XML format, 229
  remote provisioning and, 242–243
extranets, 61

F

false negatives/false positives, in troubleshooting
  mobile security, 405–406
Faraday cages, 99
FCS (frame check sequence), 13
Federal Communications Commission (FCC)
  history of Wi-Fi, 133
  Wi-Fi standards, 143
FHSS (frequency hopping spread spectrum), 88
File Transfer Protocol (FTP)
  overview of, 24
  ports used by, 19
files
  file-level encryption, 328
  object-level permissions, 341
  permissions, 249, 293–294

filters
    content filtering, 296–297
    MAC filtering for LAN security, 279
FIN flag, TCP segments, 15
firewalls
    egress and ingress rules, 47–48
    misconfigured, 403–404
    overview of, 45–46
    ruleset, 46–47
    software firewalls, 292–293
FM (frequency modulation), 87
folders
    folder-level encryption, 329
    object-level permissions, 341
    permissions, 293–294
footer, of PDU, 11
FQDNs (Fully Qualified Domain Names), 367
frame check sequence (FCS), 13
frames, of data at data link layer, 13
free space path loss (FSPL), 91
freezes, troubleshooting device issues, 430, 432
frequencies
    bandwidth of, 86
    measuring wavelength of radio waves, 83–85
    ranges for cellular communication, 110
    ranges for radio signals, 82–83, 85
    ranges for Wi-Fi, 143, 148–149
    reuse, 110
frequency hopping spread spectrum (FHSS), 88
frequency modulation (FM), 87
FSPL (free space path loss), 91
FTP-Secure (FTPS), 25
full backups, 205–206
full disk encryption, 326–327
full-duplex communication, 42–43
Fully Qualified Domain Names (FQDNs), 367

G

gain, characteristics of radio antennas, 95
gateways
    overview of, 48–49
    settings for vendor gateway servers, 258
GCM (Google Cloud Messaging). *See* Google Cloud
    Messaging (GCM)
General Packet Radio Service (GPRS), 121
generations, of cellular technologies
    3G technologies, 120–122
    4G technologies, 122–124
    overview of, 109
    summary, 124
geo-fencing, location-based services, 193–194
geo-location, location-based services, 193
gigahertz (GHz), frequency ranges for cellular
    communication, 110

gigahertz (GHz), in measuring radio frequency, 83
global positioning satellite (GPS)
    demand on battery power, 420
    location services issues, 440–441
Global System for Mobile Communication (GSM)
    in history of cellular, 108
    IMEI numbers, 239–240
    signaling technologies for cellular
        communication, 118–119
Google
    Android devices and, 176–177
    app store issues, 437
Google Cloud Messaging (GCM)
    overview of, 30
    push notification technologies, 266–267
    types of cloud services, 198
Google Play, application stores, 196–197
GPOs (Group Policy Objects), 227
GPRS (General Packet Radio Service), 121
GPS (global positioning satellite)
    demand on battery power, 420
    location services issues, 440–441
Group Policy Objects (GPOs), 227
groups
    creating device and group profiles, 229–231
    deployment to small groups in phased
        approach, 234
    mobile device, 187
    object-level permissions, 341
    permissions, 294
GSM (Global System for Mobile Communication). *See*
    Global System for Mobile Communication (GSM)

H

half-duplex communication, 42
hardware
    for connectivity testing, 364–365
    device cloning, 286
    device hardening, 297–298
    device loss, 287
    device theft, 286–287
    disabling physical ports as risk mitigation
        strategy, 298
    keystroke loggers, 285
    OEM vendors, 213
    risks, 286
    troubleshooting cellular networks, 375
    troubleshooting device crashes and hardware
        issues, 430–431
    trusted platform module (TPM) devices, 296
hardware addresses. *See* Media Access Control
    (MAC)/MAC addresses
hashing, 317–318
headers, of PDU, 11

Hertz (Hz)
    frequency ranges for cellular communication, 110
    measuring radio bandwidth, 86
    measuring radio frequency, 83
high availability, 207–208
High Speed Downlink Packet Access (HSDPA), 121
High Speed Uplink Packet Access (HSUPA), 121
hosts
    host-based intrusion detection/prevention, 294
    hosting solutions in network configuration,
       259–260
    network hosts, 6
    proxies, 49
    security associations (SAs) between, 321
hot sites, disaster recovery, 203
hotspots, Wi-Fi, 137–138
HPBW (Half Power Beamwidth), 93
HSDPA (High Speed Downlink Packet Access), 121
HSUPA (High Speed Uplink Packet Access), 121
HTTP (Hypertext Transfer Protocol). *See* Hypertext
  Transfer Protocol (HTTP)
HTTPS (Hypertext Transfer Protocol-Secure). *See*
  Hypertext Transfer Protocol-Secure (HTTPS)
hubs, 40–41
hybrid apps, 261–262
Hypertext Transfer Protocol (HTTP)
    overview of, 22
    port 80 used by, 19
    transport services for, 14
Hypertext Transfer Protocol-Secure (HTTPS)
    content security and, 249
    overview of, 23
    securing data-in-transit, 320
Hz (Hertz). *See* Hertz (Hz)

I

IANA (Internet Assigned Numbers Authority). *See*
  Internet Assigned Numbers Authority (IANA)
ICCID (Integrated Circuit Card Identifier)
    batch provisioning and, 242
    provisioning devices and, 239
ICMP (Internet Control Message Protocol). *See*
  Internet Control Message Protocol (ICMP)
IDS (intrusion detection systems), 294
IEEE 802.11 standards
    802.11a standard, 144
    802.11ac standard, 146–148
    802.11b standard, 145
    802.11g standard, 145
    802.11i (WPA2), 155
    802.11n standard, 145–146
    authentication, 154
    history of, 134

    overview of, 143–146
    security protocols, 275
    standards for data link layer, 13
    summary, 147
IEEE 802.15 (Bluetooth), 13, 149–151
IEEE 802.15.4 standard (ZigBee), 151–152
IEEE 802.1X standard, 156–157
    EAP and, 326
    in enterprise version of WPA/WPA2, 325
    overview of, 156–157
    securing wireless networks, 333
IEEE 802.3 (Ethernet), 13
IGMP (Internet Group Management Protocol), 14
IKE (Internet Key Exchange), IPsec and, 321–322
IMAP (Internet Message Access Protocol), 26
    messaging standards, 256–257
    troubleshooting email issues, 369–370
IMEI (International Mobile Equipment Identity),
  239, 242
IMSI (International Mobile Subscriber Identity), 239
incident response
    identification of incident, 303
    making policy-based response, 303–304
    outsourcing and, 302
    overview of, 299
    planning for, 300–301
    policies, 299–300
    preparing for, 299
    reporting incidents, 304–305
    responding, 303
    testing incident response plans, 301–302
incremental backups, 205–206
Independent Basic Service Set (IBSS), 142
Industrial, Scientific, and Medical (ISM) band
    802.11b and, 145
    band frequencies and channels in Wi-Fi, 148
    Bluetooth on, 150
    interference (noise) and, 161
    Wi-Fi frequencies and, 143
information, gathering to identify network issues, 354
information topology, in network configuration,
  258–259
Infrared Data Association (IrDA), 151
infrastructure
    for cellular communication, 113–115
    implementing MDM solution. *See* MDM,
      infrastructure implementation
    mobile device requirements, 186
    planning MDB solution. *See* MDM,
      infrastructure plan
    support in mobile devices, 190–191
Infrastructure as a Service (IaaS), 192
infrastructure mode, for wireless networks, 138–139

infrastructure risks
    denial-of-service (DoS) attacks, 276–277
    man-in-the-middle (MITM) attacks, 280
    overview of, 275
    rogue access points, 276
    tower spoofing, 281
    warchalking, 278–279
    wardriving, 278
    weak encryption keys, 279–280
    weaknesses in wireless LAN protocols, 279
    wireless risks, 275–276
installation, application installation issues, 434–435
Institute of Electrical and Electronics Engineers
    (IEEE). *See also* IEEE 802
    data link layer standards, 13
    Wi-Fi standards, 133, 143
Integrated Circuit Card Identifier (ICCID)
    batch provisioning and, 242
    provisioning devices and, 239
Integrated Digital Services Network (ISDN), 134
interference (noise)
    coverage planning in Wi-Fi and, 159–160
    factors in RF propagation, 91–92
    Faraday cages protecting against, 99
    in Wi-Fi site survey, 161
International Mobile Equipment Identity (IMEI),
    239, 242
International Mobile Subscriber Identity (IMSI), 239
International Telecommunications Union (ITU)
    3G cellular technologies and, 120
    X.509 standard for digital certificates, 336
Internet Assigned Numbers Authority (IANA)
    dynamic and ephemeral ports, 20–21
    port assignment, 19
    registered ports, 20
    well-known ports, 19
Internet Control Message Protocol (ICMP)
    ping utility, 361
    protocols of network layer, 14
    traceroute command, 361–362
Internet Group Management Protocol (IGMP), 14
Internet Key Exchange (IKE), IPsec and, 321–322
Internet Message Access Protocol (IMAP), 26
    messaging standards, 256–257
    troubleshooting email issues, 369–370
Internet, network architecture, 59
Internet Protocol (IP), 62
Internet Protocol security (IPsec)
    application access and, 333
    creating VPNs, 323
    protocols of network layer, 14
    securing data-in-transit, 321–322
    troubleshooting VPNs, 404

Internet Security Association and Key Management
    Protocol (ISAKMP), 321
Internet Service Providers (ISPs), 134
interoperability, support in mobile devices, 190–191
intranet, network architecture, 59
intrusion detection systems (IDS), 294
intrusion protection systems (IPS), 294
iOS (Apple)
    comparing mobile platforms, 176
    jailbreaking apps and, 284
    synchronization issues, 427
IP addresses
    classes and ranges, 64
    DNS and, 23
    NAT and, 67
    overview of, 62–64
    sockets and, 18
    special address ranges, 66–67
    subnetting, 64–65
    supernetting, 65
    troubleshooting, 367–369
IP (Internet Protocol), 62
iPad
    battery usage, 422
    comparing mobile platforms, 176
iPhone
    3G, 120
    activation issues, 376
    battery usage, 422
    cellular settings, 125
    comparing mobile platforms, 176
    configuration utility, 241
    power supply, 424
    SCEP configuration example, 244
    synchronization issues, 427
IPS (intrusion protection systems), 294
IPsec (Internet Protocol security). *See* Internet
    Protocol security (IPsec)
IrDA (Infrared Data Association), 151
ISAKMP (Internet Security Association and Key
    Management Protocol), 321
ISDN (Integrated Digital Services Network), 134
isotropic antennas, 96
ISPs (Internet Service Providers), 134
IT policies, for mobile devices, 182–183
ITU (International Telecommunications Union)
    3G cellular technologies and, 120
    X.509 standard for digital certificates, 336
iTunes
    app store issues, 438
    application stores, 196–197
    synchronization issues, 427

J

Jabber, 29
jailbreaking, 284
jamming, wireless access points, 277

K

key space, encryption, 314
keys
    compromised, 398
    corrupt, 397–398
    encryption and, 314
    key exchange process in SSL, 319
    private and public in PKI, 334–335
    risk of weak, 279–280
    troubleshooting, 409
keystroke logging, as software risk, 284–285
keywords, content filtering by, 407
kilohertz (KHz), in measuring radio frequency, 83

L

laptop computers, 174
latency issues, troubleshooting, 365–366
layer 1 (physical layer), of OSI model
    hubs operating at, 40
    overview of, 10, 12–13
layer 2 (data link layer), of OSI model. *See* data link layer (layer 2), of OSI model
Layer 2 Tunneling Protocol (L2TP)
    creating VPNs, 323
    troubleshooting VPNs, 371, 404
layer 3 (network layer), of OSI model. *See* network layer (layer 3), of OSI model
layer 4 (transport layer), of OSI model, 10
layer 5 (session layer), of OSI model, 10, 16–17
layer 6 (presentation layer), of OSI model, 10, 17
layer 7 (application layer), of OSI model, 10, 17
LDAP (Lightweight Directory Access Protocol)
    content security and, 249
    overview of, 28
Li-ion (lithium-ion), troubleshooting battery life, 420
life-cycle operations
    asset disposal, 255
    certificate expiration/renewal, 252–253
    change management, 254
    end-of-life, 255
    overview of, 252
    software/system development, 252
    updates, patches, and upgrades, 253–254
Lightweight AP Protocol (LWAPP), 50–51
Lightweight Directory Access Protocol (LDAP)
    content security and, 249
    overview of, 28
line-of-sight (LOS), in RF propagation, 89

Linux
    Secure Shell (SSH) use by, 24
    `traceroute` command, 361–362
    traffic sniffers, 363–364
listeners, port listeners, 19
lithium-ion (Li-ion), troubleshooting battery life, 420
LLC (Logical Link Control), 13
load balancing, for high availability, 207
lobes (RF), characteristics of radio antennas, 93
location, grouping devices by, 187
location services
    demand on battery power, 420
    overview of, 193–194
    remote administration and, 251
    troubleshooting, 440–442
lock ups, troubleshooting device issues, 430
log files
    documenting security incidents, 304–305
    monitoring security logs, 393–394
Logical Link Control (LLC), 13
Long-Term Evolution (LTE)
    4G cellular technologies, 122–123
    IMEI numbers, 239–240
    LTE-Advanced, 123, 239–240
LOS (line-of-sight), in RF propagation, 89
LTE (Long-Term Evolution). *See* Long-Term Evolution (LTE)
LWAPP (Lightweight AP Protocol), 50–51

M

MaaS360 cloud service, 179
MAC filtering, LAN security methods, 279
MAC (Media Access Control). *See* Media Access Control (MAC)/MAC addresses
Mac OSs. *See also* operating systems (OSs), 361–362
malware
    anti-malware solutions, 290–291
    defined, 290
    software risks, 282
    types of, 283
MAM (Mobile Application Management). *See* Mobile Application Management (MAM)
man-in-the-middle (MITM) attacks, 280
Management Information Base (MIB), 71
mandatory access control, 293–294
manual on-boarding, 240–241
MAPI (Messaging Application Programming Interface)
    messaging standards, 256
    overview of, 28
maps, location services issues, 440
MAU (Multistation Access Unit), 54–55
MD5 (Message Digest version 5), 318
MDM as a Service (MaaS), 192

MDM, infrastructure implementation
    activating devices, 237
    batch provisioning, 242
    on-boarding and provision process, 238–241
    centralizing management, 246–249
    conducting pilot testing, 232–233
    containerization of data, 228–229
    creating profiles, 229–231
    creating support and coordination for, 226–227
    digital certificates and, 227–228
    documenting, 233
    end user licensing agreements (EULA), 228
    establishing level of control, 236
    facilitating OTA access, 237–238
    in-house application requirements, 262–265
    installing and configuring, 225–226
    life-cycle operations, 252–255
    messaging standards, 256–257
    network configuration for, 257–260
    off-boarding and de-provisioning process, 243–246
    overview of, 225
    preinstallation tasks, 226
    program launch, 233
    push notification technologies, 265–267
    Q&A, 268–273
    remote administration, 249–251
    remote provisioning, 242–243
    review, 267–268
    scaling devices and users, 234–235
    secure device enrollment, 243
    self-service provisioning, 242
    setting up directory services, 227
    types of mobile applications, 260–262
MDM, infrastructure plan
    administrative privileges, 187–188
    application stores, 195–197
    backing up corporate data, 209–210
    backing up devices and servers, 203–204
    backing up personal data, 210
    backup and recovery policies, 184
    backup frequency, 204–207
    balancing security with usability, 183–184
    basic MDM concepts, 178–180
    Bring Your Own Device (BYOD) and, 181
    business continuity, 199–200
    captive portal, 189–190
    comparing mobile platforms, 176–177
    creating groups, 187
    data protection, 209
    default applications, 197–198
    device platform support, 191–192
    directory services, 204
    disaster recovery, 201–203
    high availability, 207–208
    history of, 174–175
    infrastructure requirements, 186

    interoperability and support, 190–191
    IT and security policies and, 182–183
    keeping up with technology changes, 213–215
    local backup, 210–211
    location-based services, 193–194
    Mobile Application Management (MAM)
        and, 194–195
    monitoring and reporting features, 190
    multi-instance support, 192–193
    OEM vendors, 185
    OS vendors, 184–185
    overview of, 173–174
    password strength and, 188–189
    on-premise vs. cloud-based support, 192
    pushing content, 198
    Q&A, 218–223
    recovering data, 211
    remote management (wipe/lock/unlock), 189
    restoring corporate and personal data, 212
    review, 215–218
    security requirements, 186–187
    telecommunication vendors, 185–186
    testing backups, 211–212
media
    network, 4
    organizational risks related to removable
        media, 289
    synchronization issues with media files, 426
Media Access Control (MAC)/MAC addresses
    bridges using to manage traffic, 41
    network cards and, 3
    sublayer of data link layer, 13
    switches using to manage traffic, 43
megahertz (MHz)
    frequency ranges for cellular communication, 110
    in measuring radio frequency, 83, 86
memory, troubleshooting, 430–431, 433
mesh networks
    designing network topology, 57
    network topologies, 56
Message Digest version 5 (MD5), 318
message digests, in hashing, 318
Messaging Application Programming Interface (MAPI)
    messaging standards, 256
    overview of, 28
messaging standards
    overview of, 256
    types of, 256–257
MHz (megahertz). *See* megahertz (MHz)
MIB (Management Information Base), 71
microcells
    enhancing cellular coverage, 114–115
    troubleshooting cellular networks, 375
Microsoft Point-to-Point Encryption Protocol
    (MPPE), 323
Microsoft SharePoint, 247–248

Microsoft Windows OSs. *See* Windows OSs
microwave, caution in working with EM radiation, 82
MIMO (multiple-input multiple-output), 146–147
mitigation strategies
    access levels, 293
    antivirus/anti-malware solutions, 290–291
    containerization of data, 295–296
    content filtering, 296–297
    Data Loss Prevention (DLP), 297
    device hardening, 297–298
    disabling physical ports, 298
    intrusion detection systems (IDS)/intrusion
        protection systems (IPS), 294
    overview of, 290
    permissions, 293–294
    sandboxing, 295
    software firewalls, 292–293
    trusted platform module (TPM), 296
MITM (man-in-the-middle) attacks, 280
Mobile Application Management (MAM)
    app distribution, 247
    app installation, 434
    app removal and missing apps, 437
    app store issues, 439
    compared with MDM, 194–195
    configuration change, 426
mobile applications, configuring and deploying
    in-house application requirements, 262–265
    messaging standards, 256–257
    network configuration for, 257–260
    push notification technologies, 265–267
    types of mobile applications, 260–262
mobile device management (MDM)
    app distribution, 247
    app installation, 434
    app removal and missing apps, 437
    app store issues, 439
    Bring Your Own Device (BYOD) and, 181
    configuration change, 426
    device groups and, 187
    device platform support, 191–192
    implementing solution. *See* MDM, infrastructure
        implementation
    interoperability and infrastructure support,
        190–191
    location-based services, 193–194
    MAM compared with, 194–195
    monitoring mobile devices, 392
    multi-instance support, 192–193
    overview of, 173, 178–180
    planning solution. *See* MDM, infrastructure plan
    self-service portal, 191
mobile devices
    access control, 333
    activating on cellular networks, 237–238

administrative permissions, 187–188
Android. *See* Android OS/Android devices
applications for device monitoring, 390–392
backing up, 209
Bring Your Own Device (BYOD). *See* Bring Your
    Own Device (BYOD)
in cellular communication, 114
cloning, 286
comparing mobile platforms, 176–177
containerization of data, 342–343
creating profiles, 229–231
deactivating, 245
establishing level of control over, 236
grouping, 187
hardening, 297–298
history of, 174–175
IMEI, ICCID, and IMSI numbers, 239–240
IT policies for, 182–183
life-cycle operations. *See* life-cycle operations
local backup, 210
managing with MDM. *See* mobile device
    management (MDM)
monitoring, 389–390
operations and management. *See* operations and
    management, of mobile devices
overview of, 173–174
password strength, 188–189
platform support, 191–192
provisioning, 243
reporting features, 190
risk of loss, 287
risk of theft, 286–287
risks related to unknown devices,
    289–290
security requirements, 186–187
troubleshooting activation, 376
troubleshooting client issues, 419–420, 425
troubleshooting crashes and hardware issues,
    430–431
troubleshooting mobile security. *See*
    troubleshooting mobile security
Mobile Switch Center (MSC)
    devices in cellular communication, 114
    roaming and switching between network
        types, 125
modulation, analog and digital, 87–88
monitoring
    applications for device monitoring, 390–392
    false negatives/false positives, 405–406
    features for, 190
    mobile devices, 389–390
    policy for, 388–389
    Security Information Event Management
        (SIEM), 392–393
    security logs, 393–394

MPPE (Microsoft Point-to-Point Encryption
   Protocol), 323
MSC (Mobile Switch Center)
   devices in cellular communication, 114
   roaming and switching between network
      types, 125
multi-instance support, mobile devices, 192–193
multicasting, comparing with unicasting and
   broadcasting, 14
multifactor authentication, 331
multiple access, CDMA, 116
multiple-input multiple-output (MIMO), 146–147
multiplexing/demultiplexing, 13
Multistation Access Unit (MAU), 54–55

N

N-tiered architecture, 6
NAC (network access control). *See* network access
   control (NAC)
nanocells, for enhancing cellular coverage, 114–115
NAS (network attached storage), 209
NAT (Network Address Translation), 67
National Institute of Standards and Technology
   (NIST), 200
native apps, 261
Near Field Communication (NFC), 151–152
`netstat` command, 21–22
network access control (NAC)
   devices, 209
   placing NAC devices, 259
   risk mitigation strategies, 291
   security risks related to unknown devices,
      289–290
   troubleshooting anti-malware issues, 410
Network Address Translation (NAT), 67
network architectures
   bastion host, 59
   DMZ (Demilitarized Zone), 60–61
   extranet, 61
   Internet, 59
   intranet, 59
   overview of, 58–59
   Q&A, 74–79
   review, 72–73
network attached storage (NAS), 209
network cards. *See* Network Interface Cards (NICs)
network configuration
   hosting solutions, 259–260
   information topology, 258–259
   overview of, 257–260
   settings for vendor proxy and gateway
      servers, 258
network devices
   access points, 50–51

backing up, 203–204
bridges, 41
common types, 40
firewalls, 45–48
gateways, 48–49
hubs, 40–41
managing, 71
overview of, 6, 39–40
proxies, 49
Q&A, 74–79
review, 72
routers, 43–45
specialized devices, 45
switches, 41–43
troubleshooting, 403–404
VPN concentrators, 49–50
wireless LAN controllers, 50
network hosts. *See* hosts
Network Interface Cards (NICs), 2–4
network layer (layer 3), of OSI model
   layer 3 switches, 67–69
   overview of, 10, 13–14
   routers operating at, 44
   switches operating at, 43
network management
   backhauling traffic, 69
   bandwidth, 70
   Power over Ethernet (PoE), 71
   Q&A, 74–79
   Quality of Service (QoS), 70
   review, 73–74
   Simple Network Management Protocol
      (SNMP), 71
   traffic shaping and routing, 70–71
Network Operations Centers (NOCs), 260
network technologies
   IP addressing, 62–64
   Network Address Translation (NAT), 67
   overview of, 62
   Q&A, 74–79
   review, 71–73
   special IP address ranges, 66–67
   subnetting, 64–65
   virtual LANs (VLANs), 67–69
network topologies
   ad-hoc, 56
   bus, 52–53
   designing, 57
   equipment in, 58
   information topology, 258–259
   mesh, 56
   overview of, 51
   physical environment and, 58
   Q&A, 74–79
   review, 72

network topologies (*cont*)
    ring, 54–55
    star, 54
    traditional, 51–52
network traffic sniffers, 363
networks/networking
    clients and servers, 5
    decentralized vs. centralized networks, 6–8
    devices. *See* network devices
    managing. *See* network management
    media, 4
    network cards, 2–4
    networked applications, 5–6
    OSI model. *See* Open Systems Interconnection
        (OSI)
    overview of, 1–2
    ports. *See* ports (network)
    protocols. *See* protocols (network)
    Q&A, 32–37
    review, 31–32
    technologies. *See* network technologies
    topologies. *See* network topologies
    troubleshooting. *See* troubleshooting network
        issues
NFC (Near Field Communication), 151–152
NiCd (nickel-cadmium), 420
nickel-cadmium (NiCd), 420
nickel-metal hydride (NiMH), 420
NICs (Network Interface Cards), 2–4
NiMH (nickel-metal hydride), 420
"the nines," measuring high availability, 208
NIST (National Institute of Standards and
    Technology), 200
NOCs (Network Operations Centers), 260

O

OCSP (Online Certificate Security Protocol), 397
octets, in IP addressing, 62–64
OEM vendors, 185, 213
OFB (Output Feedback mode), DES, 316
OFDM (orthogonal frequency division multiplexing),
    144–145
off-boarding
    content management, 245
    deactivating devices, 246
    decommissioning applications, 245
    migrating data, 245
    overview of, 243–244
    terminating employees, 244–245
omnidirectional antennas, 96–97
on-boarding
    activating devices, 237
    batch provisioning, 242
    facilitating OTA access, 237–238
    IMEI, ICCID, and IMSI numbers, 239

    installing profiles, 239–240
    manual on-boarding, 240–241
    provisioning devices and, 238
    remote provisioning, 242–243
    secure device enrollment, 243
    self-service provisioning, 242
on-premise support, for mobile devices, 192
Online Certificate Security Protocol (OCSP), 397
Open Shortest Path First (OPSF), 44
Open Systems Interconnection (OSI)
    application layer (layer 7), 17
    data flow, 10–11
    data link layer (layer 2), 13
    how it works, 9–10
    network layer (layer 3), 13–14
    overview of, 8–9
    physical layer (layer 1), 12–13
    presentation layer (layer 6), 17
    session layer (layer 5), 16–17
    TCP/IP and, 11–12
    transport layer (layer 4), 14–16
operating systems (OSs)
    app store issues, 438
    configuration change issues, 425–426
    keeping up with technology changes, 213
    location services issues, 440–441
    risks of unsupported OSs, 285–286
    security requirements and, 186–187
    synchronization issues, 427–428
    `traceroute` command variations by, 361–362
    troubleshooting client issues, 431–433
    vendors, 184–185
operations and management, of mobile devices. *See*
    also mobile device management (MDM)
    application management, 248–249
    centralizing management and distribution,
        246–247
    content updates and changes, 248
    distribution methods, 247–248
    life-cycle operations, 252–255
    overview of, 246
    remote administration, 249–251
    security and access control, 249
OPSF (Open Shortest Path First), 44
organizational risks
    Bring Your Own Device (BYOD) and, 288
    overview of, 287
    personal device security, 288
    removable media and, 289
    unknown devices, 289–290
    wiping personal data from mobile devices, 289
organizational units (OUs), 187
orthogonal frequency division multiplexing (OFDM),
    144–145
OSI (Open Systems Interconnection). *See* Open
    Systems Interconnection (OSI)

OSs (operating systems). *See* operating systems (OSs)

OTA (over-the-air). *See* over-the-air (OTA)

OUs (organizational units), 187

Output Feedback mode (OFB), DES, 316

outsourcing, incident response and, 302

over-the-air (OTA)
    facilitating access, 237–238
    troubleshooting OTA connectivity, 360–361, 374–378

P

PaaS (Platform as a Service), 192

PANs (personal area networks). *See* personal area networks (PANs)

parabolic dish antennas, 98–99

passwords
    authentication error message, 399
    expired, 400–401
    hashing and, 318
    non-expiring, 401–403
    single-factor authentication, 330–331
    strength of, 188–189
    troubleshooting device issues, 428–430
    troubleshooting email issues, 369, 371
    troubleshooting encryption issues, 409

patches
    app store issues, 438
    life-cycle operations, 253–254
    troubleshooting device issues, 433

PDU (Protocol Data Unit), 11

PEAP (Protected EAP), 326

peer-to-peer networks
    decentralized vs. centralized networks, 6–7
    designing network topology, 57

performance baselines, 373–374

permissions
    administrative, 187–188
    app store issues, 438
    content, 249
    object-level, 341
    risk mitigation strategies, 293–294

personal area networks (PANs)
    Bluetooth and infrared, 151
    Near Field Communication (NFC), 151–152
    Worldwide Interoperability for Microwave Access (WiMAX), 153

personal data
    containerization of, 228–229
    restoring, 212

personal identification number (PIN)
    in authentication, 330
    device access and, 333
    multifactor authentication, 331
    passwords and, 189

troubleshooting device issues, 429

troubleshooting encryption issues, 409

PGP (Pretty Good Privacy)
    encryption in, 334
    managing digital certificates, 340

phase, of radio waves, 88–89

physical addresses. *See* Media Access Control (MAC)/MAC addresses

physical environment, network topologies and, 58

physical layer (layer 1), of OSI model
    hubs operating at, 40
    overview of, 10, 12–13

picocells, for enhancing cellular coverage, 114–115

piconets. *See also* personal area networks (PANs), 375

pilot testing, 232–233

PIN (personal identification number). *See* personal identification number (PIN)

`ping` utility
    ICMP protocol and, 14
    testing connectivity, 361
    troubleshooting connectivity, 367
    troubleshooting latency, 366

PKCS (Public Key Cryptography Standards), 336

PKI (Public Key Infrastructure). *See* Public Key Infrastructure (PKI)

plaintext, encryption and, 314

planning
    for incident response, 300–301
    MDM infrastructure. *See* MDM, infrastructure plan
    Wi-Fi coverage, 159–160

Platform as a Service (PaaS), 192

platforms. *See also* operating systems (OSs)
    app development and, 263
    comparing mobile platforms, 176–177
    support in MDM, 191–192
    vendors, 184–185

PoE (Power over Ethernet), 71

point-to-multipoint networks, 51–53

point-to-point networks, 51–52

Point-to-Point Protocol (PPP), 323

Point-to-Point Tunneling Protocol (PPTP), 323, 371

polarization, of radio antennas, 95

policies
    backup and recovery, 184
    data segregation, 340–341
    establishing level of control over devices, 236
    incident response, 299–300
    making policy-based response to security incident, 303–304
    MDM, 180
    monitoring, 388–389
    password expiration, 401–403
    security, 182–183

POP3 (Post Office Protocol, version 3). *See* Post Office Protocol, version 3 (POP3)
portable hotspots, Wi-Fi, 137–138
ports (network)
    of common network protocols, 30
    disabling physical ports as risk mitigation strategy, 298
    dynamic and ephemeral, 20–22
    how they work, 18–19
    overview of, 18
    port based access control (802.1X), 156–157
    registered, 20
    SNMP, 71
    troubleshooting configuration issues, 373
    well-known, 19
Post Office Protocol, version 3 (POP3)
    messaging standards, 257
    overview of, 25
    troubleshooting email issues, 369–370
power levels, of radio antennas, 96
power outages, troubleshooting, 425
Power over Ethernet (PoE), 71
power supply, troubleshooting, 423–424
PPP (Point-to-Point Protocol), 323
PPTP (Point-to-Point Tunneling Protocol), 323, 371
pre-shared keys (PSK), in WPA2, 325
preinstallation tasks, for MDM infrastructure
    conducting pilot testing, 232–233
    containerization of personal and corporate data, 228–229
    creating device and group profiles, 229–231
    creating support and coordination, 226–227
    digital certificates and, 227–228
    documenting, 233
    end user licensing agreements (EULA), 228
    overview of, 226
    setting up directory services, 227
presentation layer (layer 6), of OSI model, 10, 17
Pretty Good Privacy (PGP)
    encryption in, 334
    managing digital certificates, 340
preventative measures, troubleshooting and, 359
printers, permissions, 293–294
privileged communications, exceptions to monitoring, 390
profiles
    creating device and group profiles, 229–231
    installing when provisioning devices, 239–240
    troubleshooting client issues, 430
Protected EAP (PEAP), 326
Protocol Data Unit (PDU), 11
protocols (network)
    Airsync, 28
    APNs, 29
    DHCP, 23
    DNS, 23

    FTP, 24
    FTPS, 25
    GCM, 30
    HTTP, 22
    HTTPS, 23
    IMAP, 26
    Jabber, 29
    LDAP/AD, 28
    MAPI, 28
    network layer (layer 3), 14
    overview of, 22
    POP3, 25
    port numbers for, 30
    RDP, 29
    rules for data communication, 8
    SFTP, 24–25
    SMTP, 25
    SRP, 29
    SSH, 24
    SSL, 22–23
    SSMTP, 27
    Telnet, 24
provisioning devices
    batch provisioning, 242
    on-boarding and, 238
    de-provisioning. *See* off-boarding
    IMEI, ICCID, and IMSI numbers, 239
    installing profiles, 239–240
    manual on-boarding and, 240–241
    remote provisioning, 242–243
    secure device enrollment, 243
    self-service provisioning, 242
proxies
    overview of, 49
    settings for vendor proxy servers, 258
PSH flag, TCP segments and, 15
PSK (pre-shared keys), in WPA2, 325
PSTN (Public Switched Telephone Network), 114
public key cryptography
    defined, 314
    overview of, 334–335
    troubleshooting encryption issues, 409
Public Key Cryptography Standards (PKCS), 336
Public Key Infrastructure (PKI)
    for authentication in wireless networking, 153–154
    certificate lifecycle and, 336–338
    certificate management infrastructure, 338–340
    content security and, 249
    digital certificates in, 227–228, 335–336
    for encryption in wireless networking, 154
    IEEE 802.1X and, 157
    overview of, 334
    public key cryptography, 334–335
    troubleshooting authentication issues, 399

Public Switched Telephone Network (PSTN), 114
publishing apps, 262–263
push notification
    Apple Push Notification service (APNs), 266
    Exchange ActiveSync (EAS), 266–267
    Google Cloud Messaging (GCM), 266–267
    overview of, 198, 265–266
pushing content, in mobile app management, 198

Q

QoS (Quality of Service). *See* Quality of Service (QoS)
Quality of Service (QoS)
    digital modulation and, 88
    in network management, 70
    troubleshooting network saturation, 374

R

RA (Registration Authority), 339–340
radio frequency (RF)
    amplitude, 85–86
    antenna characteristics, 92–96
    antenna types, 96–99
    antennas generally, 92
    bandwidth, 86
    comparing network bandwidth and RF
      bandwidth, 70
    coverage planning in Wi-Fi and, 159–160
    factors in RF propagation, 89–92
    Faraday cages, 99
    frequency, 82–83
    modulation, 87–88
    overview of, 81–82
    phase, 88–89
    Q&A, 100–105
    review, 99–100
    short-range (Bluetooth), 149–151
    signal strength and, 160–161
    spectrum analysis, 162–163
    wavelength, 83–85
RADIUS (Remote Authentication Dial-In User
    Service), 157
RC4 protocol
    overview of, 324
    weak encryption keys and, 279
    in WEP security, 155
RDP (Remote Desktop Protocol), 29
Real-Time Control Protocol (RTCP), 325
Real-time Transport Protocol (RTP), 325
receive signal strength, in Wi-Fi site survey, 161
received signal strength indicator (RSSI), 161
recovery
    data recovery, 211
    disaster recovery, 201–203
    policies, 184
    restoring corporate and personal data, 212

redundancy of data, for high availability, 207
reflection, factors in RF propagation, 90
refraction, factors in RF propagation, 90
registered ports, 20
Registration Authority (RA), 339–340
remote administration
    location services, 251
    lock/unlock, 189, 250
    overview of, 249
    provisioning devices, 242–243
    remote control, 250
    remote wipe, 189, 250
    reporting features, 251
Remote Authentication Dial-In User Service
    (RADIUS), 157
Remote Desktop Protocol (RDP), 29
Remote Procedure Call (RPC), 16–17
removable media
    encryption of, 329
    organizational risks related to, 289
repeater hubs, 40
reports
    device compliance and audit information
      reports, 388
    false negatives/false positives, 405–406
    features of mobile devices, 190
    remote administration and, 251
    security incident, 304–305
RF lobes, characteristics of antennas, 93
RF (radio frequency). *See* radio frequency (RF)
ring networks, 54–55
RIP (Routing Information Protocol), 44
risks, security. *See* security risks
roaming
    between cellular network types, 125–126
    troubleshooting cellular networks, 375
    troubleshooting OTA connectivity, 374
rogue access points, 276
root certificates, 339
rooting, software risks, 284
routers
    default gateway and, 48–49
    layer 3 switches compared with, 68
    overview of, 43–45
    Quality of Service (QoS) and, 70
    troubleshooting network saturation, 374
routing, in network management, 70–71
Routing Information Protocol (RIP), 44
RPC (Remote Procedure Call), 16–17
RSA algorithm, 325
RSA Security, Inc., 336
RSSI (received signal strength indicator), 161
RST flag, TCP segments and, 15
RTCP (Real-Time Control Protocol), 325
RTP (Real-Time Transport Protocol), 325
ruleset, firewall, 46–47, 403–404

S

SaaS (Software as a Service)
  multi-instance support for mobile devices, 193
  on-premise vs. cloud-based support for mobile devices, 192
SAN (storage area network), 209
sandboxing
  restricting access to apps, 228–229
  as risk mitigation strategy, 295
SAs (security associations), between hosts, 321
scaling devices and users, for MDM infrastructure, 234–235
SCEP (Simple Certificate Enrollment Protocol)
  iPhone example, 244
  for secure device enrollment, 243
screen (display), demand on battery power, 421
SDLC (systems development life cycles). See also life-cycle operations, 252
Secure Copy (SCP)
  copying files between hosts, 24
  securing data-in-transit, 320–321
secure device enrollment (SCEP), 243
Secure Hashing Algorithm version 1 (SHA-1), 318
Secure Real-time Transport Protocol (SRTP), 325
Secure Shell (SSH)
  overview of, 24
  securing data-in-transit, 320–321
  securing wireless networks, 333
  SSH-FTP (SFTP), 24–25
  troubleshooting encryption issues, 408–409
Secure SMTP (SSMTP), 27
Secure Sockets Layer (SSL)
  application access and, 333
  creating VPNs, 323
  FTP-Secure (FTPS) and, 25
  HTTPS and, 320
  overview of, 22–23
  securing data-in-transit, 319–320
  securing wireless networks, 333
  troubleshooting digital certificates, 373
  troubleshooting encryption issues, 409
security
  managing mobile devices, 249
  mobile device requirements, 186–187
  monitoring security logs, 393–394
  policies. See security policies
  risks. See security risks
  technologies. See security technologies
  troubleshooting. See troubleshooting mobile security
security associations (SAs), between hosts, 321
Security Information Event Management (SIEM), 392–393

security policies
  balancing security with usability, 183–184
  for mobile devices, 182–183
  monitoring policy and, 388–389
security risks
  app stores and, 281–282
  Bring Your Own Device (BYOD) and, 288
  denial-of-service (DoS) attacks, 276–277
  device cloning, 286
  device loss, 287
  device theft, 286–287
  hardware risks, 286
  incident response. See incident response
  infrastructure risks, 275
  jailbreaking, 284
  key logging, 284–285
  malware, 282
  man-in-the-middle (MITM) attacks, 280
  mitigation strategies. See mitigation strategies
  organizational risks, 287
  overview of, 275
  personal device security, 288
  Q&A, 307–312
  removable media and, 289
  review, 305–307
  rogue access points, 276
  rooting, 284
  software risks, 281
  spyware, 283
  tower spoofing, 281
  Trojans, 283
  unknown devices, 289–290
  unsupported OSs and, 285–286
  viruses, 282
  warchalking, 278–279
  wardriving, 278
  weak encryption keys, 279–280
  weaknesses in wireless LAN protocols, 279
  wiping personal data from mobile devices, 289
  wireless risks, 275–276
  worms, 283
security technologies
  block-level encryption, 327
  certificate lifecycle, 336–338
  certificate management infrastructure, 338–340
  containerization of data, 342–343
  data access controls, 341–342
  data segregation, 340
  digital certificates, 335–336
  encryption, 313–315
  file-level encryption, 328
  folder-level encryption, 329
  full disk encryption, 326–327
  implementing encryption, 326
  overview of, 313

policies for data segregation, 340–341

protecting data-at-rest, 315–318

protecting data-in-transit, 318–326

public key cryptography, 334–335

Public Key Infrastructure (PKI), 334

Q&A, 345–350

removable media encryption, 329

review, 343–345

segments, TCP, 14–15

self-service portal, in MDM, 191

self-service provisioning, 242

Server Routing Protocol (SRP), 29

servers

backing up, 203–204

backup servers, 209

in networking, 5

service-level agreements (SLAs)

distribution methods and, 248

organizational control and, 260

outsourcing considerations in incident
response, 302

Service Set Identifier (SSID)

overview of, 139–142

rogue access points and, 276

SSID hiding/cloaking, 142

session layer (layer 5), OSI model, 10, 16–17

SFTP (SSH-FTP)

overview of, 24–25

securing data-in-transit, 320

SHA-1 (Secure Hashing Algorithm version 1), 318

SharePoint, Microsoft, 247–248

Short Message Service (SMS), 266

sideloading apps, 196

SIEM (Security Information Event Management),
392–393

signal power, of radio antennas, 93

signal strength

receive signal strength, 161

troubleshooting cellular networks, 375

in Wi-Fi site survey, 160

signal-to-noise ratio (SNR)

interference and, 92

receive signal strength and, 161

signaling technologies, for cellular communication

Circuit-Switched Data (CSD), 119

Code Division Multiple Access (CDMA), 116

Global System for Mobile Communication
(GSM), 118–119

overview of, 115

Time Division Multiple Access (TDMA), 116–118

SIM (Subscriber Identity Module). *See* Subscriber
Identity Module (SIM)

Simple Certificate Enrollment Protocol (SCEP)

iPhone example, 244

for secure device enrollment, 243

Simple Mail Transfer Protocol (SMTP)

overview of, 25

SSMTP, 27

TCP providing transport services for, 15

troubleshooting email issues, 369–370

Simple Network Management Protocol (SNMP)

managing network devices, 71

messaging standards, 257

UDP and, 16

sine wave, RF signals as, 82

single-factor authentication, 330–331

Single Sideband (SSB), radio modulation, 88

site surveys, for Wi-Fi

capacity planning, 159

coverage planning, 159–160

documenting site survey, 163–164

following up after site survey, 164

interference (noise) and, 161

overview of, 157–158

preparing for, 158

signal strength and, 160–161

spectrum analysis, 162–163

troubleshooting network saturation, 374

size, in designing network topology, 57

SLAs (service-level agreements). *See* service-level
agreements (SLAs)

smartcards, in single-factor authentication,
330–331

smartphones. *See also* Android OS/Android devices;
iPhone

battery usage, 421–423

in cellular communication, 114

cellular settings, 125–126

history of mobile devices in enterprise, 174

power supply, 424

SMS (Short Message Service), 266

SMTP (Simple Mail Transfer Protocol). *See* Simple
Mail Transfer Protocol (SMTP)

SNMP (Simple Network Management Protocol). *See*
Simple Network Management Protocol (SNMP)

SNR (signal-to-noise ratio)

interference and, 92

receive signal strength and, 161

sockets, IP addresses and, 18

software

applications for device monitoring, 391

for connectivity testing, 363–364

life-cycle operations, 252

software-based container access and data
segregation, 340

software firewalls, 292–293

Software as a Service (SaaS)

multi-instance support for mobile devices, 193

on-premise vs. cloud-based support for mobile
devices, 192

software risks
    app stores and, 281–282
    jailbreaking, 284
    key logging, 284–285
    malware, 282
    overview of, 281
    rooting, 284
    spyware, 283
    Trojans, 283
    unsupported OSs and, 285–286
    viruses, 282
    worms, 283
SolarWinds, 363
source ports
    ephemeral nature of, 20
    `netstat` command showing, 21–22
    vs. destination ports, 19
spectrum analysis
    hardware tools for network troubleshooting, 365
    in Wi-Fi site survey, 162–163
spread spectrum, digital modulation and, 88
spyware, 283
SRP (Server Routing Protocol), 29
SRTP (Secure Real-time Transport Protocol), 325
SSB (Single Sideband), radio modulation, 88
SSH-FTP (SFTP)
    overview of, 24–25
    securing data-in-transit, 320
SSH (Secure Shell). *See* Secure Shell (SSH)
SSID hiding/cloaking, 279
SSID (Service Set Identifier). *See* Service Set
    Identifier (SSID)
SSL (Secure Sockets Layer). *See* Secure Sockets
    Layer (SSL)
SSMTP (Secure SMTP), 27
standards
    for cellular communication, 115
    Wi-Fi, 143–146
star networks
    designing network topology, 57
    overview of, 54
station (STA), 135
storage area network (SAN), 209
streaming algorithms, 314
subnet masks, in IP addressing, 63–65
Subscriber Identity Module (SIM)
    in GSM, 118
    ICCID and IMSI numbers and, 239–240
    troubleshooting cellular networks, 376
    troubleshooting OTA connectivity, 374
subscribers, CDMA, 116
supernetting, IP addresses, 65
supplicant, in 802.1X, 157
switches
    hubs compared with, 40
    overview of, 41–43

    troubleshooting network saturation, 374
    troubleshooting VLANs, 404
    VLANs and, 67
switching, between cellular network types, 125–126
symmetric algorithms, 314
symmetric key cryptography, 334
symptoms, identifying network issues, 354
SYN flag, TCP segments and, 15
synchronization
    missing apps, 437
    troubleshooting, 426–428
system development, life-cycle operations, 252
systems development life cycles (SDLC). *See also* life-
    cycle operations, 252

T

tablet devices. *See also* iPad, 114
TACACS (Terminal Access Controller Access-Control
    System), 157
TCP/IP (Transmission Control Protocol/Internet
    Protocol), 11–12
TCP (Transmission Control Protocol). *See*
    Transmission Control Protocol (TCP)
TDMA (Time Division Multiple Access), 116–118
TDRs (time-domain reflectometers), 364
team, in incident response
    planning and, 300–301
    policy-based response to security incidents, 303
    reporting security incidents, 304–305
technology, keeping up with changes in, 213–215
telecommunications
    keeping up with technology changes, 214
    vendors, 185–186
Telecommunications Industry Association/Electronic
    Industries Alliance (TIA/EIA), 12
Telnet, 24
Temporal Key Integrity Protocol (TKIP), 155, 324
Terminal Access Controller Access-Control System
    (TACACS), 157
testing
    hardware for testing connectivity, 364–365
    incident response plans, 301–302
    MDM infrastructure (pilot tests), 232–233
    software for testing connectivity, 363–364
    theory of probable cause in troubleshooting,
        356–357
    utilities for testing connectivity, 361–362
theft, hardware risks, 286–287
TIA/EIA (Telecommunications Industry Association/
    Electronic Industries Alliance), 12
Time Division Multiple Access (TDMA), 116–118
time-domain reflectometers (TDRs), 364
TKIP (Temporal Key Integrity Protocol), 155, 324
TLS (Transport Layer Security). *See* Transport Layer
    Security (TLS)

Token Ring, 55
tokens
    in access control, 332
    single-factor authentication, 330–331
tower spoofing, 281
TPM (Trusted Platform Module). *See* Trusted
  Platform Module (TPM)
`traceroute` command
    testing connectivity, 361–362
    troubleshooting connectivity, 367
    troubleshooting latency, 366
    troubleshooting network services, 367–368
traffic shaping
    defined, 74
    in network management, 70–71
    troubleshooting network saturation, 374
Transmission Control Protocol/Internet Protocol
  (TCP/IP), 11–12
Transmission Control Protocol (TCP)
    overview of, 14–16
    ports and, 18–19
    SNMP ports, 71
    transport protocol for network protocols, 30
    troubleshooting port configuration, 373
transport layer (layer 4), of OSI model
    overview of, 10, 14–16
    Transmission Control Protocol (TCP), 14–16
    User Datagram Protocol (UDP), 16
Transport Layer Security (TLS)
    FTPS and, 25
    HTTPS and, 320
    securing data-in-transit, 319–320
    troubleshooting digital certificates, 373
    troubleshooting encryption issues, 409
transport mode, IPsec, 321
Triple DES (3DES), 316
Trojans
    key logging and, 284
    software risks, 283
troubleshooting clients
    app issues, 436–437
    app store issues, 437–439
    application installation issues, 434–435
    authentication and password issues, 428–430
    battery life issues, 420–423
    configuration change issues, 425–426
    device crashes and hardware issues, 430–431
    device issues, 419–420, 425
    email issues, 442–443
    location services issues, 440–442
    OS issues, 431–433
    overview of, 419
    power outages, 425
    power supply issues, 423–424
    profile authentication and authorization issues, 430

Q&A, 444–449
review, 443–444
synchronization issues, 426–428
upgrade issues, 435
troubleshooting methodology
    creating theory of probable cause, 356
    documenting the issue, 359–360
    establish plan of action, 357–358
    identifying the problem, 353–355
    implementing the solution, 358
    overview of, 351–352
    steps in, 352–353
    testing theory of probable cause, 356–357
    verifying solution, 359
troubleshooting mobile security
    anti-malware issues, 410
    applications for device monitoring, 390–392
    authentication issues, 399–403
    certificate issues, 394–398
    content filtering issues, 407–408
    crossover error rate, 406
    encryption issues, 408–410
    false negatives and false positives, 405–406
    monitoring devices, 389–390
    monitoring policy, 388–389
    monitoring security logs, 393–394
    network device issues, 403–404
    overview of, 387
    Q&A, 412–417
    review, 411–412
    Security Information Event Management
      (SIEM), 392–393
troubleshooting network issues
    certificate issues, 372–373
    connectivity issues, 366–367
    email issues, 369–371
    hardware for connectivity testing,
      364–365
    latency issues, 365–366
    network saturation issues, 373–374
    network service issues, 367–369
    OTA connectivity, 360–361, 374–378
    overview of, 351
    port configuration issues, 373
    Q&A, 380–385
    review, 378–380
    software for connectivity testing, 363–364
    utilities for connectivity testing, 361–362
    VPN issues, 371–372
TrueCrypt application, for containerization
  of data, 342
Trusted Platform Module (TPM)
    block-level encryption, 327
    full disk encryption, 326
    overview of, 296

tunnel mode, IPsec, 322
tunneling protocols
    L2TP and PPTP, 323
    troubleshooting VPNs, 404
    VPNs and, 322
Twofish, encryption algorithm, 317

U

U-NII (Unlicensed National Information
  Infrastructure)
    interference (noise) and, 161
    Wi-Fi frequencies and, 143
UDP (User Datagram Protocol). *See* User Datagram
  Protocol (UDP)
Ultra High Frequency (UHF)
    frequency ranges for cellular communication, 110
    measuring wavelength of radio waves, 83
UMTS (Universal Mobile Telephone System), 121
unicasting, comparing with multicasting and
  broadcasting, 14
Uniform Resource Locators (URLs)
    content filtering by, 407–408
    DNS and, 23
Universal Mobile Telephone System (UMTS), 121
Unix
    Secure Shell (SSH) use by, 24
    `traceroute` command, 361–362
Unlicensed National Information Infrastructure
  (U-NII)
    interference (noise) and, 161
    Wi-Fi frequencies and, 143
updates
    app store issues, 438
    content, 248
    life-cycle operations, 253–254
    synchronization issues, 426
upgrades
    life-cycle operations, 253–254
    troubleshooting, 435
URG flag, TCP segments, 15
URLs (Uniform Resource Locators)
    content filtering by, 407–408
    DNS and, 23
usability, balancing security with, 183–184
USB drives, full disk encryption and, 327
User Datagram Protocol (UDP)
    overview of, 16
    ports and, 18–19
    SNMP ports, 71
    troubleshooting port configuration, 373
users
    creating group profiles, 231
    deployment to small groups in phased
      approach, 234

    establishing level of control over, 236
    permissions, 294, 341
    questions in process of identifying network
      issues, 354–355
    scaling MDM infrastructure during deployment
      phases, 234–235
utilities, for connectivity testing, 361–362

V

Validation Authority (VA), in certificate management
  infrastructure, 339
vendors
    app development and, 263–264
    OEM vendors, 185
    OS vendors, 184–185, 213
    role of third-party vendors in keeping up with
      technology changes, 214
    telecommunication vendors, 185–186
Virtual Local Area Networks (VLANs)
    overview of, 67–69
    security risks related to unknown devices, 290
    switches and, 43
    troubleshooting, 404
Virtual Private Networks (VPNs)
    securing data-in-transit, 322–323
    troubleshooting, 371–372, 404
    tunneling protocols, 322
    VPN concentrators, 49–50
virtual sandboxing, 295
virtualization, for high availability, 208
viruses
    antivirus solutions, 290–291
    software risks, 282
VLANs (Virtual Local Area Networks). *See* Virtual
  Local Area Networks (VLANs)
Voice over IP (VoIP), 325
VPNs (Virtual Private Networks). *See* Virtual Private
  Networks (VPNs)

W

W (watts), measuring radio signal power in, 93
WANs (wide area networks), 69
WAPs (wireless access points). *See also* access
  points (APs)
    infrastructure mode and, 139
    jamming, 277
warchalking, 278–279
wardriving, 278
warm sites, disaster recovery, 202
watts (W), measuring radio signal power in, 93
wavelength
    frequency ranges for cellular communication, 110
    of radio waves, 83–85
Web apps, 261

web of trust model, 334, 340
well-known ports, 19
WEP (Wired Equivalent Privacy). *See* Wired
  Equivalent Privacy (WEP)
Wi-Fi
    access points, 137
    ah-hoc and infrastructure modes, 138–139
    authentication, 153–154
    capacity planning, 159
    coverage planning, 159–160
    documenting site survey, 163–164
    encryption, 154
    following up after site survey, 164
    history of, 134
    IEEE 802.11 standards, 143–146
    IEEE 802.15 standard (Bluetooth), 149–151
    IEEE 802.1X standard, 156–157
    interference (noise) and, 161
    location services, 251
    overview of, 133
    PANs, 151–153
    portable hotspots, 137–138
    preparing for site survey, 158
    Q&A, 166–171
    review, 164–166
    service sets options, 139–142
    signal strength and, 160–161
    site surveys and, 157–158
    spectrum analysis, 162–163
    standards, 13, 143
    WEP, 154–155
    wireless channels and frequencies, 148–149
    wireless clients, 135–136
    WPA, 155–156
Wi-Fi protected access (WPA)
    CCMP and, 325
    DoS attacks and, 276
    encryption, 155–156
    overview of, 324–325
    securing wireless networks, 333
    weak encryption keys and, 279–280
    WPA2. *See* WPA2
wide area networks (WANs), 69
WiMAX (Worldwide Interoperability for Microwave
  Access), 124, 153
Windows OSs
    BitLocker Drive Encryption, 327, 329
    comparing mobile platforms, 177
    file-level encryption, 328
    folder-level encryption, 329
    history of mobile devices in enterprise, 174
    synchronization issues with Windows
      devices, 427
    `tracert` command, 361–362
    traffic sniffers, 363–364

wiping personal data, from mobile devices, 289
Wired Equivalent Privacy (WEP)
    DoS attacks and, 276
    encryption, 154–155
    securing data-in-transit, 323–324
    securing wireless networks, 333
    weak encryption keys, 279
wired media, 4
wireless access points (WAPs). *See also* access
  points (APs)
    infrastructure mode and, 139
    jamming, 277
wireless channels, 148–149
wireless clients, 135–136
wireless LANs (WLANs)
    security protocols, 279
    wardriving and, 278
    weak encryption keys, 279–280
    weaknesses in security protocols, 279
    wireless LAN controllers, 50
wireless media, 4
wireless metropolitan area networks (WMANs), 124
wireless networks, access control, 333
wireless personal area networks (WPANs), 149–151
wireless risks, 275–276
Wireshark traffic sniffer, 364
wiring/cabling
    cable testers, 364–365
    Power over Ethernet (PoE) and, 71
    T568A and T56B wiring standards, 12
WLANs (wireless LANs). *See* wireless LANs (WLANs)
WMANs (wireless metropolitan area networks), 124
Worldwide Interoperability for Microwave Access
  (WiMAX), 124, 153
worms, 282–283
WPA (Wi-Fi protected access). *See* Wi-Fi protected
  access (WPA)
WPA2
    overview of, 155, 324–325
    securing wireless networks, 333
    weak encryption keys and, 279
WPANs (wireless personal area networks), 149–151

X

X.509 standard for digital certificates, 336
XML (eXtensible Markup Language). *See* eXtensible
  Markup Language (XML)

Y

Yagi antennas, 98

Z

ZigBee (IEEE 802.15.4 standard), 151–152